ENVIRONMENTAL HEALTH

DADE W. MOELLER

ENVIRONMENTAL HEALTH

Revised Edition

HARVARD UNIVERSITY PRESS
Cambridge, Massachusetts
London, England
1997

Printed in the United States of America

Library of Congress Cataloging-in-Publication Data

Moeller, D. W. (Dade W.)
Environmental health / Dade W. Moeller. — Rev. ed.
p. cm.
Includes bibliographical references and index.
ISBN 0-674-25859-2 (alk. paper)
1. Environmental health. I. Title.
RA565.M64 1997
616.9'8—dc20 96-43287

To Betty Jean, who for almost fifty years has been the joy of my life, and to Rad, Mark, Kehne, Matt, and Anne, who never cease to make us proud

CONTENTS

PREFACE TO THE REVISED EDITION

This book continues to be an outgrowth of a course, "Principles of Environmental Health," that I taught at the Harvard School of Public Health for twenty-seven years and for shorter periods of time, in modified form, at Harvard College and the Harvard University Extension School.

One of my primary objectives in preparing this revised edition was to incorporate new developments in the field. In the legislative arena, these included passage of the Food Quality Protection Act, which, among other things, repealed the Delaney Clause (Chapter 6), and passage of the Safe Drinking Water Act Amendments (Chapter 7), which require that drinking-water supplies be analyzed for a wide range of specific contaminants.

As with the first edition, my goal has been to write a book that provides comprehensive coverage of the field. In seeking to achieve this, I have taken care to present topics from both local and global perspectives, and in relation to both short- and long-range impacts. At the same time I have sought to provide perspective, as, for example, in summarizing data on the major causes of cancer (Chapter 1). As will be noted, diet and personal living habits (most importantly, the use of tobacco products) are estimated to be the sources of about 65 percent of current cancer cases. This demonstrates that many of the factors affecting our health are within our control; they do not necessarily arise through noncontrollable sources such as environmental pollutants.

Also incorporated into the revised edition are discussions of a number of emerging and/or controversial issues in environmental and public health. These range from considerations of environmental justice, deforestation, and the protection of endangered species (Chapter 19) to topics such as multiple chemical sensitivity (Chapter 2), the applicability of the threshold concept in evaluating the effects of toxic chemicals (Chapter 4) and radiation (Chapter 12), and the uncertainties in extrapolating laboratory data obtained through studies with animals, such as mice, to estimate related potential health effects in humans (Chapter 2).

Care has also been taken to ensure that the reader understands the limitations associated with techniques, such as epidemiology (Chapter 3) and risk assessment (Chapter 16), that are commonly applied in evaluating the impacts of various environmental stresses. In the discussions on epidemiology, for example, I point out that though this technique can be used to show an association between a given environmental stress and a specific health effect, it cannot be used to demonstrate causality.

Another feature of the revised edition is the effort to ensure that the reader understands the differences among clinical medicine, public health, and environmental health (Chapter 1). Equally important is the emphasis on using a total systems approach in assessing environmental problems. Although all of us recognize the need to manage and control various pollutants within individual segments of the environment (air, water, food), we also must understand and take into account potential interrelationships of these segments, one with another. Within this context, care has been taken to ensure that the reader is aware of the need to protect both humans and our natural resources. This is exemplified by a review of the concept of ecological risk assessment, and the discussion of primary and secondary standards for airborne contaminants—primary to protect the health of people; secondary to protect the environment. It is also exemplified by the discussion of acid precipitation (Chapter 5) and ozone depletion and global warming (Chapter 19).

As would be expected for an undertaking of this magnitude, I am grateful to a host of fellow environmental and public health professionals for sharing their talents and expertise with me. Special thanks are due my former colleagues at the Harvard School of Public Health, William A. Burgess, Melvin W. First, John D. Graham, David Hemenway, John B. Little, Richard R. Monson, Jacob Shapiro, Robert Schlegel, Andrew Spielman, and Jay A. Winsten. Other associates who provided invaluable support include John B. Garrick, William E. Kennedy, Matthew P. Moeller, Paul M. Newberne, and Cynthia Palmer. I also want to express my appreciation to Janet Francoeur, who prepared many of the figures used in the book. Finally, I deeply appreciate the editorial suggestions of Vivian Wheeler and Christine Thorsteinsson at Harvard University Press.

And God pronounced a blessing upon Noah and his sons and said to them, be fruitful and multiply and fill the earth.

And the fear of you and the dread and terror of you shall be upon every beast of the land, every bird of the air, all that creeps upon the ground, and upon all the fishes of the sea. Into your hands they are delivered.

Genesis 9:1–2

ABBREVIATIONS

AAEE	American Academy of Environmental Engineers
ACGIH	American Conference of Governmental Industrial Hygienists
AEA	Atomic Energy Act
AIDS	Acquired Immune Deficiency Syndrome
AIHA	American Industrial Hygiene Association
ALARA	As Low As Reasonably Achievable
ALI	Annual Limit on Intake
AMA	American Medical Association
ASME	American Society of Mechanical Engineers
ATSDR	Agency for Toxic Substances and Disease Registry, U.S. Department of Health and Human Services
BACT	Best Available Control Technology
BEIs	Biological Exposure Indices
BEIR	Committee on the Biological Effects of Ionizing Radiation, National Research Council
BOD	Biochemical Oxygen Demand
BST	Bovine Somatotropin
BTI	*Bacillus thuringiensis israeliensis*
BTK	*Bacillus thuringiensis kurstaki*
BWR	Boiling-Water Reactor
CDC	Centers for Disease Control and Prevention, U.S. Department of Health and Human Services
CEQ	Council on Environmental Quality
CERCLA	Comprehensive Environmental Response, Compensation, and Liability Act (Superfund)
CFC	Chlorofluorocarbon
CO	Carbon Monoxide
$C0_2$	Carbon Dioxide
COD	Chemical Oxygen Demand

CRCPD	Conference of Radiation Control Program Directors
DAC	Derived Air Concentration
DDT	Dichlorodiphenyltrichloroethane
DEET	Diethyltoluamide
DNA	Deoxyribonucleic Acid
DO	Dissolved Oxygen
DOE	U.S. Department of Energy
EIS	Environmental Impact Statement
EMAP	Environmental Monitoring and Assessment Program
EPA	U.S. Environmental Protection Agency
EPRI	Electric Power Research Institute
eV	Electron Volt
FDA	Food and Drug Administration, U.S. Department of Health and Human Services
FEMA	Federal Emergency Management Agency
FIFRA	Federal Insecticide, Fungicide, and Rodenticide Act
GI Tract	Gastrointestinal Tract
GRAS	Generally Recognized As Safe
HHS	U.S. Department of Health and Human Services
HVAC	Heating, Ventilating, and Air Conditioning
Hz	Hertz (cycles per second)
IAEA	International Atomic Energy Agency
ICNIRP	International Commission on Non-Ionizing Radiation Protection
ICRP	International Commission on Radiological Protection
IIHS	Insurance Institute for Highway Safety
INPO	Institute of Nuclear Power Operations
IRIS	Integrated Risk Information System
IRPA	International Radiation Protection Association
IVHS	Intelligent Vehicle Highway Systems
LASER	Light Amplification by Stimulated Emission of Radiation
LD_{50}	Lethal Dose for 50 percent of the exposed population
LLRW	Low-Level Radioactive Waste
MCL	Maximum Contaminant Level
MRS Facility	Monitored Retrievable Storage Facility
MTD	Maximum Tolerated Dose
NAAQS	National Ambient Air Quality Standards
NAFTA	North American Free Trade Agreement
NASA	National Aeronautics and Space Administration
NCRP	National Council on Radiation Protection and Measurements
NEPA	National Environmental Policy Act
NHEXAS	National Human Exposure Assessment Survey
NIOSH	National Institute for Occupational Safety and Health, U.S. Department of Health and Human Services
NO_2	Nitrogen Dioxide
NPDES	National Pollution Discharge Elimination System
NPL	National Priorities List
NRC	National Research Council

NRPB	National Radiological Protection Board (UK)
NSC	National Safety Council
OECD	Organization for Economic Cooperation and Development
OSHA	Occupational Safety and Health Administration, U.S. Department of Labor
OTA	Office of Technology Assessment, U.S. Congress
PAHO	Pan American Health Organization
PC	Personal Computer
PCB	Polychlorinated Biphenyls
PST	Porcine Somatotropin
PWR	Pressurized-Water Reactor
RACT	Reasonably Available Control Technology
RCRA	Resource Conservation and Recovery Act
SARA	Superfund Amendments and Reauthorization Act
SO_2	Sulfur Dioxide
SUMA	Program Supply Management Program
TLVs	Threshold Limit Values
TSCA	Toxic Substances Control Act
UNESCO	United Nations Educational, Scientific, and Cultural Organization
USDA	U.S. Department of Agriculture
USNRC	U.S. Nuclear Regulatory Commission
UV Radiation	Ultraviolet Radiation
WHO	World Health Organization

1

THE SCOPE

MANY aspects of human well-being are influenced by the environment, and many diseases can be initiated, promoted, sustained, or stimulated by environmental factors. For that reason, the interactions of people with their environment are an important component of public health.

In its broadest sense, environmental health is the segment of public health that is concerned with assessing, understanding, and controlling the impacts of people on their environment and the impacts of the environment on them. Still, environmental health is defined more by the problems faced than by the approaches used. These problems include the treatment and disposal of liquid and airborne wastes, the elimination or reduction of stresses in the workplace, purification of drinking-water supplies, the impacts of overpopulation and inadequate or unsafe food supplies, and the development and use of measures to protect hospital and medical workers from being infected with diseases such as acquired immune deficiency syndrome (AIDS). Environmental health professionals also face long-range problems that include the effects of toxic chemicals and radioactive wastes, acidic deposition, depletion of the ozone layer, global warming, resource depletion, and loss of forests and topsoil. The complexity of these issues requires multidisciplinary approaches. Thus a team coping with a major environmental health problem may include scientists, physicians, epidemiologists, engineers, economists, lawyers, mathematicians, and managers. Input from all these experts is essential to the development and success of broad strategies that take into account both lifestyles and the environment.

Just as the field of public health involves more than disease (for example, health care management, maternal and child health, epidemiology), the field of environmental health encompasses the effects of the environment on animals other than humans, as well as on trees and vegetation and on natural and historic landmarks. While many aspects of public health deal with the "here and now," many of the topics addressed within the subspecialty of environmental health are concerned with the previously cited impacts of a long-range nature.

Defining the Environment

To accomplish their goals effectively, environmental health professionals must keep in mind that there are many ways to define the environment. Although no single definition is without its deficiencies, each offers benefits in terms of perspective and understanding.

The inner versus outer environment. From the standpoint of the human body, there are two environments: the one within the body and the one outside it. Separating them are three principal protective barriers: the skin,

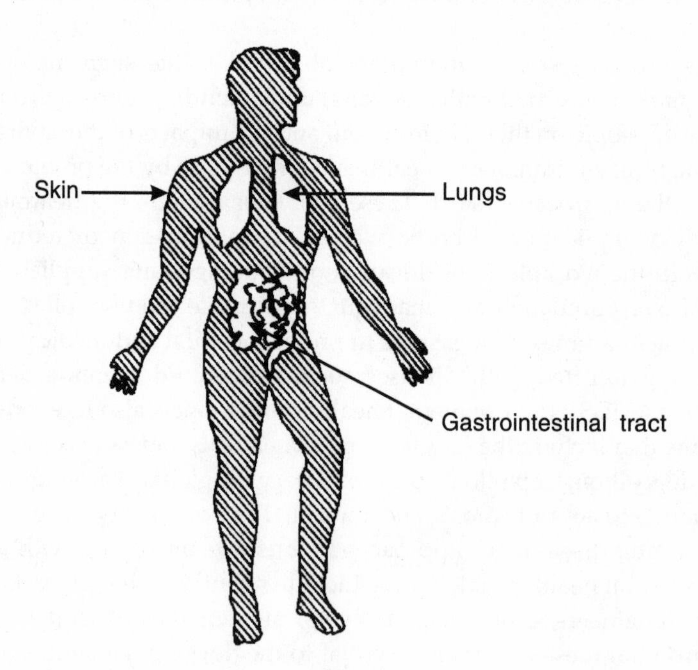

Figure 1.1 Barriers between the inner and outer environments

Table 1.1 Characteristics of the principal barriers between the outer and inner body

	Area		Thickness		Weight		Daily exposure	
Barrier	ft^2	m^2	in	μm	lb	kg	lb	kg
Skin	21	2	4×10^{-3}	100	30	12–16	Variable	
GI tract	2,150	200	4×10^{-4}	10–12	15	7	4–6	2–3
Lungs	1,500	140	1×10^{-5}	0.2–0.4	2	0.8–0.9	50	24

which protects the body from contaminants outside the body; the gastrointestinal (GI) tract, which protects the inner body from contaminants that have been ingested; and the membranes within the lungs, which protect the inner body from contaminants that have been inhaled (Figure 1.1, Table 1.1).

Although they may provide protection, each of these barriers is vulnerable under certain conditions. Contaminants can penetrate to the inner body through the skin by dissolving the layer of wax generated by the sebaceous glands. The GI tract, which has by far the largest surface area of any of the three barriers, is particularly vulnerable to compounds that are soluble and can be readily absorbed and taken into the body cells. Fortunately, the body has mechanisms that can protect the GI tract: unwanted material can be vomited via the mouth or rapidly excreted through the bowels (as in the case of diarrhea). Airborne materials in the respirable size range may be deposited in the lungs and, if they are soluble, may be absorbed. Mechanisms for protecting the lungs range from simple coughing to cleansing by macrophages that engulf and promote the removal of foreign materials. Unless an environmental contaminant penetrates one of the three barriers, it will not gain access to the inner body. And even if a contaminant is successful in gaining access, the body still has mechanisms for removing it. For example, materials entering the circulatory system can be detoxified in the liver or excreted through the kidneys.

Although an average adult ingests about 1.5 kilograms of food and 2 kilograms of water every day, he or she breathes roughly 20 cubic meters of air per day. This amount of air weighs more than 24 kilograms. Because people usually cannot be selective about what air is available, the lungs are the most important pathway for the intake of environmental contaminants into the body. The lungs are also by far the most fragile and susceptible of the three principal barriers.

The personal versus ambient environment. In another definition, people's "personal" environment, the one over which they have control, is contrasted with the working or ambient environment, over which they may have essentially no control. Although people commonly think of the working or ambient environment as posing the greater threat, environmental health experts estimate that the personal environment, influenced by hygiene, diet, sexual practices, exercise, use of tobacco, drugs, and alcohol, and frequency of medical checkups, often has much more influence on well-being.

Table 1.2 summarizes the estimated contributions of these various factors to cancer deaths in an industrialized society. As may be noted, the personal environment is seen as accounting for 75 percent or more of such deaths. Cigarette smoking leads to increased deaths not only from lung cancer but

Table 1.2 Proportion of cancer deaths attributable to various factors, England and Wales, 1995

	Percentage of all cancer deaths	
Agent or class of agents	Best estimate	Range of estimates
Diet	35	20–60
Tobacco	31	29–33
Natural hormones	15	10–20
Infections	10	5–15
Electromagnetic radiation		
Ionizing 4.5		
Ultraviolet 2.5	8	5–10
Lower frequency <1		
Alcoholic beverages	5	3–7
Occupational exposure	3	2–6
Environmental pollution	2	<1–4
Medicines and medical procedures	1	0.5–2
Industrial products	<1	<1–4
Other	?	?

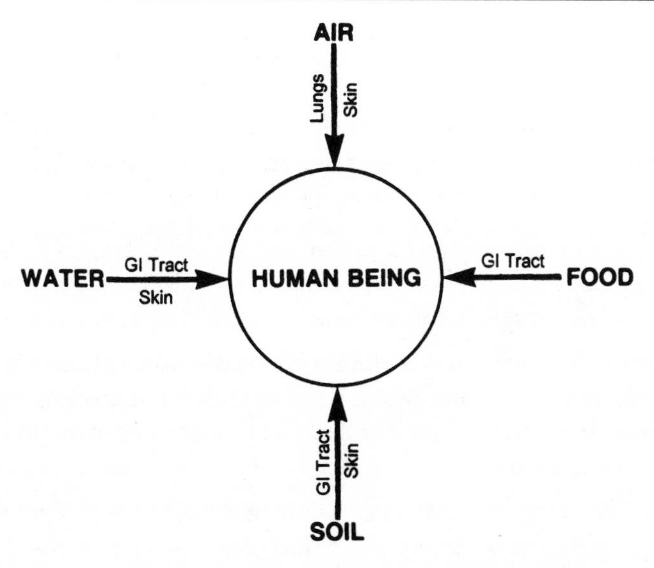

Figure 1.2 Routes of human exposure through the gaseous, liquid, and solid environments

also from heart disease. As a result, this single factor is estimated to account for 15 to 20 percent of all deaths in the United States (Surgeon General, 1989). The Centers for Disease Control and Prevention estimate that cigarette smoking is responsible annually for more than 400,000 deaths nationwide. The associated medical-care costs for 1993 were estimated at $50 billion, more than 40 percent of the total annual medical-care expenditures (Anonymous, 1994).

The amount of pollution taken into a smoker's lungs as a result of inhaling the various products from cigarettes is several orders of magnitude greater than the amount normally inhaled due to industrial airborne pollution. Unless it is controlled, cigarette smoke can account for a significant fraction of the fine-particle content of air inside buildings. In fact, it accounts for more than 1 percent of the fine-particle content of the outdoor air in Los Angeles (*New York Times*, 1994).

The gaseous, liquid, and solid environments. The environment can also be considered as existing in one of three forms—gaseous, liquid, or solid. Each of these is subject to pollution, and people interact with all of them (Figure 1.2). Particulates and gases are released into the atmosphere, sewage and liquid wastes are discharged into water, and solid wastes, particularly plastics and toxic chemicals, are disposed of on land.

The chemical, biological, physical, and socioeconomic environments. Another perspective considers the environment in terms of the four avenues or mechanisms by which various factors affect people's health.

1. *Chemical* constituents and contaminants include toxic wastes and pesticides in the general environment, chemicals used in the home and in industrial operations, and preservatives used in foods.
2. *Biological* contaminants include various disease organisms that may be present in food and water, those that can be transmitted by insects and animals, and those that can be transmitted by person-to-person contact.
3. *Physical* factors that influence health and well-being range from injuries and deaths occurring as a result of accidents, to excessive noise, heat, and cold, to the harmful effects of ionizing and nonionizing radiation.
4. *Socioeconomic* factors, though perhaps more difficult to measure and evaluate, significantly affect people's lives and health (Graham, Chang, and Evans, 1992). Statistics demonstrate compelling relationships between morbidity and mortality and socioeconomic status. People who live in economically depressed neighborhoods are less healthy than those who live in more affluent areas.

Clearly, illness and well-being are the products of community, as well as of chemical, biological, and physical, forces. Factors contributing to the differences range from the unavailability of jobs, inadequate nutrition, and lack of medical care to stressful social conditions, such as substandard housing and high crime rates.

The contributing factors, however, extend far beyond socioeconomics. Studies have shown that those without political power, especially disadvantaged groups who live in lower-income neighborhoods, often bear a disproportionate share of the risks of environmental pollution. One common example is increased air and water pollution due to nearby industrial and toxic waste facilities. In many cases it appears that personnel in various governmental agencies, including those at the federal level, have intentionally selected lower-income communities as sites for the more hazardous types of industrial operations, including waste disposal facilities and waste incinerators (Easterling, 1994). Disadvantaged groups also suffer more frequent exposure to lead paint in their homes and to pesticides and industrial chemicals in their work.

Recognizing these and related problems, the President, on 11 February 1994, signed an Executive Order on environmental justice, with the goal of ending this form of environmental inequity and discrimination (Clinton, 1994). Included among the objectives was a reaffirmation that all communities and individuals, regardless of economic status or racial makeup, are entitled to a safe and healthful environment and that, in the future, the risks associated with hazardous industrial facilities will be distributed equitably across population groups. As part of the siting of any potentially hazardous operation, regulators will be required to identify and critically examine all potentially adverse impacts on the health and environment of minority and low-income populations. The Order also required that disadvantaged populations have an opportunity to participate fully in decisions that affect their health and environment.

None of the above definitions of the environment is without its deficiencies. Classification in terms of inner and outer environments, or in terms of gaseous, liquid, and solid environments, for example, fails to take into account the significant socioeconomic factors cited above, or physical factors such as noise and ionizing and nonionizing radiation. Consideration of the full range of existing environments is essential to understanding the complexities involved and to controlling the associated problems.

Assessing the Problems

In the course of their work, medical and public health personnel have achieved remarkable success in decreasing human morbidity and mortality. One major benefit during the past century has been a significant increase in the average human life span. One important consequence has been a dramatic growth in the world's population. This, in turn, has led to environmental degradation and accompanying threats to human health, and has clearly demonstrated that the environment has a limited capacity to support life (Morris and Hendee, 1992). Fortunately, this problem is being recognized and addressed, as exemplified by the third United Nations International Conference on Population and Development, held in Cairo in September 1994.

A large share of the social, economic, and environmental decline in many parts of the world today results from increased production of materials and wastes, greater consumption of resources as a result of the expanding expectations of a growing population, and the use of ever more sophisticated technologies to satisfy continually increasing demands for goods and services. Many of these practices have global ecological effects, and the combination of local and global effects will inevitably affect human health.

One of the primary goals of environmental health professionals is to understand the various ways in which humans interact with their environment. A primary step is to study the process or operation that leads to the generation of an environmental problem and to determine how best to achieve control. Components of such an analysis include (1) determining the source and nature of each environmental contaminant or stress; (2) assessing how and in what form that contaminant comes into contact with people; (3) measuring the resulting effects; and (4) applying controls when and where appropriate. Instead of focusing on air pollution or water pollution facility by facility, environmental health professionals should gather data on all the discharges from a given facility, all the sources of a given pollutant, and all the pollutants being deposited in a region regardless of their nature, origin, or pathway (Train, 1990).

Even though tracing the source and path of a contaminant is important, an essential part of the process is to determine the effects on human health. Working with an interdisciplinary team, environmental health professionals must establish quantitative relationships between the exposure, the resulting dose, and its effects. On the basis of such data standards can be recommended for acceptable limits of exposure to the contaminant or stress.

To assess the effects of exposures correctly, environmental health workers must take into account not only the fact that exposures can derive from multiple sources and enter the body by several routes, but also that elements in the environment are constantly interacting. In the course of transport or degradation, agents that were not originally toxic to people may become so, and vice versa. If the concentration of a contaminant in the ambient (outdoor) environment (for example, a substance in the air) is relatively uniform, local or regional sampling may yield data adequate to estimate human exposure. If concentrations vary considerably over space and time (as is true of certain indoor pollutants) and the people being exposed move about extensively, it may be necessary to measure exposure of individual workers or members of the public by providing them with small, lightweight, battery-operated portable monitoring units. Development of such monitors and the specifications for their use requires the expertise of air pollution engineers, industrial hygienists, chemists and chemical engineers, electronics experts, and quality control personnel. Once the levels of exposure are known, they can be compared to existing standards, and controls can be applied when and where warranted.

At the same time, environmental health professionals must recognize that advances in technology have produced highly sophisticated and sensitive analytical instruments that can measure many environmental contaminants at concentrations below those that have been demonstrated to cause harm to health or the environment. For example, techniques capable of measuring contaminants in parts per billion are common. The mere act of measuring and reporting the presence of certain contaminants in the environment often leads to concern on the part of the public, even though the reported levels may be well within the acceptable range. The accompanying fears, justified or not, can lead to expenditures on the control of environmental contaminants instead of on other, more urgent problems. Those responsible for protecting people's health must be wary of demands for "zero" pollution: it is neither realistic nor achievable as a goal in today's world. Rather, given the host of factors that are an integral part of our daily lives, the goal should be an optimal level of human and environmental well-being.

The Systems Approach

Attempts to control pollution in one segment of the environment can often result in the transfer to or creation of a different form of pollution. Such interactions can be immediate or they can take place over time; they can occur in the same general locality or at some distance. On a short-term basis, the incineration of solid wastes can cause atmospheric pollution; the application of scrubbers and other types of air-cleaning systems to airborne effluents can produce large amounts of solid wastes; and the chemical treatment of liquid wastes can produce large quantities of sludge. On a longer-term basis, the discharge of sulfur and nitrogen oxides into the atmosphere can result in acidic deposition at some distance from the point of release; the discharge of chlorofluorocarbons can lead to the destruction of the ozone layer in the upper atmosphere; and the discharge of carbon dioxide can lead to global warming.

At the same time, it must be recognized that many uses of chemicals have brought major benefits to humankind. The chlorination of drinking water, for example, has led to significant reduction in the rates of many infectious diseases. In a similar manner, the use of chlorofluorocarbons has led to low-cost refrigeration and longer-term storage and transportation of milk and food, as well as of vaccines and antibodies (Train, 1990). In other cases, however, widespread and indiscriminant uses of chemicals, most notably as insecticides and pesticides and in various types of industrial operations, have led to a global legacy of enormous chemical contamination (Canadian Public Health Association, 1992). Unless environmental health professionals recognize the severity and widespread nature of these problems, attempts to deal with them will be inadequate, piecemeal, and destined to fail.

Clearly, what is done to the environment in one place will almost certainly affect it elsewhere. A systems approach ensures that each problem is examined not in isolation, but in terms of how it interacts with and affects other segments of the environment and our daily lives.

Intervention and Control

Because the complexity of the problems in environmental health requires multidisciplinary approaches to their evaluation and control, the techniques for addressing environmental problems often differ from those applied in medical practice. Physicians traditionally deal with one patient at a time, whereas environmental health specialists must consider entire populations. To the extent possible, they must also try to anticipate problems to prevent them from developing. As depicted in the clinical intervention model (Figure 1.3a), the goal of the physician is to prevent a specific disease from leading to death (Morris and Hendee, 1992). The public health intervention model (Figure 1.3b), in contrast, calls for preventing the development of disease. Far superior to either is the environmental stewardship model (Figure 1.3c), where the goal is to protect humans by preventing environmental degradation and its resulting impacts on health.

Even after a problem is understood, environmental health personnel need strong support from other groups if their goals are to be achieved. A prime necessity is the assurance of legislators that the requisite laws and regulations, as well as financial resources, are available. Public health educators need to ensure that the public participates in development of the programs, and that the associated regulations and requirements are fully understood by the industrial organizations and other groups who are expected to comply. Also needed is the input of program planners and economists to assure that available funds, invariably limited in quantity, are spent

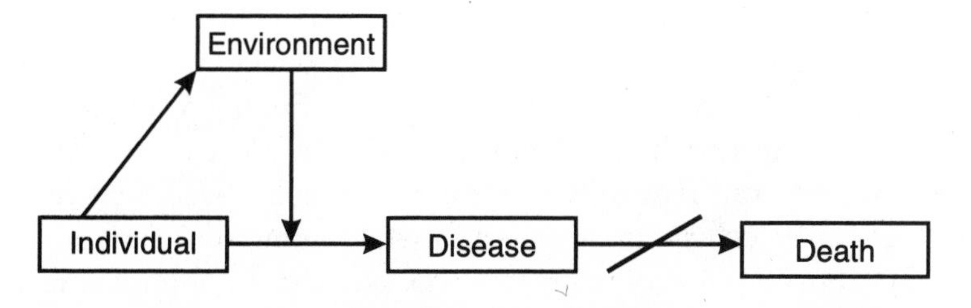

a. Clinical intervention model

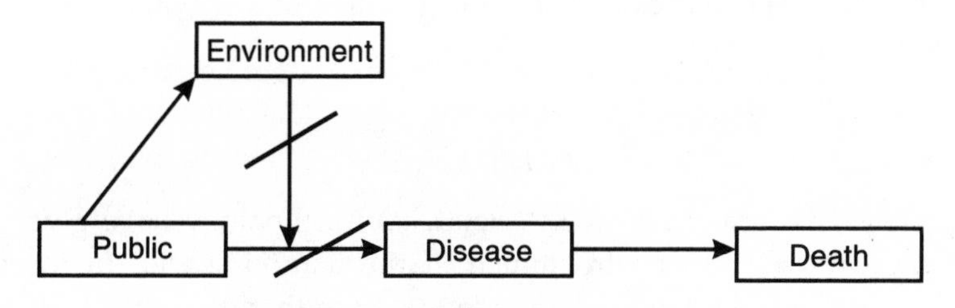

b. Public health intervention model

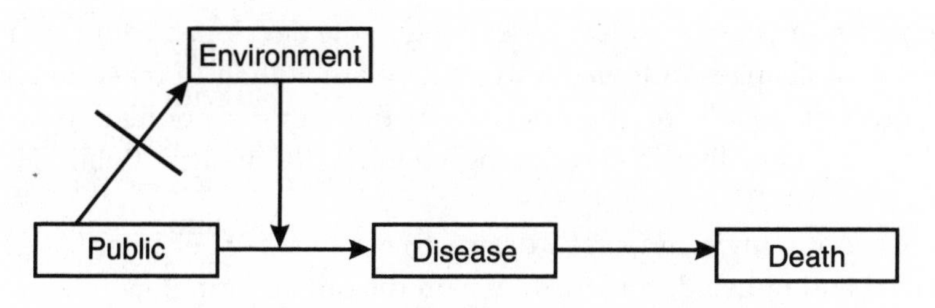

c. Environmental stewardship model

Figure 1.3 Various models for improving the state of human health and the environment

in the most effective manner. Far too often, decisions on where and how monies should be spent to improve the environment are based on emotions, not science (Swogger, 1992). With current programs on environmental protection in this country necessitating expenditures of more than $100 billion annually, it is imperative that the funds be directed to the most pressing situations.

Regardless of how competent they may be, environmental health professionals cannot be expected to solve these problems alone. In the long run, the strong commitment and support of individual members of the public is essential to success. A prime example is the personal environment, whose control is largely dependent on individual action. But the responsibilities of individuals do not end here. Members of society must be constantly reminded that they can reduce the production of solid wastes by recycling newspapers, plastics, glass bottles, and metal (aluminum) cans. They can reduce the consumption of energy by car pooling, by minimizing home heating and cooling costs through the installation of storm windows and other weatherproofing measures, and by conserving water through the use of low-flow showerheads and the installation of low-water-consuming flush toilets.

The General Outlook

Fortunately, advances in modern science and technology have given humans the capability to control much of the natural world. In the end, however, choices will have to be made to assure that the controls that are applied result in an optimal level of health for both the environment and the public. The overall goal should be to achieve the greatest good for the maximum number of people. Those living in the developed countries must decide what changes in their lifestyles they are willing to make to ensure that "greatest good" for the majority of the world's population, a vast number of whom live in developing countries (Canadian Public Health Association, 1992).

Once specific environmentally destructive patterns of behavior have been identified and targeted, constructive patterns can be formulated as alternatives. The goal can best be described as application of the principles of *environmental stewardship* (Morris and Hendee, 1992) and *global bioethics* (Potter, 1992). The overall objective should be to achieve both sustainable development and a sustainable environment (Toman and Darmstadter, 1996). This will involve many types of tradeoffs and, in some instances, could well entail the exchange of one set of environmental problems for another.

The concept of a sustainable environment is based on the premise that renewable resources should only be used at a rate that ensures their continued existence (sustained yield); nonrenewable resources should be used sparingly and recycled wherever possible (conservation); and natural systems should not be polluted to the point where they are no longer able to cope with the resulting damage (pollution prevention). As defined by the

World Commission on Environment and Development, sustainable development "meets the need of the present without compromising the ability of future generations to meet their own needs" (Canadian Public Health Association, 1992).

The problems of the environment are enormous. Solutions will require the cooperation of government, industry, and commerce, as well as the concern and dedication of individuals throughout the world.

2

TOXICOLOGY

It has been estimated that, worldwide, approximately 70,000 chemicals are in common use, and the chemical industry markets an additional 200–1,000 new synthetic chemicals each year (Stone, 1993b). As a result, humans are exposed to a host of chemicals in the home, in the workplace, and in the general environment. Trace quantities of toxic chemicals are present in our food, our air, and our drinking water.

To ensure that the public is being adequately protected, environmental and public health officials need constant information on the biological effects (or toxicity) of a multitude of chemicals. One source of such information is provided through epidemiological studies of human populations known to be exposed to certain agents. But this kind of research is not easy to conduct, the data are difficult to interpret, and the results are available only after the exposures and effects have occurred. Therefore, such studies are not preventive or predictive in nature. A further difficulty is accurate estimation of the level of exposure that caused the effects, and determination of whether the suspected exposures were accompanied by exposures to other toxic agents.

Faced with these challenges and the need to estimate the health effects of a wide range of toxic agents, scientists have increasingly turned to laboratory studies of animals. Those who conduct such studies are referred to as toxicologists, and their efforts involve both science and art. The science lies in the observational or data-gathering aspects, and the art is in the projection of these data to situations where there is little or no information (Doull and Bruce, 1986). When the evaluations are directed at the presence of chemicals in the environment, the situation is far more complicated. Here the environmental health professional must expand the work of the toxicologist, who traditionally deals with the effects of a single chemical in a single animal species, to include assessment of the effects, both direct and indirect, of combinations of chemicals on total ecosystems. This is what is known as *environmental toxicology.*

Pathways of Exposure and Excretion

Although protection of other species is important, the initial discussion here will be of the impact of toxic chemicals on humans. From the standpoint of occupational and environmental health, the major routes of intake are the lungs (inhalation), the gastrointestinal tract (ingestion), and the skin (absorption). In the case of the respiratory tract, the primary site of uptake is through the alveoli in the lungs—especially for gases such as carbon monoxide, nitrogen oxides, and sulfur dioxide and for vapors of volatile liquids such as benzene and carbon tetrachloride. Their ready absorption is related to the large alveolar area, high blood flow, and proximity of the blood to the aveolar air. Liquid aerosols and airborne particles may also be absorbed through the lungs. In contrast, the deposition of airborne particles is heavily influenced by their size, the particles of primary interest being those in the size range 0.001–0.1 micrometer.

Once a chemical is absorbed, the nature and intensity of its effects depend on its concentration in the target organs, its chemical and physical form, what happens to it after it is absorbed, and how long it remains in the tissue or organ in question (following the central tenet that "the dose makes the poison"). After being taken up in the blood, a toxic chemical will be rapidly distributed throughout the body. As part of this process, the chemical may be translocated from one organ or tissue to another, and it may be converted into a new compound or metabolite. This process is known as biological transformation. Metabolic processes in the cytoplasm, for example, can alter toxic substances through various chemical reactions, including oxidation and reduction. In the main, these reactions tend to result in new products that are less absorbable and more polar (charged) chemically, and thus are more readily excreted in the urine. The removal of toxic chemicals from the body is thereby enhanced. In certain cases the new product or metabolite may be more toxic than the parent compound; such reactions are known as bioactivation (Lu, 1991). In most cases, however, the newly formed compounds tend to be less toxic (Smith, 1992).

The principal means of excretion of chemicals from the human body is the urine, but the liver (via reabsorption from the bile into the blood and excretion through the bowels) and the lungs (via various clearance mechanisms described in Chapter 5) are also important excretory organs for certain types of chemicals. Among the less significant routes of excretion are the sweat glands. In general, the GI tract is not a major route of excretion of toxicants (Lu, 1991).

Toxic chemicals may cause injuries at the site of first contact, or they may be absorbed and distributed to other parts of the body where they exhibit their effects. Those effects may be reversible (that is, appear to cease after exposure terminates) or they may be considered irreversible (that is, continue long after exposure concludes). In general, reversible effects are observed for short-term exposures at low concentrations; irreversible effects are more commonly observed following long-term exposures at higher concentrations. Toxic agents may also produce either immediate or delayed effects. A notable example of the latter is carcinogenesis; many cancers do not appear in humans until a decade or more after exposure to a toxic agent. The effects of a toxic agent may be influenced by previous sensitization of the exposed person to the same or a similar chemical. Such effects are often classified as allergic reactions (Lu, 1991).

Other factors that can modify the response of animals to toxic chemicals include the species and strain of animal being affected, its age and sex, and its nutritional and hormonal status. Because young animals have less effective mechanisms for biotransforming and detoxifying certain chemicals, they may be more susceptible to certain toxic agents. In similar manner, people with diseases of the liver, which is a major detoxifying and biotransforming organ within the body, are more susceptible to a variety of chemicals.

Physical factors too can alter the effects of toxic chemicals. For example, a rise in ambient temperature will increase the toxicity of dinitrophenol, occasionally used as a herbicide, in adult male workers. Usually, however, the duration of the response will be shorter when the temperature is higher, apparently because of the temperature-dependent biochemical reactions responsible for biotransformation of the chemical. Alterations in response have also been observed with humidity, higher levels tending to increase the acute toxicity of certain chemicals. The effects of toxic chemicals also show a diurnal pattern that is mainly related to the light cycle. Social factors also can affect toxicity. Those that have been shown to be important, particularly in laboratory testing, include the types of cages in which the animals are kept, whether they are housed singly or in groups. and the bedding materials provided.

Table 2.1 Approximate concentrations of various chemicals required to produce death in 50 percent of exposed animals

Chemical	LD_{50} (mg/kg of body weight)
Ethyl alcohol	10,000
Sodium chloride	4,000
Ferrous sulfate	1,500
Morphine sulfate	900
Phenobarbital sodium	150
Picrotoxin	5
Strychnine sulfate	2
Nicotine	1
d-Tubocurarine	0.5
Hemicholinium-3	0.2
Tetrodotoxin	0.10
Dioxin (TCDD)	0.001
Botulinum toxin	0.00001

Source: C.D. Klaasen, "Principles of Toxicology," in C.D. Klaasen, Mary O. Amdur, and J. Doull, *Casarett and Doull's Toxicology: The Basic Science of Poisons,* 3rd ed. (New York: MacMillan Publishing Company, 1986), table 2-1, p. 12. Adapted with permission of The McGraw-Hill Companies.

Individual chemicals vary widely in their toxicity. Some, such as botulism toxin, produce death in humans at concentrations of only nanograms per kilogram (10^{-9} gram per kilogram) of body weight. Others, such as ethyl alcohol, may have relatively little effect even after doses of several grams per kilogram (Table 2.1). Data of this sort are often used to rank chemicals according to their toxicity (Table 2.2). Under this categorization, botulism toxin would be classified as supertoxic, whereas ethyl alcohol would be classified as slightly toxic. Although primarily qualitative, such a classification scheme serves a useful purpose in providing laypeople with answers to the question, How toxic is this chemical? (Klaassen, 1986). Toxic chemicals can also be classified in terms of their target organ (liver, kidney), their use (pesticide, food additive), their source (animal or plant toxin), and their effects (cancer, mutations).

The presence of toxic chemicals in various media within the environment, and their uptake by different species can lead to a variety of interesting situations. As a result of biological and chemical processes, the concentrations of certain chemicals in aquatic species will be much higher than those in the water in which they live. The concentrations of mercury in

Table 2.2 Toxicity ratings

	Probable lethal dose for humans	
Toxicity rating	Dosage	For average adult
Practically nontoxic	>15 g/kg	More than 1 quart
Slightly toxic	5–15 g/kg	Between 1 pint and 1 quart
Moderately toxic	0.5–5 g/kg	Between 1 ounce and 1 pint
Very toxic	50–500 mg/kg	Between 1 teaspoon and 1 ounce
Extremely toxic	5–50 mg/kg	Between 7 drops and 1 teaspoon
Supertoxic	<5 mg/kg	A taste (less than 7 drops)

Source: C.D. Klaasen, "Principles of Toxicology," in C.D. Klaasen, Mary O. Amdur, and J. Doull, *Casarett and Doull's Toxicology: The Basic Science of Poisons,* 3rd ed. (New York: Macmillan Publishing Company, 1986), table 2–2, p. 13. Adapted with permission of The McGraw-Hill Companies.

plankton, for example, will be higher than in the surrounding water, and the concentrations in small fish and larger fish will be higher still. The concentrations in birds that feed on the fish will be even higher, perhaps by as much as several hundredfold. This phenomenon is known as biological magnification (Moriarty, 1988). Such magnification through the food chain led, for example, to the harmful effects of DDT on pelicans, via a thinning of the shells of their eggs. In similar manner DDT will concentrate in a human mother's milk to the extent that her baby's intake of this pesticide per unit of body weight may be more than 20 times that in the mother's diet.

Conventional Tests for Toxicity

The effects of toxic chemicals on animals may range from rapid death to sublethal effects to situations in which there are apparently no effects at all. Often the first step in the prediction of effects is to conduct a series of laboratory studies involving a single chemical and a single animal species. Generally, the animals are exposed to a range of doses and/or concentrations and over different periods of time. Because of legal and ethical limitations, most such studies are conducted using rats or mice rather than humans. To examine the effects associated with exposure over various time periods, toxicological studies have generally been divided into three categories (Lu, 1991):

Acute toxicity studies—either a single administration of the chemical being tested or several administrations within a 24-hour period;

Short-term (also known as subacute and subchronic) toxicity studies—repeated administrations, usually on a daily basis, over a period of about 10 percent of the life span of the animal being tested (for example, about three months in rats and one to two years in dogs); however, shorter durations such as 14-day and 28-day treatments have also been used by some investigators;

Long-term toxicity studies—repeated administrations over the entire life span of the test animals (or at least a major fraction thereof). For mice, the time period would be about 18 months; for rats, 24 months; for dogs and monkeys, 7 to 10 years.

ACUTE TOXICITY STUDIES

Analyses of the data derived from acute toxicity studies generally begin with the plotting of a curve that shows the relationship between the dose or concentration of a toxic chemical and the number or percentage of test animals that demonstrate an effect. Such curves often exhibit the distribution of sensitivities shown in Figure 2.1.

This type of curve is representative of that observed for a large number of variables, including death, change in body weight, or size of the animals at a given age. Where death is used as an endpoint, such tests fall into the category of acute toxicity studies. The peak of the curve indicates the dose that produces effects in 50 percent of the animals. Either an increase or a decrease in dose will result in proportionately fewer animals responding. This is another way of saying that some animals will exhibit the response at a lower dose, whereas for other animals higher doses are required to demonstrate the same effect. The portion of the curve between "minimum" and point *B* represents the response of the most susceptible animals; the portion between *B′* and "maximum" represents the response of the most resistant animals. Since the curve follows a normal or Gaussian distribution, statistical procedures can be used to evaluate the resulting data (Loomis, 1968).

Although the Gaussian distribution is interesting, data resulting from toxicological studies are generally plotted in the form of a curve relating the dose or concentration to the *cumulative* percentage of animals exhibiting the given response. The curves in Figure 2.2 show this type of plot for two different chemicals, A and B.

The curve to the left represents the more toxic of the two compounds,

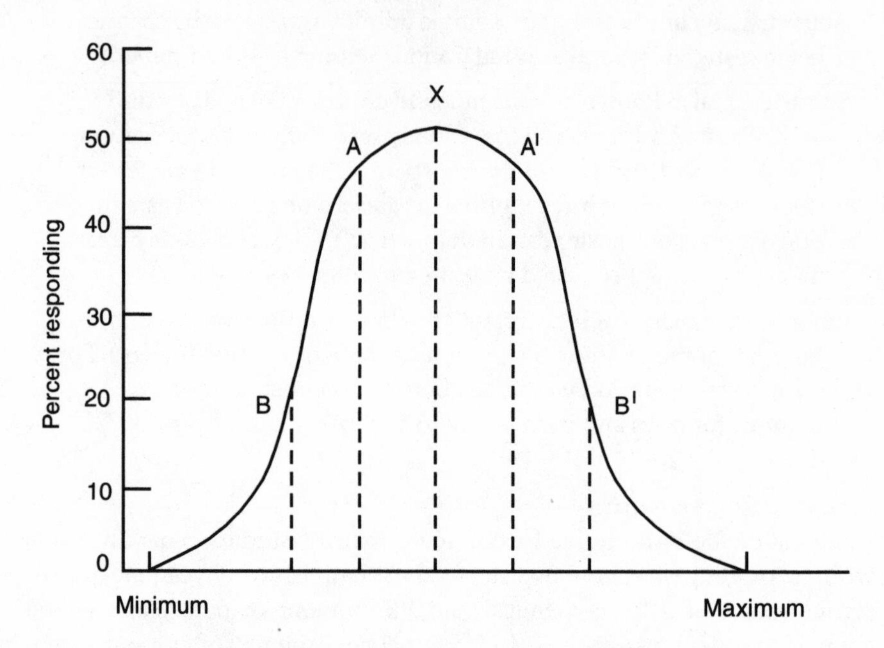

Figure 2.1 Distribution of animal responses to a toxic chemical as a function of dose. The midpoint (peak) of the curve (denoted on the graph as *X*) indicates the dose that produces effects in the largest percentage of the animals.

since the dose (or concentration) required to produce death in 50 percent of the exposed population is lower. Such graphs are commonly referred to as dose-response curves and are plotted using an arithmetic scale on the vertical axis and a logarithmic scale on the horizontal axis. One advantage of this format is that a major portion of the curve is linear: for this portion the response (in this case, death) is directly related to the dose or concentration of the chemical agent (Smith, 1992).

Figure 2.2 also illustrates the approach for determining the lethal dose for half (the so-called LD_{50}) of the exposed animal population within a certain period of time. This endpoint is easily measurable; it either occurs or it does not. In previous years, determination of the LD_{50} was one of the primary goals of many acute toxicity studies. This is far less true today, particularly in light of the diminished need for this type of information for the regulation of toxic chemicals. Another contributing factor is the increased interest in both cancerous and noncancerous diseases, as well as possible behavioral effects, that may be caused by chemical exposures.

Other benefits of acute toxicity studies are that they can provide information on the probable target organs for the chemical and its specific toxic effect, as well as guidance on the doses to be used in the more prolonged (long-term) studies. Acute toxicity studies can also provide information on the synergistic and antagonistic effects of certain combinations of chemicals. Such information is very important in the evaluation of environmental exposures, which typically include simultaneous exposures to more than one chemical.

SHORT-TERM AND LONG-TERM TESTS

Humans are most often exposed to chemicals in amounts much lower than those that are acutely fatal, and they are exposed over much longer periods of time. Short-term and long-term toxicity tests are designed to determine responses in humans in these more realistic situations. Under the conditions of *short-term* studies, generally two or more species of animals are used, the objective being to have them biotransform the chemical in a manner essentially identical to the process in humans. It cannot be assumed, however, that other animals will biotransform chemicals in *exactly*

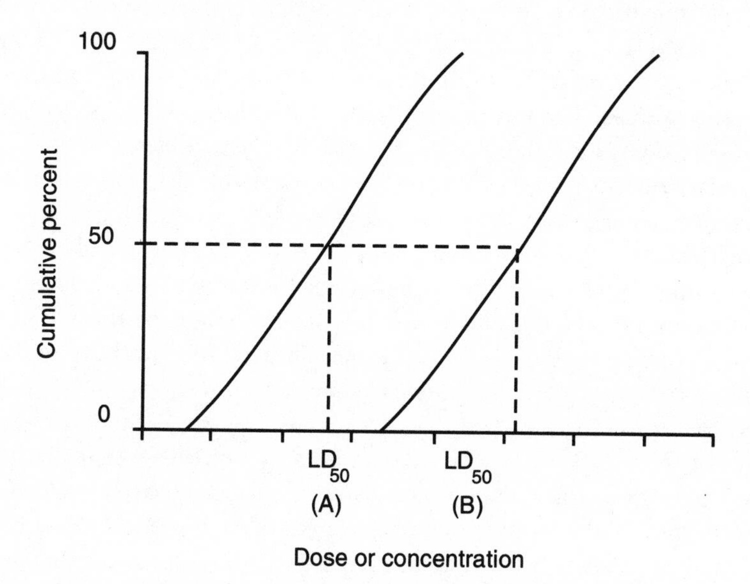

Figure 2.2 Cumulative percentages of animals showing responses to toxic chemicals. The LD_{50} designates the dose that is lethal to 50 percent of the exposed animals; the curve to the left represents the more toxic of the two chemicals.

the same manner as humans. In fact, differences in the abilities of various species to biotransform chemicals are the basis for the effectiveness of many of the pesticides that have been developed to be selectively toxic to only one insect, plant, or animal (Smith, 1992).

Under normal circumstances, the animals selected are the rat and the dog because of their appropriate size, ready availability, and the preponderance of toxicologic information on their reactions to a wide range of chemicals (Lu, 1991). Differences in response by gender require that equal numbers of male and female animals be used, and that a control group be maintained for comparison purposes. In addition, the chemical should be administered by the same route of exposure that is anticipated for humans. To assure that the studies encompass the full range of anticipated outcomes, most investigators select three dose ranges—one sufficiently high to elicit definite signs of toxicity but not high enough to kill many of the animals; one sufficiently low that it is not expected to induce any toxic effects; and an intermediate dose (Lu, 1991).

One of the outcomes of such studies is guidance on the "acceptable intake" of a chemical. Another piece of information that is needed, however, is an indication of the "no-effect level," or the "no observed adverse effect level." For this purpose long-term studies are generally employed. As with short-term studies, generally one or more species of animals are used, with the rat being preferred. The routes of administration are similar to those in short-term studies. The information collected includes body weight, body size, and food consumption, supplemented by general observations, laboratory tests, and postmortem examinations. General observations include appearance, behavior, and any abnormalities; laboratory tests generally include hematologic examinations, supplemented by analyses of blood and urine; postmortems include gross pathological examinations, including histologic examinations, supplemented by determination of the weights of individual organs such as the liver, kidneys, heart, brain, and thyroid.

OUTCOMES

Studies of these types provide data on toxicity with respect to the target organs, the effects on these organs, and the associated dose-effect and dose-response relationships. One determination that is often made on the basis of *acute toxicity studies* of suspected carcinogens is the maximum tolerated dose (MTD), the highest dose just below the level at which toxic effects other than cancer can occur. The basic reason for using the MTD as an endpoint is to estimate the carcinogenic potential of a chemical in the shortest possible time, using the fewest exposed animals. Again, acceptable levels of intake for humans would be extrapolated from the MTD, taking into account appropriate safety factors. The concept of the MTD has been criticized by some toxicologists who believe that the high doses introduce artifacts that exaggerate carcinogenicity in humans. Because of the controversy, scientists do not currently agree on the usefulness and applicability of this test (National Academy of Sciences, 1993).

A variety of complicating factors make it unlikely that the establishment of an acceptable level of intake of a chemical for humans will provide adequate guidance in setting a corresponding limit for the environment. Exposed population groups may include some members who are unusually susceptible. In addition, detrimental effects may have occurred but not been observed. These include changes in reproduction, increased susceptibility to disease, and decreased longevity. Furthermore, there is no justifiable reason to assume a constant relationship, for different chemicals or different species, between the dose required to kill and that needed to impair an organism (Moriarty, 1988).

Endpoints for Toxicological Evaluations

While acute and short-term tests were a mainstay during earlier toxicological studies, with only death or tissue damage as recognized endpoints, today the evaluation of human exposures tends to be directed at studies encompassing a full range of effects, including those on behavior (Weiss, 1990) and other noncancer endpoints. According to the National Research Council (1986) and Lu (1991), the more prominent endpoints include the following.

1. *Carcinogenesis.* Chemical carcinogenesis is recognized today as a multistage process, involving at least three steps: initiation, promotion, and progression. Although formerly it appeared that various chemical compounds and physical agents were either purely initiators or purely promoters, more recent interpretations suggest that some chemicals and agents are both initiators and promoters. Current theory posits that the development of cancer involves the activation or mutation of oncogenes, or the inactivation of suppressor genes, and that this causes a normal cell to develop into a cancerous cell.

Because of the expense and time required for related tests using animals, toxicologists have for years experimented with the development of short-term, in vitro tests (experiments conducted outside the body) as an alternative. One of the most widely applied is the Ames test (Ames, 1971). This test for mutagenicity in bacteria is based on evidence that deoxyribonucleic acid (DNA) is the critical target for most carcinogens, and on the observation that mutagenic chemicals are often also carcinogenic.

2. *Reproductive toxicity.* Toxic effects on reproduction may occur anywhere within a continuum of events ranging from germ cell formation and sexual functioning in the parents through sexual maturation in the offspring. The relationship between exposure and reproductive dysfunction is highly complex because exposure of the mother, father, or both may influence reproductive outcome. In addition, critical exposures may include maternal exposures long before or immediately prior to conception as well as exposure of the mother and fetus during gestation (National Research Council, 1986).

3. *Developmental toxicity (teratogenesis).* The type of illness involving the formation of congenital defects has been known for decades and is an important cause of morbidity and mortality among newborns. Developmental effects encompass embryo and fetal death, growth retardation, and malformations, all of which can be highly sensitive to chemical exposures. For some years no connection was suspected between such effects and chemicals; toxicologists had a tendency to assume that the natural protective mechanisms of the body, such as detoxication, elimination, and the placental barrier, were sufficient to shield the embryo from maternal exposure to harmful chemicals. These concepts changed dramatically after the clinical use of thalidomide, a sedative first employed in Germany in the late 1950s to relieve morning sickness in pregnant women (Smith, 1992).

4. *Neurotoxicity.* Although fewer than 10 percent of the approximately 70,000 chemicals in use have been tested, almost 1,000 have been identified as known neurotoxicants in humans and other animals (Stone, 1993b). The multitude of impacts on humans range from cognitive, sensory, and motor impairments to immune system deficits. For this reason classification of chemical neurotoxic action is constantly evolving, and the application of data from studies in animals to estimation of the risks of neurologic disease in humans is very complicated. Often there are major differences between the degree of neurotoxic response observed in animals and that found in humans.

5. *Immunotoxicology.* Various toxic substances are known to suppress the immune function, leading to reduced host resistance to bacterial and viral infections, and to parasitic infestation, as well as to reduced control of

neoplasms. The importance of these effects is well illustrated by the concern about AIDS, in which the infected person often dies owing to inability to resist an organism that would not be a problem in a healthy individual. Certain toxic agents can also provoke exaggerated immune reactions leading to local or systemic reactions.

In recent years some scientists have postulated that certain people have "multiple chemical sensitivity," which can lead to a type of "chemical AIDS" (Cullen, 1991). The supposition is that exposures to trace concentrations of multiple chemicals present in the environment may impair the body's immune system. Other scientists have noted that industrial society is producing hormone-like pollutants that could interfere with human reproduction (Kaiser, 1996). Still other scientists are skeptical of the scientific basis for these claims (Whelan, 1993). Such controversies and the wide span of opinions are typical of the early stages of emerging scientific questions. This is especially true when concerns are being raised but existing information, as well as scientific consensus about the meaning of that information, is insufficient to resolve whether the expressed fears are well founded (Rhomberg, 1996).

Extrapolation of Animal Data

In order to use animal bioassay data to predict human responses to environmental exposures, two kinds of extrapolations are necessary. One is determination or estimation of the relative responsiveness of humans and the animal species used in the bioassays—the so-called extrapolation from small animals to humans; the other is extrapolation from the biological effects observed as a result of exposures at the relatively high doses administered in the laboratory to the much lesser effects anticipated at the lower-level, but longer-duration, exposures expected within the environment. The second of these generates by far the greater uncertainty (Lippmann, 1992).

To cope with interspecies extrapolations, researchers must consider a variety of factors. These include differences in the pathways of exposure and the uptake of chemicals from various environmental media, the rates at which these materials are metabolized, the lengths of time they are retained in the target tissues, and the sensitivities of these tissues. The goal of the assessment is to predict what response, if any, might occur in the range of exposures anticipated for humans, recognizing that such exposures are often tenfold to a thousandfold below the lowest dose administered in the toxicity tests. Since the slope of the dose-response curve becomes increasingly uncertain as one extends it beyond the range of experimental data, the estimate of the extrapolated effects may be in error by a large factor.

In making such extrapolations, one generally assumes a linear relationship between dose and response within the range under consideration. This premise is based on present understanding of the cancer process as derived from studies involving ionizing radiation and genotoxic chemicals. While the relationship between dose and response in these cases may be linear (or nearly so) in the low-dose region, promoters and cytotoxicants would be expected to produce relationships that would be very nonlinear in the low-dose region. In fact, such effects may exhibit a genuine, or practical, threshold (a dose below which no response would be anticipated). As a result, the linear multistage model may be inappropriate for evaluation of the biological effects of chemicals such as dioxin, thyroid-type carcinogens, mitrolotriacetic acid, and, presumably, similar nongenotoxic chemicals. For these types of chemicals, other models may be more appropriate to avoid being overly protective (Lippmann, 1992).

In the absence of human data the weight of evidence that an animal carcinogen will be a human carcinogen increases with the number of animal studies and the number of different strains and species showing a positive response, and with the number of different tissues in the body that develop tumors following exposure to the compound. The weight of evidence also increases in the presence of a definitive dose-response relationship for all, but especially for malignant, tumors, and for those cases in which there is a shortening of life because of the induced tumors (Rosenthal, Gray and Graham, 1992).

Establishment of Exposure Limits

Two basic principles should be applied in setting health-based exposure limits for human populations. The first is use of human data whenever possible; the second is use of surrogate chemicals or surrogate species only when scientific evidence indicates that they provide an appropriate basis for the prediction (Doull, 1992).

In many cases, applicable human data for the establishment of exposure guidelines are not available for either the candidate chemical or the appropriate surrogate. Thus it is frequently necessary to use data obtained through studies with other animals. Application must be based on information concerning the chemical and its adverse effects in the test species, and on the exposure conditions in the target population. As explained above, the first step in this process is usually identification of all the adverse effects that can be produced by either acute or chronic exposure to the agent; the second step is establishment of dose-response relationships for each of the adverse effects (Doull, 1992).

Ideally, these studies would also provide information on the effects of administration by various routes, on different rates and duration of exposure, and on other test species. This information, together with data on the chemical and physical-chemical properties of the agent, kinetic data in various species, genotoxic studies, teratology and reproductive studies, and other types of target organ and mechanistic studies, constitutes what is referred to as the database for the target chemical.

The next step is to determine whether the information in the database is relevant and appropriate for making the prediction in the target species using actual exposure conditions. In situations where the information is adequate and relevant, and where there is a threshold or no-effect level for the specific adverse effect on which the prediction will be based, the final step in the process is to use these data to establish an appropriate exposure limit for humans. In setting such a limit, scientists and regulators generally incorporate a safety factor into the threshold or no-effect level observed in animals. Selection of this factor should not only reflect the confidence of the evaluator in the quality and relevance of the data, but also account for differences in the susceptibility and kinetics between test and target species and between individual members of the exposed population (Doull, 1992).

The magnitude of these safety factors is illustrated by the values used by the Safe Drinking Water Committee of the National Research Council to establish suggested no-response levels for various toxic agents in drinking water (NRC, 1983):

A factor of 10 was used when valid chronic exposure data existed on humans and supportive chronic data were available on other species; the factor was added to assure protection of the more sensitive individuals;

A factor of 100 was used when there were no data on humans but satisfactory chronic toxicity data existed for one or more other animal species; the 100 includes a factor of 10 to protect sensitive individuals and a factor of 10 to account for interspecies extrapolations;

A factor of 1,000 was used when the chronic toxicity data were limited or incomplete.

Regardless of how sound these safety factors are thought to be, a basic principle of health protection is to keep all exposures as low as is reasonably achievable.

Applying Toxicological Data to the Environment

Whereas the laboratory toxicologist is primarily concerned with the effects of toxic chemicals on individual organisms, evaluation of the effects of these same chemicals in the environment is far more complicated. The complications arise from several sources (Moriarty, 1988):

1. Different species, and different groups and individuals within a single species, may react differently to identical exposures to the same chemical.
2. Some pollutants may occur in more than one form, and the determination of either the details of exposure or the resulting biological effects may be difficult.
3. The presence of combinations of toxic chemicals may lead to synergistic or antagonistic effects.
4. The indirect effects of the toxic chemical may be of equal or greater importance than the direct effects.

Although data on the biological effects of toxic chemicals on single organisms or individual plants and animals are important, equally significant are the effects—either direct or indirect—of toxic agents on population groups. The fact that a chemical kills half of the individuals in a species population may be of little or no ecological significance, whereas a chemical that kills no organisms but retards development may have considerable ecological impact. Nonlethal chemicals that impair the ability of an organism to respond to its environment may very well have serious effects, such as increasing susceptibility to disease. This, in turn, can lead to a shortening of life.

When multiple species are involved, additional complications arise. Often it is easier to observe certain species than others. Even though predictions of biological effects may be correct for these species, other species may be significantly more vulnerable and/or susceptible. If environmental health specialists decide to be concerned only with the effects of a given chemical on a selected species, who is to judge whether the degree of protection afforded will be adequate to protect other species? To some extent, the concern should not be with how most species will be affected, but with identification of the first-affected species. Further complicating assessment of the effects of chemicals within the environment is (1) the fact that the effects of many pollutants on wildlife may pass unnoticed, and (2) the difficulty, even when an obvious effect is observed, of identifying the chemicals that are responsible.

The problems of assessing the effects of chemicals within the environment do not end here. Some chemicals may have no apparent direct effects on living organisms, but they may have very serious indirect effects. For example, alterations in the physical and chemical characteristics of the environment may have an impact on the ability of a species to survive: witness the releases of sulfur dioxide into the atmosphere that result in acid rain, and airborne discharges of carbon dioxide and other chemicals that affect global temperatures. Lakes and streams may be enriched through the release of sewage and agricultural chemicals, which in turn leads to eutrophication and detrimental impact on the survival of certain types of animal life within these waters. The analysis of indirect effects must take into account not only the realization that the impacts of certain airborne emissions may be global in nature, but also that their concentrations and resulting impacts can vary significantly from one region of the world to another.

As a result of these complications, it is quite probable that precise predictions of the effects of chemicals within the environment are unlikely to be achieved in the foreseeable future. Nonetheless continuing guidance is needed to make sound judgments relative to the introduction and use of chemicals, and environmental toxicologists will undoubtedly continue to direct their attention to these problems (Moriarty, 1988).

As a general guide, the chemicals that will be most important in terms of the environment are those that have known toxic effects, that are persistent, and that are biologically concentrated by various animals and/or plants.

The General Outlook

Most human carcinogens have been identified by epidemiological studies or by physicians who have noticed clusters of cancer cases resulting from specific exposures. As indicated above, however, classic human epidemiological studies suffer from a number of disadvantages, including the fact that they reveal a problem only after it has occurred. They also suffer from low sensitivity and large uncertainties in the accompanying estimates of associated health effects. Now being explored is the examination of populations at a molecular level (Garner, 1992; Marshall, 1993). This approach is based on the hypothesis that organic chemical carcinogens are activated in the body to interact with biological molecules. Such interactions lead to DNA-bound moieties, chromosomal aberrations, sister chromatid exchanges, and oncogene activation. These changes can be detected in samples of blood, urine, even exhaled breath, making it possible to understand events that precede a clearly identifiable toxic response. Instead of studying animals stressed to the maximum with near-lethal doses of toxic compounds, toxicologists can now concentrate on relatively normal biological processes in an effort to understand the biomechanisms by which damage is done. While this approach appears promising, some toxicologists caution that the data that result must be related to the whole animal.

As part of this same trend, other cellular and molecular approaches are being explored. These include an increasing use in toxicological tests of isolated organs, tissue and cell cultures, and lower forms of life. The thrust is to develop tests that will be faster and cheaper. An example is the report that toxicologists are turning to earthworms to complement their rodent-based assays. One goal of these studies is to develop assays to indicate how substances such as PCBs and metals impair the immune system (Holden, 1993a).

Adequate toxicological data do not exist for most chemicals in commerce; the available information seldom includes noncancer effects such as neurotoxicity or reproductive toxicity. Even when data are available, quantitative risk estimates still rely on assumptions concerning the mechanisms of disease, dose-response relationships, and the effective dose in relation to exposure. Obviously, such estimates cannot be expected to be completely accurate. Regulatory agencies are obliged to make assumptions that they believe are conservative, to avoid underestimating the risk to the human population. Unfortunately, overestimating a risk can unnecessarily eliminate jobs or the use of what otherwise would be commercially important materials, thereby decreasing our standard of living.

In addition, it is important to recognize that the resources available to control or remove toxic agents are limited. Overestimating the health effects of chemicals in the environment can result in underregulation or inadequate attention being directed to what may be more hazardous situations. The more that empirical assumptions can be replaced with experimentally validated procedures, the better will be the standards developed for protecting both human health and economic well-being (NRC, 1986). Unfortunately, until such time as toxicologists better understand the fundamental mechanisms by which chemicals cause adverse effects, controversies of the type that have accompanied the evaluation of chemicals such as ALAR (Smith, 1990; Holden, 1993b) and Dioxin (Stone, 1993a) will continue.

Recognizing the widespread need for toxicological data, Congress included in the Superfund Amendments and Reauthorization Act (SARA, 1986) the requirement that the Agency for Toxic Substances and Disease Registry (ATSDR) prepare toxicological profiles for a wide range of hazardous substances. These were to include those substances most commonly found at facilities on the National Priorities List (Chapter 9) that pose the most significant potential threat to human health, as determined by ATSDR and the Environmental Protection Agency (EPA).

Many of these profiles are now available and give detailed information on the toxicological and adverse health effects of substances ranging from aldrin and asbestos to zinc (HHS, 1993, 1994, 1995). For each substance the profile identifies and reviews the key peer-reviewed literature and describes its toxicological properties, then presents information on levels of significant human exposure and, where known, significant human health effects. Included is a determination of whether adequate information on the health effects is available; if not, the additional testing needed to provide the required information is identified.

These profiles are proving an excellent source of information for environmental and public health officials (Walker, 1993). Much of this and related information is available through the Integrated Risk Information System (IRIS), an electronic data base established and maintained by the EPA to aid practitioners in performing risk assessments. Also proving helpful is the "Pocket Guide to Chemical Hazards" (HHS, 1990).

The need for information on a wide range of chemicals will continue to expand. Toxicologic evaluations of carcinogenicity are available for less than 20 percent of the chemicals in use today, and epidemiologic assessments for less than 1 percent. Animal studies for teratogenesis have been performed on less than 10 percent of the chemicals currently in commercial use (Green, 1993). As supplies of fossil fuels disappear, new synthetic fuels will undoubtedly have to be developed. These and other needs will necessitate the development and use of many new chemicals, and toxicologists will bear a major share of the responsibility for assessing their safety (Lu, 1991). This will require increasing input from the companies that manufacture and sell these products. The care demonstrated by the chemical industry in recent years not only to develop and produce chemicals that can be manufactured, transported, used, and disposed of safely, but also to recognize and respond to related community concerns, is an encouraging sign (Abelson, 1992).

By way of final comment, it is important to keep in mind that the environment in its natural state contains a wide range of toxic chemicals, including microbial toxins, such as those produced by bacteria, blue-green algae, and dinoflagellates; phytotoxins, such as those produced by higher plants; and zootoxins, such as those produced by animals. Even though a multitude of toxic substances have been introduced into this same environment as the result of industrial and other human activities, some toxicologists estimate that more than 99 percent of the current intake of toxic substances by human populations is of natural origin.

3

EPIDEMIOLOGY

FOR WELL over a hundred years, epidemiologic studies have played an important role in the investigation of the ways in which infectious diseases spread through the community. With the growing awareness of environmental pollution and its potential effects on health, the techniques of epidemiology have been expanded to examine the effects of a variety of chemical and physical agents within the environment. The result has been the science of "environmental epidemiology," defined by the National Research Council (NRC, 1991) as "the study of the effect on human health of physical, biologic, and chemical factors in the external environment, broadly conceived. By examining specific populations or communities exposed to different ambient environments, it seeks to clarify the relationship between physical, biological or chemical factors and human health."

In this definition it is important to note that the techniques of environmental epidemiology are not designed—nor should they be expected—to prove that a given environmental agent *causes* a given disease or health effect; the best outcome that can be anticipated is that the methods of environmental epidemiology will demonstrate a *relationship* or *association* between a given agent and one or more specific health effects. The basic difficulty is that few of the nonbiologic agents have unique effects on health; conversely, the effects considered may often be related to a wide range of factors (WHO, 1983). Thus, when decisions have to be made about the need to control suspected agents within industry or the community at large, many aspects of the situation must be taken into account—the strength and consistency of the association, toxicologic and clinical findings, and the economic and social implications of control measures. An important consideration is whether there is a plausible mechanism through which the given chemical or physical agent can cause the suspected effect.

As contrasted to the field of toxicology, which is experimental in nature and generally involves the conduct of laboratory studies using animals, the field of environmental epidemiology is nonexperimental and involves studies of existing human population groups who have been inadvertently exposed to one of more chemical and/or physical agents. This chapter will outline the general principles of environmental epidemiology and highlight some of the precautions that must be taken in the design of such studies and in the analysis and interpretation of the collected data.

A Classic Example

Often recognized as the founder of epidemiology is John Snow who conducted what is regarded today as a classic study of the transmission of cholera in London in the mid-1800s (Monson, 1990). The case illustrates many of the principles of a valid environmental epidemiologic study.

Snow, a practicing physician, observed that people working with cholera patients did not always contract the disease, and that people who did not have contact with infected patients often did contract the disease. He postulated the existence of some vehicle that transmits the disease and, with support from other physicians and local laypeople, hypothesized that one possibility was the presence of sewage (fecal) contamination in drinking water. Snow conducted a study of population groups in different parts of the city who obtained their drinking water from different suppliers. Recognizing that other factors could influence the spread of the disease, he analyzed the mortality rates in a single subdistrict, where the only observable difference was that one portion of the population obtained its drinking water from one supplier and the other obtained its water from a second supplier. Using a chemical test that took advantage of a difference in the chloride content of the two water supplies, he was able to identify the supplier of each individual household. Using these data, he confirmed that the disease was transmitted by sewage in the drinking-water supplied by one of the companies (Goldsmith, 1986; Monson, 1990).

As pointed out by Monson, several factors make Snow's study a model of environmental epidemiology.

1. Snow recognized an association between exposure and disease—that is, between the source of the drinking-water supply and the incidence of cholera.

2. He formulated a hypothesis—that fecal contamination of drinking water was the specific agent of transmission of the disease.
3. He collected information to substantiate his hypothesis—in subdistricts where the drinking water was supplied by only one company, the association was stronger.
4. He recognized that there could be an alternative explanation for the association—that social class or place of residence might influence transmission of the disease.
5. He applied a method to minimize the effects of the alternative explanation—he compared cholera rates within a single district or neighborhood, rather than between neighborhoods, on the basis of their water supply.
6. He effectively minimized the collection of biased or false information—since most residents were not aware of the name of the company that supplied their water, he applied a chemical test to make this determination in a positive manner (Goldsmith, 1986).

These criteria have withstood the test of time. In fact, investigators today assess the associations between environmental agents and disease based on the design of environmental epidemiologic studies that incorporate many of the principles established by John Snow.

Modern Environmental Epidemiology

As exemplified by the work of Snow, early epidemiologic studies were "disease centered," and the diseases primarily involved were infectious in nature. As a result, investigators during this time period relied primarily on laboratory investigations with little attention to study design. Their basic principles were that a microorganism should be considered as causally related to a disease when it was present in all subjects affected and when it was absent, or found as a fortuitous parasite, in other diseases. The implication that a given agent was the source of the disease was then confirmed by isolating it in the laboratory, inoculating it into animals, and demonstrating that the animals developed the disease (Terracini, 1992).

Today the trend is to employ epidemiologic studies that are "exposure centered." This approach is an outgrowth of the realization of a multitude of factors. One is that in the developed countries of the world the degenerative diseases such as cancer, whose etiology is multifactorial, have become the prevailing pathology. Furthermore, many environmental hazards have been found to be associated with more than one disease, and many ill effects are thought to be due to interactions (addition, synergism, or antagonism) among a variety of environmental agents. The result has been an increasing awareness of the need for a rational, systematic, explicit, and reproducible approach to evaluating the associations between various diseases and environmental agents. Meeting this need requires the consideration of certain basic criteria, enumerated by Hill (1965):

1. The strength and specificity of the association;
2. The consistency of the findings in different studies;
3. The existence of a dose-response gradient between the exposure and the occurrence of the disease;
4. The biological plausibility of the proposed association;
5. The coherence of the evidence with the natural history of the disease;
6. The supporting experimental, or quasi-experimental, evidence.

Although subsequent investigators have expanded on these criteria, they have served as one of the foundations of modern epidemiology, much as Snow's principles did during the early years. The primary changes have been to emphasize the control of confounding variables and to improve study design (Terracini, 1992).

DESIGN OF AN EPIDEMIOLOGIC STUDY

One of the first considerations in the design of an environmental epidemiologic study is the definition of its objectives and scope. As an extreme, one might consider monitoring the health records of the whole population and linking that information with as many data on environmental factors as possible. Basic to such a study would be national death statistics and records on morbidity. To extend this type of study to include inquiries into the "health and habits" of individual members of the population on a national scale, however, might be considered an intrusion on privacy—and the financial costs would be prohibitive. Nonetheless, if success is to be achieved, some form of additional data gathering may be required (WHO, 1983).

An alternative approach is to focus on small groups of people considered to be at risk. The objective would be to consider a specific disease or effect, and to compare the available information on exposures in this group to those in a control group. Because it is unethical to expose people to potentially hazardous environmental agents solely for purposes of epidemiologic study, essentially all such studies are nonexperimental. As a result, it may be difficult to define or quantify the exposures received by the population group being evaluated.

Two of the multitude of ways in which environmental epidemiologic studies can be classified are the following:

Cohort study. A group of persons who has received unusual exposures is followed over time to determine what diseases they develop and whether there is an increase in the incidence of those diseases that might be presumed to have been caused by the exposures. The epidemiologic studies of the survivors of the World War II atomic bombings in Japan exemplify the cohort study (Radiation Effects Research Foundation, 1994).

Case-control study. People who are known to have a specific disease are examined to determine what if any exposures that they are receiving now, or have received with unusual frequency in the past, might have been the source of the disease. Early epidemiologic studies of the relationship between cigarette smoking and lung cancer (Doll and Hill, 1950) are one example of a case-control study. It has also been used to evaluate various diseases in occupational settings, one example being the associations between certain illnesses and pesticide exposures (Cantor et al., 1992).

A cohort study may be either prospective, in which case the disease has not yet occurred, or retrospective, in which case the disease has already occurred, at the time the exposed and nonexposed groups are defined. Case-control studies, in contrast, are frequently termed retrospective because the investigator is looking backward from disease to exposure. That is to say, individuals are included on the basis of whether they have or do not have the disease being evaluated. Simply referring to a study as retrospective or prospective, however, leads to confusion, especially in the case of retrospective cohort studies, where the investigator is looking forward from exposure to disease but is basing the analysis on data collected in the past. As might be anticipated, various combinations of approaches are often included within a single study.

The basic differences in the various types of epidemiologic studies can be summarized as follows (Monson, 1990).

In a cohort study, individuals are included on the basis of whether they have been exposed; in a case-control study, individuals are included on the basis of whether they have the disease being evaluated.

In a prospective cohort study, the disease has not occurred at the time the exposed and nonexposed groups are defined; in a retrospective cohort study, the disease has already occurred.

In a prospective cohort study, the investigator usually compares the disease rates of two or more groups (for example, smokers and nonsmokers); in a retrospective cohort study, mortality rates among the exposed group are compared to mortality rates of some general population (no formal comparison group).

Since in a case-control study the past history of exposure is the primary information that is collected, such studies can be completed in a relatively short period; because time must pass in order for the disease to develop, completion of a prospective cohort study often requires a relatively long interval.

In a case-control study, the general approach is to evaluate a number of exposures in relation to one disease; in a cohort study, one exposure is evaluated in relation to a number of diseases.

In the conduct of current environmental epidemiologic studies, the general approach is not to compare an exposed and a presumably unexposed group; rather, it is to compare the incidence of a given disease as a function of the degree, extent, or amount of exposure. This approach is taken because it is often difficult to identify persons who have not been exposed at all to a given physical or chemical agent.

Major Challenges

A variety of challenges face environmental epidemiologists. Some of the more important are enumerated below.

EXPOSURE ASSESSMENT

As mentioned above, assessment of the exposures to which a population study group has been subjected is a crucial, often inadequately addressed component of environmental epidemiology. The challenge of monitoring the environment and collecting the necessary data for estimating such exposures, especially in the case of toxic chemicals and radioactive materials, is sufficiently complex that Chapter 15 of this book is devoted in its entirety to this subject. Regardless of the complexity, valid environmental monitoring measurements and accurate estimates of exposures are essential if confidence is to be placed in the associations that are developed between exposures and observed adverse consequences to human health (NRC, 1991).

Exposures to physical agents, such as noise or vibration, may be transitory and the resulting assessments must therefore be made on a real-time basis at the sites of exposure. Unfortunately, in some cases (as with electric and magnetic fields), assessment personnel do not yet fully understand which parameters are indicative of exposure. Nor do they know whether it is average or peak exposures that are important. In the case of chemical and radioactive contaminants, the field of environmental monitoring and exposure assessment requires consideration of the source of the contaminant, its associated media or pathways of exposure, its avenues of transport through each medium, its routes of entry into the body, the intensity and frequency of contact with the contaminant of the persons exposed, and its spatial and temporal concentration patterns. The importance of such movement and interactions is exemplified by the fact that the composition as well as the physical form of chemical contaminants can be readily altered thereby. The progression from the release of a contaminant, its movement through the environment and uptake by humans, to the production of associated health effects, is depicted in Figure 3.1.

Accurate assessment of exposures from airborne particulates requires, for example, not only identifying what the contaminant is, its physical form (amorphous, crystalline, discrete particulate, or fibrous), and particle-size distribution, but also, in many cases, its physicochemical surface properties. Especially complex is the assessment of exposures necessary for cross-sectional studies of the effects of air pollution. In earlier times, such studies typically involved a comparison of the health of populations in communities having different ranges of specific contaminants in the outdoor air. It is now recognized that people spend 90 percent or more of their time indoors. Thus, the assessment of their exposures must include determination of the concentrations of airborne contaminants inside their homes and places of work.

Important factors that can contribute to airborne exposures within a home include personal or family eating habits, the type of cooking facilities (natural gas or electricity), personal hobbies and recreational activities, pesticide applications within the home and garden, and the nature of the domestic water supply. Contaminants released into the air during shower-

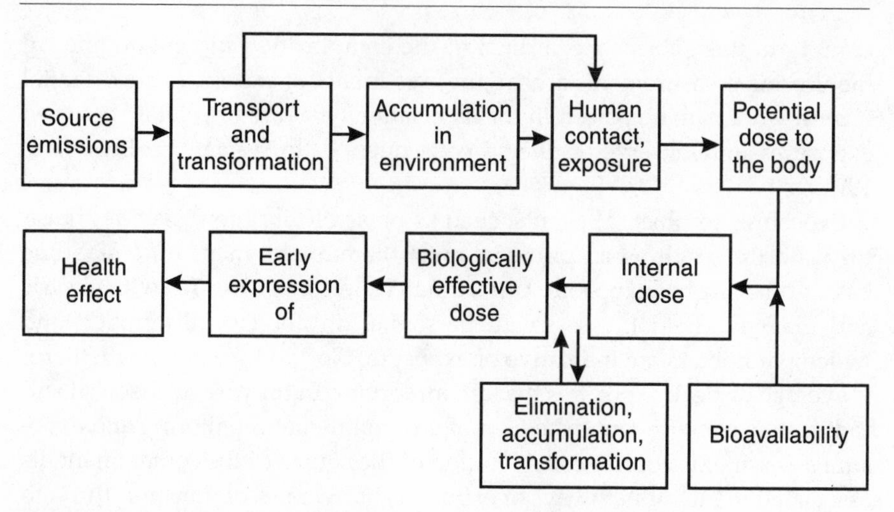

Figure 3.1 Progression of factors that influence the behavior of a contaminant within the environment, its uptake by humans, and the resulting health effects

ing, bathing, and cooking may become sources of exposure through inhalation (NRC, 1991). The hierarchy of exposure data or surrogates, ranging from quantified measurements of individual exposures to simply knowing the person's residence or place of employment, is depicted in Table 3.1.

Because environmental epidemiology often involves studies that have a retrospective component, the accompanying exposure assessments have proved difficult to accomplish. Records of ambient pollutant concentrations sometimes provide a surrogate for exposure but are not always available. In addition, direct measures of past exposures have not usually been recorded and must be estimated using environmental models (National Research Council, 1991). Because it is recognized that exposures do not always have a direct correlation with the dose an exposed person receives, in many environmental epidemiologic studies today no attempt is made to convert exposure into dose. Instead, the goal is simply to determine if there is an association between an exposure and a health effect.

HEALTH ENDPOINTS

A second major challenge in the design and implementation of an environmental epidemiologic study is selection of the health endpoints to be evaluated. Formerly it might have been adequate simply to determine whether the chemical or physical agent in question was causing an increase in the number of deaths (mortality) or hospital admissions (morbidity) among the exposed population. Subsequently, the potential increase in the incidence of cancer became the health indicator of primary importance. Today environmental and public health officials, as well as the public at large, are concerned with the possible impacts of environmental agents on the quality of life. They are demanding that a variety of possible pathological conditions—biochemical, physiological, and neurological dysfunctions—also be considered. These include effects on the respiratory and cardiovascular, central nervous, and musculoskeletal systems, as well as behavioral, hemopoietic, growth, and reproductive effects.

Such considerations add enormous complexity to the studies. If, for example, the only effect of an agent at a given intensity is a small change in bodily function, well within an individual's normal physiological range of variation, then its importance in comparison with other factors affecting health must be carefully weighed. Competing factors that must be considered include the duration of the effects and the number of persons likely to be affected. Also to be considered is the relative importance of a minor immediate effect versus a potentially more serious but delayed effect (WHO, 1983). A key criterion is whether the chemical or physical agent being evaluated has been demonstrated to be capable of causing the suspected effect. Unless it has, successful conduct of the proposed study may be seriously impaired.

Assessment of any of these endpoints requires some standardized measure of effects. The indicators that have been developed for measuring be-

Table 3.1 Hierarchy of exposure data or surrogates

Types of data	Approximation to actual exposure
Quantified personal measurements	Best ↑
Quantified area or ambient measurements in vicinity of residence or other sites of activity	
Quantified surrogates of exposure (e.g., estimates of drinking water and food consumption)	
Residence or employment in proximity to site of source of exposure	
Residence or employment in general geographic area (e.g., county) of site of source of exposure	↓ Poorest

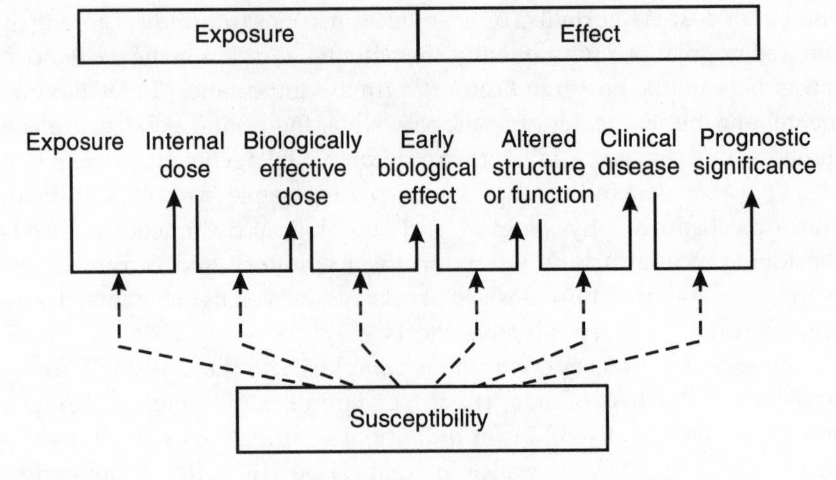

Figure 3.2 Relationship of susceptibility, exposure, and effect

havioral effects of noxious environmental agents, for example, fall into two broad groups: (1) measures of psychological and psychophysiological functioning and (2) measures of mental state and behavior. Psychological tests have proved effective in the detection and measurement of organic brain damage, and relatively simple techniques, such as Raven's progressive matrices and vocabulary and memory tests, can be both reliable and practical in field studies involving the screening of large numbers of individuals (WHO, 1983).

In recent years the range of indicators of exposure to various environmental agents has been expanded to include biologic markers (Figure 3.2). They are generally divided into three groups and are defined as any cellular or molecular indicator of (1) toxic exposure, (2) adverse health effects, or (3) susceptibility. Markers are not a new tool; blood lead, urinary phenol levels in benzene exposure, and liver function assays after solvent exposure have been used for some time in occupational and public health research and practice to indicate recent exposures to these compounds. What distinguishes the current generation of markers is their increased analytical sensitivity and their ability to describe events all along the continuum between exposure and clinical disease. Health events are less likely to be viewed as binary phenomena (presence or absence of disease) than as a series of changes through homeostatic adaptations and dysfunction to disease and death (NRC, 1991).

POTENTIAL BIASES

Another challenge in the conduct of environmental epidemiologic studies is the variety of potential biases that can cause the outcome to be in error. As noted above, measurement errors occur because of the assumption that all subjects within a given population group receive the same exposure. There can be undetected differences among communities in risk factors such as illness, tobacco use, or occupational exposure. Also contributing to biases is the lack of standardization in the equipment used to measure exposures at different locations (NRC, 1991). As indicated below, "recall bias" can also be present.

Monson (1990) has separated the biases that can influence epidemiologic studies into three categories:

Selection bias occurs as a result of deficiencies in the study design. If two groups of persons, one exposed and one not exposed, are identified today and followed forward in time until disease occurs (as in a prospective cohort study), no selection bias is possible. If, however, a group of people with disease is identified and a group of controls (for example, in a case-control study) is selected, selection bias is possible owing to the fact that the disease had occurred prior to the initiation of the study. Once selection bias has occurred, no amount of data manipulation can correct its effects—the two groups are forever noncomparable. In essence, selection bias cannot be controlled; it must be prevented.

Observation bias is also a result of deficiencies in the study design. In a cohort study, for example, observation bias occurs when information on disease outcome is obtained in a noncomparable manner from exposed and nonexposed groups. In case-control studies, observation bias occurs when information on exposure is obtained in a noncomparable manner from cases and controls. An obvious way to prevent observation bias in a cohort study is for the data collectors or interviewers not to know the exposure status of study individuals when they gather information. Likewise, in a case-control study no observation bias is possible if neither the patient nor the data collector knows the diagnosis. Again, this type of bias must be considered in the design of the study and efforts made to minimize its effects. Underlying the prevention of observation bias is the need for all concerned with a study to be impartial. Also influencing the types of data obtained through interviews is what is called "recall bias." Often the input data are based on past experiences of people, which in turn depend on the accuracy of their memories and the information they are willing to share (Taubes, 1995).

Confounding bias is inevitable in all types of studies. For example, an evaluation of two variables (exposure and disease) is influenced by a third variable that is a cause of the disease and also associated with the exposure. Specifically, cigarette smoking is a cause of lung cancer. Cigarette smoking is also associated with high alcohol consumption. If one examines the relationship between high alcohol consumption and lung cancer, it will be observed that the rate of lung cancer in high alcohol consumers is higher than in nonconsumers. All that can be done is to collect information on known or suspected confounding factors, so as to be able to measure any bias that is introduced. Confounding bias does not result from any error of the investigator; it is a basic characteristic of existence.

Studies of geographic variations in disease rates illustrate the potential biases in environmental epidemiologic studies. One of the principal goals of such studies is to formulate hypotheses about the etiology of disease by taking into account spatial variations in environmental factors. Testing such hypotheses on the basis of geographic variations, however, is generally not possible. The hypotheses need to be tested by more rigorous methods, using cohort and case-control studies, the primary reason being that in geographic studies the exposure to a particular environmental agent (for example, water containing a specific contaminant) and the suspected effect (an increase in cancer) are not measured in the same individuals. Nonetheless, because they take advantage of large differences in both the frequency of disease and the prevalence of exposure, geographic studies at an international level have been successful in identifying a number of possible risk factors for disease (English, 1992).

One of the basic applications of geographic studies in the field of epidemiology is in conjunction with simple descriptive studies of geographic variation, where the goal is to determine if variations in disease rates are associated with variations in the accompanying levels of exposure. The place and time of residence of the affected populations are often used as surrogates for the exposures of interest. Although the relationship between surrogate and exposure may be direct (say between cosmic radiation and altitude, or between ultraviolet radiation and latitude), in most cases it is indirect. The validity of the geographic approach depends on how well the surrogate serves as a measure of the actual exposure of an individual who develops the disease.

Conduct of an Environmental Epidemiologic Study

The many practical problems in the organization of an environmental epidemiologic study include both the level of study to be conducted (simple to complex) and the resources required.

One of the first objectives is to identify the population group to be examined. It is often helpful to consider the conduct of an initial study among workers who may be exposed to the same agent. One advantage of this approach is that exposures in occupational settings are often higher than in the general environment. At the same time, it should be borne in mind that a working population is preselected: it excludes children, the elderly, and those whose health is already impaired, as well as individuals who may be hypersensitive to certain agents. The working population also frequently includes a disproportionately low number of women. Furthermore, exposures of workers are limited in most cases to eight hours a day. As a result, caution must be exercised in extrapolation of the resulting observations to the general population (WHO, 1983).

Once the study group has been identified, contacts need to be established with individuals within the group to guarantee their interest and cooperation. Where individuals decline to participate, care must be taken that their response does not bias the results of the study. If a number of people are engaged in collecting information, joint training sessions are required to ensure uniformity of approach, and it may be necessary and beneficial to interchange the teams periodically during the data collection period. Experience has shown that the most effective approach, where the effects of common environmental agents on individuals within the general population are being studied, is to have the data collectors visit the subjects in their homes. Although this technique is labor intensive, it is often justified by the improved quality of the results. Any instruments used to collect data need to be calibrated on a regular basis, and all related methods should be standardized. If biologic indexes of effects are used, it may be necessary to have all measurements performed in a single laboratory.

Ethical problems may also arise. If some tests are intrusive (for example, the collection of blood samples), prior permission will be required. In addition, confidentiality is an issue. Thus it is common practice to include the names and addresses of those interviewed only on the original survey form.

As indicated by Monson (1990) and the World Health Organization (1983), the computer has had a revolutionary impact on the conduct of environmental epidemiologic studies. In fact, the dramatic increase in the number of such studies since the 1950s is directly related to the development and wide availability of computers. As Monson says, the ability to collect large amounts of data, to store them, and to conduct extensive analyses is "the hallmark of epidemiology today." This is especially true of data showing weak associations between exposure and effects. Still, the computer has separated many epidemiologists from the data they are analyzing. They may not be familiar with weaknesses inherent in the collection of the data or with limitations in the computer programs being used.

Even when a significant correlation is determined to exist between an index of health and one or more environmental factors, the relevance of this finding must be carefully considered, including whether the observation has a plausible biologic explanation. Above all, it must be recognized that the indication of a correlation does not imply causation.

Case Studies

A wide range of environmental epidemiologic studies have been conducted in the past that can serve as examples of the beneficial uses and applications of this methodology. Summarized below are several examples.

FLUORIDE IN DRINKING WATER AND DENTAL CARIES

One of the earliest case-control studies was the determination of an association between fluoride in drinking water and the condition of the teeth of those who consumed the water. Having observed reports of mottled enamel on the teeth of people who drank water from certain sources, scientists in the U.S. Public Health Service in the late 1920s began a series of epidemiologic explorations to try to identify the cause of the problem. Many of these early studies were similar to those of Snow, in that the investigators noted differences in mottling depending on which spring served as the source of drinking water. Analyses of the water in the various springs highlighted differences in the fluoride concentrations. Subsequent studies showed that children who lived in areas where mottled teeth were endemic had far less tooth decay than those who lived where mottled enamel was nonexistent. Associated studies demonstrated that the concentration of fluoride in drinking water needed to prevent tooth decay was far below that which would cause mottling. This work was pioneering in that it demonstrated that epidemiologic studies could be used to promote better health (in this case, less tooth decay), as well as to determine the sources of various types of detrimental health effects, such as mottling (Williams, 1951; Terracini, 1992; WHO, 1983).

CIGARETTES AND LUNG CANCER

Cigarettes have unfortunately become a routine part of the personal environment of many people in the world. The determination of a definitive association between cigarette smoking and lung cancer is a classic example of the useful application of environmental epidemiology. It is also an example of how the personal choices of individuals can have an extremely detrimental effect on their health and of how difficult it is, even when a relationship has been thoroughly demonstrated, to implement effective control measures.

In the middle to late 1940s, physicians in several of the industrialized countries of the world, including the United States and the United Kingdom, noted an increasing number of diagnoses of men with lung cancer. A decade earlier such cancer had been a medical curiosity. Although cigarette smoking was immediately suspected as a cause, obviously the presumption had to be confirmed. Two types of studies were undertaken—case-control studies in which persons with and without lung cancer were asked about past habits, including smoking; and cohort studies in which smokers and nonsmokers were followed and the rates of development of a variety of diseases, including lung cancer, were measured (Monson, 1990).

One of the leading epidemiologists who conducted such studies was Richard Doll, working first with A. Bradford Hill and later with Richard Peto. On the basis of an initial case-control study, Doll and Hill (1950) concluded that "smoking is a factor, and an important factor, in the production of carcinoma of the lung." They admitted, however, that they had no evidence about the nature of the carcinogen. As a result of a subsequent series of longer-term cohort studies, Doll and Peto (1976) concluded that the death rate from lung cancer in smokers was ten times the rate in nonsmokers.

These studies, and related research by a multitude of other investigators, led to growing awareness not only of a definite association between cigarette smoking and lung cancer but to the conclusion, on the part of many public health officials, that cigarette smoking was a cause of lung cancer. These concerns culminated with the issuance in 1964 of the Surgeon General's report on *Smoking and Health* (USPHS, 1964) and the subsequent development of a wide range of antismoking campaigns, including the banning in this country of cigarette commercials on television stations (Surgeon General, 1989). Interestingly enough, one of the principal actions that finally brought about a noticeable reduction in cigarette smoking in the United States was publication of the results of epidemiologic studies which showed that nonsmokers were harmed by "secondhand" (side-stream) smoke (Trichopoulos, 1994). Today, smoking is prohibited on commercial passenger airplanes, and in buildings of the federal government and numerous state and local agencies, as well as in many workplaces, restaurants, and public buildings.

IONIZING RADIATION AND CANCER

Shortly after the discovery of X rays in 1895, reports of radiation injuries, including the subsequent development of cancers, began to appear in the published literature. It was soon recognized that ionizing radiation had harmful health effects and that standards were needed for the control of related exposures. This evidence was followed several decades later by observations of bone sarcomas in a number of the young women who ingested radium in conjunction with their work in applying luminous compounds to watch dials (Rowland, 1994). Shortly after detonation of the two atomic bombs in Japan near the end of World War II, researchers saw that these events, while tragic for the people who were exposed, offered a unique opportunity to quantify the relationship between exposures to ionizing radiation and health effects. Almost immediately after the cessation of hostilities, a joint Japanese-U.S. program was established to conduct long-range cohort studies of the survivors of the bombings (Radiation Effects Research Foundation, 1994).

The resulting studies had the benefit of detailed health information maintained by Japanese health officials on the inhabitants of the two bombed cities. As in all such programs, estimation of the exact doses received by the people exposed has proved to be difficult (Straume et al., 1992). Nonetheless, the data generated over succeeding years have served as a principal source for establishing quantitative dose-effect relationships for ionizing radiation exposures among human populations. Based on a careful analysis of the data generated from 1945 through 1970, the Committee on the Biological Effects of Ionizing Radiation (BEIR, 1972) concluded that the dominant effects of ionizing radiation among the Japanese survivors were the solid tumors—cancers of the lungs, breast, thyroid, and so on—and that, as long as the dose limits were set so as to avoid an excessive number of cases of these tumors, the problems of leukemia and genetic effects would likewise be adequately controlled. That conclusion still holds today.

ELECTRIC AND MAGNETIC FIELDS AND LEUKEMIA

For some years various scientists, public health officials, and others have expressed concern about the possible health effects of human exposure to electric and magnetic fields. A major impetus was a case-control study of leukemia in children living in the Denver area (Wertheimer and Leeper, 1979). The investigators concluded that the relative cancer risk for children living in homes located near high-current power lines was double that for children living elsewhere. They also concluded that the stronger magnetic fields in the homes of the children near the high-current lines was an underlying factor in their elevated cancer risk.

Although some studies conducted by other investigators have supported the findings of Wertheimer and Leeper, questions have been raised about the validity of their observations. In a case-control study designed to replicate the Denver study, Savitz and colleagues (1988) estimated the children's exposures on the basis of the configuration of nearby power lines, as well as by conducting point-in-time measurements of the electric and magnetic fields in the residences of each of the cases and controls. No statistically significant correlation was found between exposures (based on measurements of the magnetic fields) and childhood cancer. As in the study by Wertheimer and Leeper, the cancer risk was found to be correlated with exposure levels estimated on the basis of power-line configurations.

Subsequent reviews of several of the studies of the health effects of electric and magnetic fields revealed a number of methodological deficiencies. In nearly all instances, no quantitative assessment of power-frequency field exposures was made; that is, most of the estimates of exposure were based on surrogate measures. In many studies, the sample populations were small; thus the observed increases in cancer could have been due either to chance or to some unidentified factor.

For these and other reasons, a number of organizations and groups have conducted independent reviews of possible associations between electric and magnetic fields and effects on health (Bierbaum and Peters, 1991; Oak Ridge Associated Universities, 1992). The authors of the Oak Ridge reviews, conducted under the auspices of the Federal Committee on Interagency

Radiation Research and Policy Coordination, concluded that "there is no convincing evidence in the published literature to support the contention that exposures to extremely low frequency electric and magnetic fields (ELF-EMF) generated by sources such as household appliances, video display terminals, and local powerlines are demonstrable health hazards." Of interest also is the committee's note that "no plausible biological mechanism is presented that would explain causality."

The General Outlook

In terms of many environmental agents and suspected carcinogens, the literature contains one or more so-called epidemiologic studies that can be cited to either support or negate essentially any conclusion that a person wants to draw—witness daily articles in newspapers and magazines, as well as items reported on radio and television network news. To help understand how such a situation has developed, one needs to examine the differences in the characteristics of those epidemiologic studies for which the results have been widely accepted and those for which they have not.

Among the most significant of the studies are those linking cigarette smoking with lung cancer. On the basis of the latest studies, the risk of developing lung cancer, for moderate to heavy smokers, has been estimated to be as much as 3,000 percent (thirty times) greater than for nonsmokers. The evidence is overwhelming! Another agent in our daily lives that has been determined to have a definite relationship to health is exposure to sunlight and the resulting increase in skin cancer. In the field of occupational epidemiology, classic examples include the role of airborne asbestos in the development of lung cancer, and the association of benzene exposures with leukemia (Taubes, 1995). In terms of the personal environment, increased physical activity has benefits that relate to cardiovascular disease, and a high intake of fruits and vegetables protects against many cancers (Willett et al., 1995).

Among the more controversial areas of epidemiology are the previously cited studies that have shown alleged associations between exposures to electric and magnetic fields and certain health effects. While details of any possible relationship have not been completely settled, the associations that exist are so weak as to make the observations highly questionable.

Similar controversies exist on the quantitative relationship between low-level exposures to ionizing radiation and the number of excess cancers that may be induced. The difficulty of establishing such a relationship is exemplified by the following data on the survivors of the World War II atomic bombings in Japan. If the "normal" proportional mortality from cancer in the exposed population is assumed to be 17 percent, it would have been anticipated that perhaps 17,000 of the 100,000 survivors would have developed and ultimately died from cancer. To date, the excess cancers that have been identified due to radiation exposures by this group number less than 500. The statistical challenges alone are enormous!

For these and other reasons, some epidemiologists caution that, in cases where the finding stands in a biological vacuum or has little or no biomedical credibility, small relative risks should not be taken seriously (Trichopoulos, 1995). Similarly, the findings of individual studies should not be accepted as factual until they have been confirmed by other investigators (Willett et al., 1995). It is vital that in these cases epidemiologists share their skepticism with the public and the press, particularly as they search for subtler links between diseases and environmental agents. Some epidemiologists have even suggested that they may be pushing the edge of what can be accomplished through the application of epidemiologic techniques (Taubes, 1995).

4

THE WORKPLACE

As early as the fourth century B.C., Hippocrates apparently observed adverse effects on miners and metallurgists caused by exposure to lead. In 1473, Ulrich Ellenbog recognized that the fumes of some metals were dangerous and suggested preventive measures. In 1556, Georg Bauer (known as Georgius Agricola), a physician and mineralogist, attributed lung disease among miners in the Carpathian Mountains to the inhalation of certain kinds of mineral dusts, observing that so many miners succumbed to the disease that some women there married as many as seven times. In 1700, Bernadino Ramazzini published the first complete treatise on occupational diseases, *De Morbis Artificum Diatriba.* In the mid-1880s, Karl Bernhard Lehmann, whose work continues to serve as a guide on the effects of exposure to airborne contaminants, conducted experiments on the toxic effects of gases and vapors on animals (Patty, 1978). During the same period the first occupational cancer, scrotal cancer in chimney sweeps, was observed in England.

In the United States, occupational health problems received little attention until the twentieth century. The U.S. Bureau of Labor was created in 1885, but even when it became the Department of Labor in 1913 its stated goals included no mention of workers' health beyond "promoting their material, social, intellectual, and moral prosperity" (U.S. Congress, 1913). And Alice Hamilton's classic work, *Exploring the Dangerous Trades,* now perhaps the most widely quoted book on the field in the world, was not published until 1943 (Hamilton, 1943).

Protective Legislation

Protective legislation came piecemeal and slowly (Table 4.1). In 1908, the federal government provided limited compensation to civil service employees injured on the job. In 1911, New Jersey became the first state to enact a workers' compensation law; although many other states rapidly followed suit, it was not until 1948 that all the states required such compensation (Patty, 1978). Today, each of the fifty states has a workers' compensation law.

Workers' compensation laws, passed in France, Germany, and the United Kingdom in the nineteenth century, were one of the earliest forms of social insurance on a prepaid basis, with no direct monetary contribution from workers. By removing from the courts the determination of compensation for occupational injuries, these laws revolutionized the approach to control of workplace hazards and did more than any other measure to reduce occupational injuries in the United States. Eventually some states expanded their laws to include a full range of occupational diseases and required that compensation for occupational injuries be paid on a no-fault basis, that settlements be reached promptly through administrative tribunals, and that payments be made in accordance with a system of scheduled benefits.

The Walsh-Healey Public Contracts Act of 1936 established safety and health standards in industries conducting work under contract to the federal government. This forerunner of modern occupational health regulations stimulated research on occupational disease and the development of occupational health programs by state and local agencies, insurance companies, foundations, management, and unions (Cralley and Konn, 1973). The first significant federal legislation for workers outside government projects did not come until 1969, with the Federal Coal Mine Health and Safety Act. This legislation was followed by the landmark Occupational Safety and Health Act of 1970, whose announced principal purpose was to "assure so far as possible every working man and woman in the Nation healthful working conditions and to preserve human resources." Among the provisions of this act was the establishment of the Occupational Safety and Health Administration (OSHA) and the creation of the National Institute for Occupational Safety and Health (NIOSH). The stipulated purposes of OSHA were to encourage the reduction of workplace hazards, to provide for occupational and health research, to establish separate but dependent responsibilities and rights for employers and employees, to maintain a

Table 4.1 Significant federal legislation pertaining to occupational health and safety

Year	Act	Content
1908	Federal Workers' Compensation Act	Granted limited compensation benefits to certain U.S. civil service workers for injuries sustained during employment
1936	Walsh-Healey Public Contracts Act	Established occupational health and safety standards for employees of federal contractors
1969	Federal Coal Mine Health and Safety Act	Created forerunner of Mine Safety and Health Administration; required development and enforcement of regulations for protection of mine workers
1970	Occupational Safety and Health Act	Authorized federal government to develop and set mandatory occupational safety and health standards; established National Institute for Occupational Safety and Health to conduct research for setting standards
1976	Toxic Substances Control Act	Required data from industry on production, use, and health and environmental effects of chemicals; led to development of "right- to-know" laws, which provide employees with information on nature of potential occupational exposures
1990	Pollution Prevention Act	Established policy to ensure that pollution is prevented or reduced at source, recycled or treated, and disposed of or released only as last resort; led to substitution of less toxic substances in wide range of industrial processes, with significant reductions in worker exposure

reporting and record-keeping system to monitor job injuries, to establish training programs, to develop mandatory safety and health standards, and to provide for development and approval of state occupational safety and health programs. The responsibilities of NIOSH were for conducting research on which new occupational health standards could be based and for implementing education and training programs to provide adequate numbers of people to implement and enforce the Occupational Safety and Health Act.

In a later development, Congress incorporated "right-to-know" provisions in amendments to the 1976 Toxic Substances Control Act to require employers to furnish workers with information about the health hazards of their occupational environment. This stipulation has made it much easier for workers to be aware of the hazards they face and to raise questions about the protection being provided.

One recent law that has significantly reduced occupational exposures is the Pollution Prevention Act of 1990 (U.S. Congress, 1990). This law, which established a national policy to encourage the prevention of pollution at the source, with disposal to the environment acceptable only as a last resort, has led to the substitution of less toxic substances for those used in a wide range of industrial processes. In turn, significant reductions in worker exposures have resulted.

Identification of Occupational Health Problems

Today well over 100 million men and women are gainfully employed in the United States, and increasing numbers of people, particularly women, are entering the workforce. To some degree, all these workers are exposed to occupational hazards, and all risk job-related adverse health effects. Compounding these problems is the fact that more than 25 percent of Americans are employed in businesses that have fewer than 20 employees, and more than 50 percent in companies with fewer than 100 employees. These smaller companies often lack the knowledge to identify occupational health hazards and the funds to finance associated control programs; moreover, many are exempt from state and federal occupational health and safety regulations.

The effects of occupational exposures range from lung diseases, cancer, hearing loss, and dermatitis to more subtle psychological effects, many of which are only now beginning to be recognized (Table 4.2). Workplace exposures include airborne contaminants, ionizing radiation, ultraviolet and

Table 4.2 The ten leading work-related diseases and injuries, United States, 1990

Type of disorder or injury	Examples
Occupational lung diseases	Asbestosis, byssinosis, silicosis, coal workers' pneumoconiosis, lung cancer, occupational asthma
Musculoskeletal injuries	Disorders of the back, trunk, upper extremity, neck, lower extremity; traumatically induced Raynaud's phenomenon
Occupational cancers (other than lung)	Leukemia, mesothelioma, cancers of the bladder, nose, and liver
Severe occupational injuries	Amputations, fractures, eye loss, lacerations, traumatic deaths
Cardiovascular diseases	Hypertension, coronary artery disease, acute myocardial infarction
Reproductive disorders	Infertility, spontaneous abortion, teratogenesis
Neurotoxic disorders	Peripheral neuropathy, toxic encephalitis, psychoses, extreme personality changes
Noise-induced loss of hearing	
Dermatologic conditions	Dermatoses, burns, chemical burns, contusions
Psychological disorders	Neuroses, personality disorders, alcoholism, drug dependency

visible light, electric and magnetic fields, infrared radiation, microwaves, heat, cold, noise, extremes of barometric pressure, and stress. Each of these may also interact with other chemical, physical, or biological agents. For example, cardiovascular diseases may be related to a combination of physical, chemical, and psychological job stresses. The workplace can also be the source of a wide range of infectious diseases. Hospital workers in particular must be concerned with protection against hepatitis B, tuberculosis, influenza, and other viral infections, including acquired immune deficiency syndrome (AIDS).

The National Safety Council (1996) estimates that almost 500,000 cases of job-related illnesses occur annually in the United States. Industrial accidents account for another 3.2 million disabling injuries and some 9,000 deaths. The costs associated with accidental deaths and injuries—loss of wages, medical expenses, insurance administrative costs, fire losses, and other indirect expenses—are estimated to exceed $110 billion per year. Insurance data indicate that people in certain relatively high-risk occupations, such as agriculture, construction, mining, and quarrying, have three to four times the average death rate for all industries.

Large though these numbers are, the true magnitude of the health and economic impacts of occupational disease and injury remains unknown. First of all, the recording of data on workers' illnesses and deaths is often incomplete or erroneous. Physicians frequently fail to relate observed diseases to occupational exposures. This is particularly true for neurologically based illnesses (Weiss, 1990) and for chronic degenerative diseases such as atherosclerosis and chronic obstructive respiratory ailments. In other cases, the diagnosed cause of death may not be coded onto the death certificate. Even when the required information is available, it may not be used to promote worker protection. Second, because the health effects of chronic exposures to various toxic agents in the workplace are delayed, and because many workers change jobs frequently, by the time a disease manifests itself it may be difficult to relate it to a specific exposure or combination of exposures. Third, even if an association between a specific disease and a given toxic agent is known, it is often difficult to quantify the concentration of the toxic agent to which the worker was exposed and to estimate the intake and the accompanying dose.

Economic considerations also tend to delay or reduce attempts to solve occupational health problems. In large corporations the directors and officers acting on behalf of the stockholders may insist on operating industrial facilities with an emphasis on profits rather than on occupational health. For example, they may insist that a refinery be kept in operation, with minimum downtime for maintenance or overhaul, at the expense of worker health and safety. Moreover, workers themselves frequently object to controls designed to enhance health and safety when such measures slow production or interfere with comfort. This is especially true in times of economic recession, when many people fear losing their jobs. In addition, there is a chronic shortage of people qualified to investigate and control exposures in industry. Federal funding for training professional occupational health personnel has been declining, and the overall shortage of funds for regulatory organizations has reduced the number and frequency of OSHA inspections. In fact, records show that three-quarters of the U.S. workplaces that suffered serious accidents in 1994 had not been inspected during the nineties (Associated Press, 1995).

Another problem is that the patterns of occupational disease are constantly changing, requiring ever more refined methods to uncover the subtle injuries and disabilities resulting from low-level, on-the-job psychological stress and other nonphysical or chemical hazards. Conducting more dose-response studies entails not only training more professionals in the necessary disciplines but also developing better record-keeping and health data systems to facilitate epidemiologic studies. Also needed are more reliable animal models for predicting human effects, earlier indicators of diseases as they develop, and improved environmental and biological monitoring procedures.

Unless these problems are solved, it will continue to be difficult to identify occupational hazards, determine their magnitude, and judge the adequacy of control measures. The success of these efforts has implications far beyond the occupational environment: because concentrated exposures to hazardous agents frequently occur first in the workplace, and the associated health effects are initially identified and observed among workers, the monitoring of occupational exposures can, and often does, provide the first warning of the presence of potential hazards in the general environment. Clearly, occupational diseases and injuries have consequences not only for workers but also for their families and their communities.

Types and Sources of Occupational Exposures

Years ago, most of the people who were classified as workers were employed in manufacturing. Over the past several decades, however, this situation has changed significantly. Today only about 20 million of the workers in the United States are employed in manufacturing; the remainder are in service industries. Both types of employment have associated occupational health problems and, as would be expected, many problems are common to both. One of the most common problems in manufacturing is the presence of contaminants in the air that result from various industrial processes. Other problems include noise, vibration, and ionizing radiation. Common problems in the service industries include inadequate indoor air quality, low-back pain, and cumulative trauma disorders. In certain situations, problems not heretofore recognized are assuming importance. These include the need to protect workers from potential exposures to biological agents and to provide them with safe (nonslip) floors and stairs and comfortable, employee-friendly work-station environments.

Summarized below are the types and sources of exposures in a variety of activities that involve workers today. The sources of exposure are divided into three principal groups: toxic chemicals, biological agents, and physical factors.

TOXIC CHEMICALS

Toxic chemicals used or generated in industry play a major role in occupationally related diseases. The two primary portals of entry for such agents are the skin and the respiratory tract. Once inside the body, toxic agents can affect other organs, such as the liver and kidneys. Obviously, the ideal way to assure that exposures to toxic chemicals are properly controlled is to assess their toxicologic risk before they are introduced into the workplace (Burgess, 1995). Despite advances in "predictive" toxicology, much work needs to be done to develop practical and reliable screening systems to identify chemicals that have a potential for harming human health.

Typical of the toxic chemicals that can gain access to the body through the respiratory tract are airborne releases associated with metal fabrication, machining, welding, and brazing, followed by cleaning, electroplating, or painting of the finished product. Specific toxic agents include mineral dusts, metal fumes, carbon monoxide, and resin systems used in sand bonding agents. The operations involved generate a host of physical hazards, among them noise, vibration, and heat stress, as well as dermal exposures to cutting fluids and coolants. Such exposures result in almost a half-million cases of dermatitis in this country each year (Burgess, 1991).

BIOLOGICAL AGENTS

The presence of biological agents (bioaerosols) in the air of the workplace is increasingly recognized as a common problem. This is especially true in the healthcare industry, where studies have demonstrated that blood-containing, respirable aerosols are routinely produced in the operating room during surgical procedures (Jewett et al., 1992). Similar exposures have been observed in dental offices. In like manner, flax dust in the linen industry has been shown to contain microbial contaminants. Compounding the problem is the discovery that fungi may grow in respirators designed to protect workers from inhaling airborne contaminants. This is particularly a problem for units containing cellulose and fiberglass filters (Pasanen et al., 1994). Closely related is the presence of airborne allergenic dusts in the workplace that can cause respiratory allergies, such as asthma and allergic rhinitis. These dusts are common in the agriculture and food industries (Virtanen and Mantyjarvi, 1994).

Related instances of exposures to infectious disease agents include health-care workers who are exposed to bloodborne pathogens, such as hepatitis B virus and the human immunodeficiency virus, which causes AIDS. In contrast to the examples cited above, the primary sources of these exposures are accidental punctures of the skin with contaminated needles. Overall it is estimated that almost 6 million workers are subject to these types of workplace hazards. In addition to workers in the health-care industry, these include people employed in funeral services, linen services, medical equipment repair, correctional facilities, law enforcement, and hazardous waste sites. The accompanying exposures are estimated to lead to over 9,000 infections and more than 200 deaths in the United States each year.

PHYSICAL FACTORS

Many health and safety problems in the workplace are caused by physical stresses originating from a variety of sources. Such stresses can be imposed on workers by improperly designed equipment that leads to repetitive motions, forceful motions, static or awkward postures, mechanical stresses, and local vibration (Figure 4.1). Nearly 60 percent of the illness cases reported among workers are associated with this group of factors. The annual cost in compensation expenses for U.S. workers exceeds a half-billion dollars.

Studies show that 25 percent of all injuries in the workplace occur in the

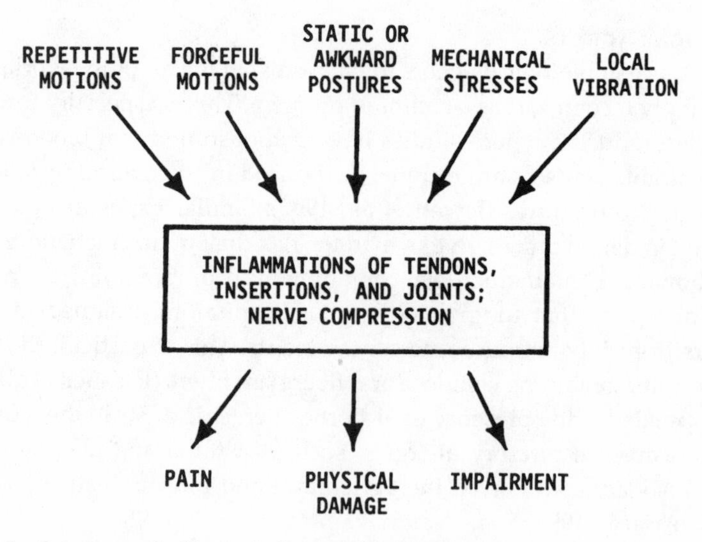

Figure 4.1 Sources and effects of physical factors in the occupational environment

process of lifting and moving objects. Another 15–20 percent of workplace injuries occur as a result of slips and falls. Thus, these two categories represent almost half of all such injuries. Data also indicate that if a worker who has suffered a low-back injury has not returned to work within six months, he or she will probably never return (Snook, 1989). Additional data on the extent of the impacts of physical factors are exemplified by recent studies showing that carpal tunnel syndrome was reported as being a problem within the previous 12 months by 1.4 percent of a sample of currently employed persons 18 years of age or older; repeated neck, back, or spine trouble was reported by 19.1 percent (Schneider, 1994). For these and other reasons evaluation and control of the physical factors in the workplace have been designated as the occupational safety and health issue of the 1990s.

Another major source of physical stress is noise, one of the most common occupational problems since the Industrial Revolution. Because noise-induced hearing loss occurs gradually, invisibly, and often painlessly, many employers and employees do not recognize the problem early enough to provide protection; indeed, for years hearing loss was considered a "normal" hazard of employment. Today most people recognize that noise can interfere with communication, can disturb concentration, and can cause stress. In fact, people subjected to excessive noise have elevated blood pressure, an increased pulse rate, and a higher respiratory rate. By-products of these effects are increased levels of fatigue and higher rates of injuries. Some 28 million people in this country have impaired hearing, and at least 1 million U.S. workers employed in manufacturing have sustained a hearing handicap from noise, meaning that the hearing loss is serious enough to interfere with the activities of daily life (Suter, 1994).

Another pervasive problem is heat stress, especially among workers who wear protective clothing. Heat stress affects as many as 6 million U.S. workers (Grubb, 1990). As body temperature rises, the circulatory system seeks to cool the body by increasing the heart's pumping rate, dilating the blood vessels, and increasing blood flow to the skin. If these mechanisms do not provide sufficient cooling, the body perspires; the evaporation of sweat will cool the skin and the blood and reduce body temperature. Because sweating causes a loss of both water and electrolytes, some form of heat stress, including heat stroke, may develop if body temperature is not decreased.

Other problems include inadequate lighting and the wide range of musculoskeletal disorders that can be caused or aggravated by improper design of equipment and work stations. These problems are discussed in more detail below.

Occupational Exposure Standards

The late 1930s brought the first organizations of occupational health professionals, and with them the first occupational health and safety standards. The American Conference of Governmental Industrial Hygienists

Table 4.3 Threshold limit values for selected chemical substances in the air

Substance	Typical industrial uses or sources	Time-weighted average[a]		Short-term exposure limit[b]	
		ppm[c]	mg/m³	ppm[c]	mg/m³
Ammonia	Coke ovens	25	17	35	24
Benzene[d]	Gasoline refining, organic chemical synthesizing	10	32	—	—
Carbon monoxide	Blast furnaces, coal mines	25	29	—	—
Chlorine	Fabric bleaching, water purification	0.5	1.5	1	2.9
Formaldehyde	Embalming, pathology	—	—	0.3	0.37
Lead (elemental and inorganic compounds)	Battery manufacturing	—	0.05	—	—
Toluene (skin absorption)	Lacquer manufacturing, petroleum refining	50	188	—	—
Trichloroethylene	Metal degreasing	50	269	100	537
Vinyl acetate	Artificial leather manufacturing	10	35	15	53

a. For normal 8-hour day, 40-hour workweek.

b. Not to exceed 15 minutes more than four times per day.

c. Parts per million.

d. A pending proposal would reduce the time-weighted average limits to 0.5 ppm and 1.6 mg/m^3 and assign short-term exposure limits of 2.5 ppm and 8 mg/m^3.

Table 4.4 Biological exposure indices for selected chemicals

Chemical	Sampling time	Biological exposure index
Acetone		
Acetone in urine	End of shift	100 mg/l
Arsenic and soluble compounds including arsine		
Inorganic arsenic metabolites in urine	End of workweek	50 μg/g creatinine
Cadmium and inorganic compounds		
Cadmium in urine	Not critical	5 μg/g creatinine
Cadmium in blood	Not critical	5 μg/liter
Carbon monoxide		
Carboxyhemoglobin in blood	End of shift	3.5% hemoglobin
CO in end-exhaled air[a]	End of shift	20 ppm
Lead		
Lead in blood	Not critical	30 μg/100 ml
Mercury		
Total inorganic mercury in urine	Preshift	35 μg/g creatinine
Total inorganic mercury in blood	End of shift at end of workweek	15 μg/l
Trichloroethylene		
Trichloroacetic acid in urine	End of workweek	100 mg/g creatinine
Trichloroacetic acid and trichloroethanol in urine	End of shift at end of workweek	300 mg/g creatinine
Free trichloroethanol in blood	End of shift at end of workweek	4 mg/l

a. Usually represents alveolar air from lower respiratory system.

(ACGIH), established in 1938, has played a major role in developing limits for exposures in the workplace. Early on, it established "threshold limit values" (TLVs) to provide guidance on permissible concentrations of airborne contaminants in the workplace (Table 4.3; ACGIH, 1996). These TLVs now exist for more than 750 chemical substances. The American Industrial Hygiene Association (established in 1939), OSHA, and NIOSH have essentially followed the ACGIH in developing workplace exposure limits. In the course of these activities, NIOSH has issued health standards for a variety of chemicals used in industry, and it has also issued a series of "Alert" bulletins requesting the assistance of industrial personnel in the control of specific problems. These range from lead poisoning, organic dust, and silicosis to the development of a strategy for prevention and research on homicide in the workplace (NIOSH, 1992).

More recently the ACGIH has supplemented its TLVs by providing biological exposure indices (BEIs) for about three dozen chemicals. Recommended limits and indices for selected chemicals are shown in Table 4.4.

By establishing both TLVs and BEIs, the ACGIH offers a two-step approach to assessment of the importance of chemicals in the workplace: first, monitoring of the air being breathed; second, monitoring of the chemicals themselves or their metabolites in biological specimens (such as urine, blood, and exhaled air) collected from workers at specified intervals. The first step provides data on exposures of workers; the second step provides data on the resulting doses. Although the correlation between these two sets of measurements is generally close, as explained in Chapter 5 the resulting data are not synonymous. Exposures are indicators of conditions in the working environment; doses are indicators of how much of a given contaminant has been taken into the body.

The ACGIH has also established TLVs for physical agents, including heat and cold, noise (Table 4.5), vibration, lasers, radiofrequency/microwave ra-

Table 4.5 Threshold limit values for noise in the workplace

Typical industrial source	Exposure time (hours per day)	Sound level (decibels)
Textile plants, forge shops, machine shops, jackhammer operators	1/8	103[a]
	1/4	100
	1/2	97
	1	94
	2	91
	4	88
	8	85
	16	82
	24	80

a. No exposure should be permitted to continuous, intermittent, or impact noise in excess of 140 decibels.

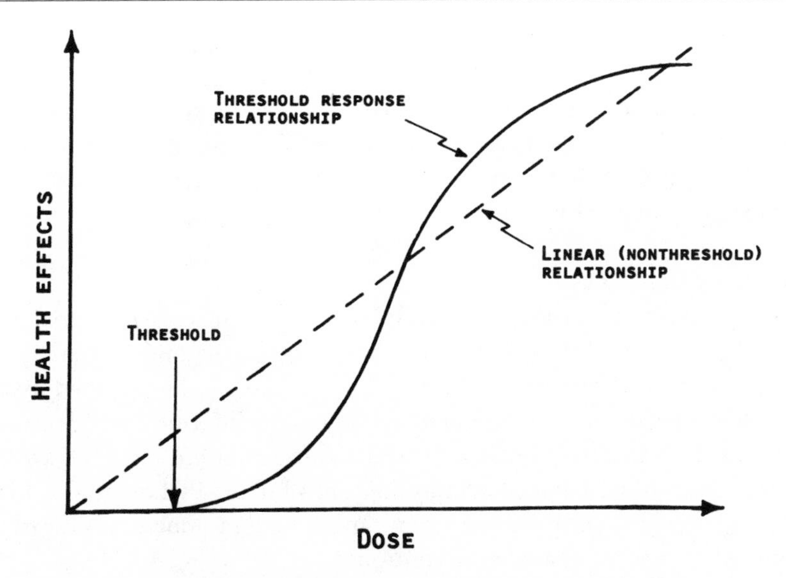

Figure 4.2 Possible relationships between dose and health effects

diation, magnetic fields, and ultraviolet and ionizing radiation (ACGIH, 1996).

The TLVs are based on the best available information from industrial experience, from experiments involving humans and other animals, and, when possible, from a combination of both. The basis on which the values are established may differ from substance to substance. Protection against impairment of health may be a guiding factor for some, whereas reasonable freedom from irritation, narcosis, nuisance, or other forms of stress may be the basis for others. Threshold limit values, however, do not always represent valid thresholds for adverse effects on health. A small percentage of workers, because of age, genetic factors, personal habits (such as cigarette smoking or the use of alcohol or other drugs), medication, or previous exposures, may be affected by some substances at concentrations at or below the threshold limit. For most workers, however, maintaining exposures below the threshold should provide protection. If the relationship between exposure (or dose) and the associated health effects is linear, there will be some effect, however small the exposure (Figure 4.2). For this reason, TLVs cannot be considered the final word on the line between safe and dangerous concentrations, nor should they be used as a definitive index of the relative toxicity of various substances (ACGIH, 1996).

Monitoring the Workplace

Workplace monitoring can be done to assess exposures of workers under routine conditions, to alert workers to abnormal (accident) situations, or to design a control strategy. The type of monitoring program depends to a large extent on the nature of the stress being evaluated.

AIRBORNE CHEMICALS

If the problem is an airborne contaminant, monitoring can be restricted to collecting samples of air. The goal here is to provide data for estimating the concentrations and quantities of contaminants likely to be breathed by the exposed workers. If the contaminant is particulate, information should also be collected on size distribution, since size affects not only how effectively the particles will be retained within the lungs but also where they will be deposited. Another purpose of air sampling is to determine how effectively the given contaminants are being controlled.

Essentially all air samplers consist of a filter or sorbent collector, an air mover or fan to pull the air and associated contaminants through the collector, and a means of controlling the rate of flow. The system selected depends on the purpose of the monitoring program and the type (particulate or gas) and concentration of the airborne contaminant. The collection medium depends on the physical and chemical properties of the materials to be collected and analyzed. Particles are generally collected by means of various types of filters. Gases and vapors are generally collected via solid sorbents and liquid reagents. The air mover may be small and serve only one sampler, or it may be a central vacuum system that serves a number of air-sampling stations. Once a sample has been collected, its identification and quantification commonly require laboratory chemical or physical analysis.

A variety of samplers are in use. They include small, lightweight units that are battery powered and can be worn by individual workers to obtain what are called personal air samples (Figure 4.3). The small size of such samplers permits them to be positioned on the lapel or collar of the workers, so they collect samples representative of the air being breathed. Passive personal samplers that do not require an air mover have also been developed. In the main, the active component of these units is a material that collects the contaminant through diffusion or direct absorption (see also Chapter 15).

A second major type of air sampler is the fixed-location sampler. The

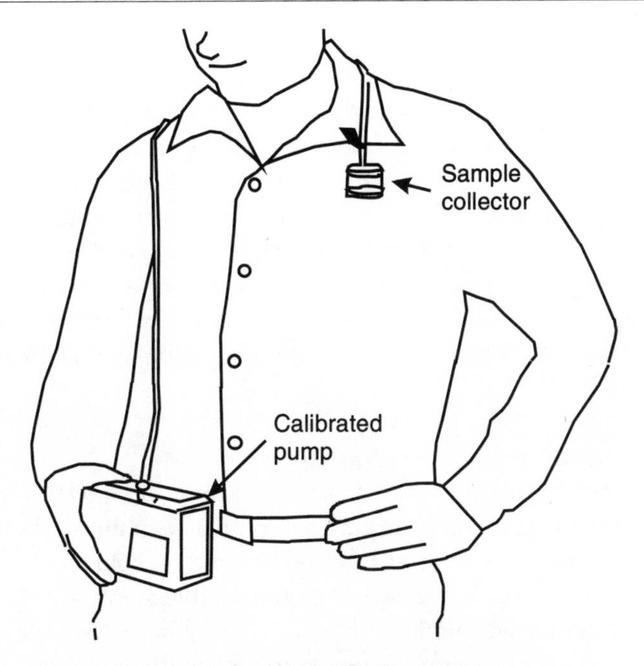

Figure 4.3 Personal air sampler

objective in this case is to determine the concentrations of airborne contaminants within a given space or area. Information provided by such samplers can be used for several purposes—to assess the ongoing effectiveness of efforts to control airborne releases; to determine whether it is permissible for workers to enter a given area; and to warn of excessive airborne contaminants in case of an unexpected or accidental release (Hickey et al., 1993).

As indicated above, an airborne monitoring program may be supplemented by a variety of measurements of biological indicators of contaminants within the bodies of the exposed workers. Generally, this method of monitoring requires the collection of prescribed samples of urine, blood, hair, and body fluids and/or tissues that are analyzed for specific contaminants or their metabolites. The primary disadvantage of such techniques is that they provide information only after an exposure has occurred; the primary advantage is that they provide information on the actual dose to the workers and can be used to confirm the adequacy of workplace monitoring programs.

BIOLOGICAL AGENTS

For those cases in which the transmission of biological agents is through the air, associated monitoring techniques closely parallel those developed for airborne gases and particulates. Because of the many different types of bioaerosols that must be evaluated, no single sampling method or analytical procedure is optimal. Once a sample is collected, the contaminants must be identified. Usually, culturing is required in the case of microorganisms, and microscopic examination in the case of contaminants such as pollen grains, fungal spores, and house dust mites. In the future, such techniques will undoubtedly be expanded to include gene probes and DNA amplification (Macher, 1993).

PHYSICAL AND PSYCHOLOGICAL FACTORS

For certain physical factors such as heat and noise, a variety of measuring instruments is available for collecting real-time data in the workplace. Ergonomic hazards obviously present a different type of problem. Complicating the assessment of ergonomic factors is the multitude of settings in which workers are employed, the large number of interfaces between them and the equipment they use, and the increasing recognition that organizational and psychological factors may be as important as physical factors in terms of the resulting impact on health. Further complicating the situation is the scarcity of data that quantify dose-response relationships for the physical factors and the total lack of data to quantify dose-response relationships for the associated psychosocial or organizational factors (AIHA, 1994). The seriousness of these deficiencies is illustrated by studies that have shown that jobs that place high psychological demands on workers, and give them little control over the work process, are causally related to atherosclerosis of the coronary arteries (Fine, 1996).

Control of Occupational Exposures

A complete and effective control program requires process and workplace monitoring systems and the education and commitment of both workers and management to appropriate occupational health practices. Ideally, protection is provided not only under normal operating conditions but also under conditions of process upset or failure, particularly in systems for controlling airborne contaminants (Gideon et al., 1979). Although a majority of the problems associated with toxic chemicals can be controlled by ventilation, those associated with biological agents, particularly in the case of health-care workers, often require personal protective equipment. The situation is similar when protecting workers against physical stresses, such as noise and heat. To assure that occupational health specialists apply the best methodologies, one must keep in mind the full range of possibilities available.

TOXIC CHEMICALS

In the control of exposures from toxic chemicals in the workplace, emphasis in recent years has been on designing each element in the process to eliminate generation of the contaminant. If this aim proves impossible to achieve, the second defense is to prevent dispersal of the contaminant into the workplace. If this goal cannot be met either, the backup is to collect and remove the contaminant by exhausting the air into which it is released. Care must be taken to remove the contaminant prior to either discharging the air to the environment or recycling it within the workplace. The basic philosophy is similar to that being exercised today in the management and control of solid waste.

In a generic sense, there are six basic approaches for controlling toxic airborne chemicals. The first four involve controlling the chemical at its source by preventing it from gaining access to the working environment (Burgess, 1991). The last two are designed for situations in which the first four approaches have failed or cannot be used. In these cases airborne toxic chemicals must be assumed to be present in the working environment. Usually control involves a combination of these approaches (Burgess, 1994).

Elimination or substitution—This approach involves control at the source by completely eliminating the use of a toxic substance or substituting a less toxic one. Examples include discontinuing the use of mercury in Leclanche-type batteries and using toluene or xylene instead of the more toxic benzene in paint strippers.

Process or equipment modification—Older processes that do not meet existing or proposed occupational health standards can be modified and upgraded. The goal is to design processes so that, as far as practical, the hazardous materials involved are contained within sealed or enclosed equipment and maintenance requirements and associated exposures are minimized.

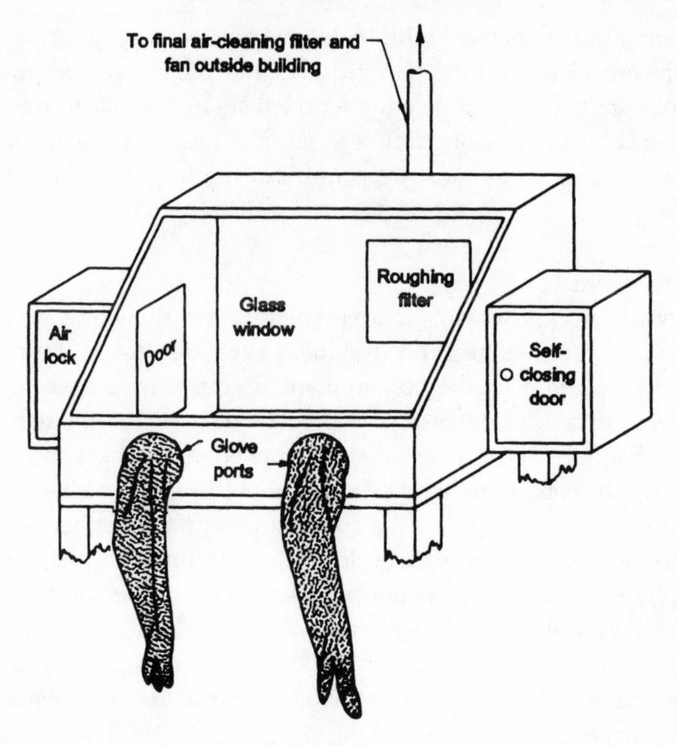

Figure 4.4 Glove box for handling highly toxic or radioactive materials

Isolation or enclosure—Operations involving highly toxic materials can be isolated from other parts of the facility by constructing a barrier between the source of the hazard and the workers who might be affected. The barrier can be a physical structure or a pressure differential. A common approach is to place toxic or radioactive materials in an enclosure with a negative pressure, or to cause the space occupied by workers to be at a positive pressure. Often the isolation of a process from a worker is made possible by the use of robots (Burgess, 1994).

Local exhaust ventilation and air cleaning—Airborne gases or particulates produced by essentially all industrial operations can be captured at the point of generation by an exhaust ventilation system. Two possible types of equipment are a glove box (Figure 4.4) and a laboratory hood (Figure 4.5). Before the exhaust air is released to the environment, however, it should be passed through an air-cleaning device (such as a filter, adsorber, or electrostatic precipitator) to remove any contaminants present.

Personal protective equipment—Controls can also be applied to individual workers, the fundamental concept being to isolate the worker rather than the source of exposure. People working with heavy equipment, for example, should be provided with protective helmets, goggles, and safety shoes. Those working with corrosive and toxic chemicals should be provided with face shields and protective clothing, the choice of clothing being based on the exposure hazard, the amount of body coverage required, and the permeability of the hazardous agent. Because protective garments are generally designed to be impervious, they impede heat loss through evaporation and workers wearing them are frequently subject to heat stress. For these and other reasons, personal protective equipment should generally be considered a last resort, particularly when controlling the inhalation of airborne contaminants through the use of respirators (Figure 4.6). Unless such equipment is used only when

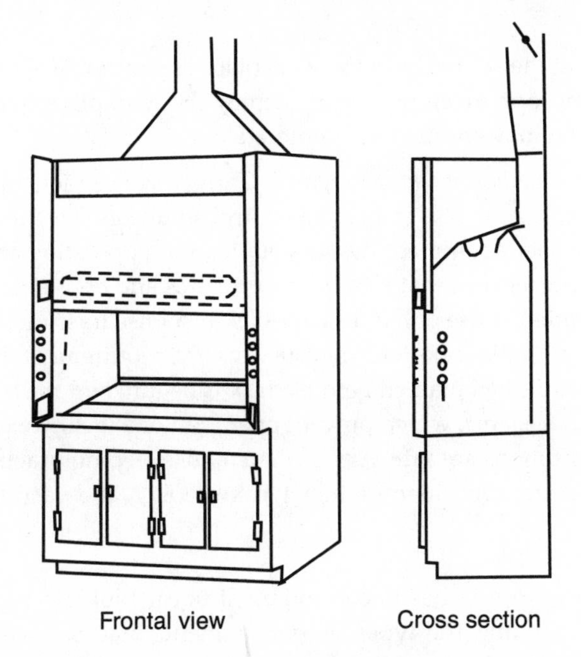

Figure 4.5 Typical laboratory hood for use when handling toxic chemical or radioactive materials

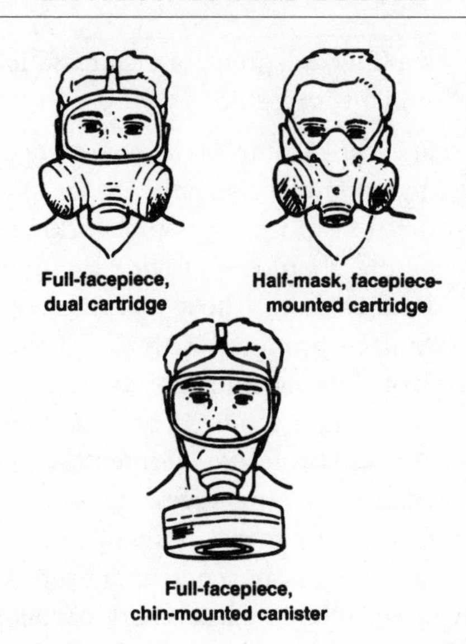

Figure 4.6 Several types of respirators used to protect workers from airborne contaminants

controls at the source or in the workplace are impractical or have failed, appropriate emphasis on keeping the workplace free of contamination may end up being ignored.

Proper work practices and housekeeping—Correct work practices and housekeeping are also important control strategies. The first step involves proper equipment design coupled with operating and maintenance procedures that minimize exposures and emissions. Examples include the use of hand-held quick-response instruments to conduct periodic leak-detection surveys, the requirement that safe-work permits be obtained before a task is begun, and the use of "lockout" systems, which prevent operation of a facility except when conditions are safe. Appropriate housekeeping practices include chemical decontamination, wet sweeping, and vacuuming.

BIOLOGICAL AGENTS

One of the best approaches to controlling airborne biological agents in the workplace is limiting the types of environments, namely, wet spots and pools of water, that promote the growth of organisms. Another key step is proper maintenance of the air handling system, especially the humidifier.

When exposures to biological agents arise primarily through puncture wounds from contaminated needles, as in the health-care setting, the principal controls are puncture-resistant containers for used needles, hand washing to reduce contamination, and appropriate personal protection such as gowns, gloves, and goggles. Control in this case also is dependent on careful housekeeping, with specific requirements for discarding contaminated needles and other sharp instruments, and proper handling of the accompanying wastes (OSHA, 1992).

PHYSICAL FACTORS

Control techniques for physical factors differ from those for toxic chemicals or biological agents. People who apply these techniques must take into account the human/machine interface and promote the use of industrial equipment designed to reduce both physical stresses and accidents. One specific example is heat stress. The degree to which heat affects workers depends on their level of physical activity, the velocity of air movement, the dry-bulb air temperature, and the relative humidity (which influences the effectiveness of perspiration as a mechanism for cooling the body). Control measures include reducing humidity to improve evaporative cooling, increasing air movement via natural or mechanical ventilation, providing radiant-reflecting shields between workers and the heat source, reducing demands in terms of workload and duration, or some combination of these elements.

As with various airborne contaminants, noise can be controlled at the source by damping, reducing, or enclosing the vibrating surface that produces it. For instance, low-speed, high-pitch fan blades can be substituted for high-speed, low-pitch ones; sound absorbers can be placed between the source and the employees; and hearing protection can be provided to individual workers. One of the recent innovations in noise control is the development of headsets that contain a small computer capable of analyzing incoming noise. The information is fed to an electronic controller, which then generates an opposing sound wave that in essence cancels out a portion of the incoming noise. Although primarily effective in controlling low-frequency sounds, such systems have been able to reduce incoming noise by 50 to 95 percent, while permitting the wearer of the headset to hear desired sounds, such as speech and music. Or the electronic controllers can be incorporated into the source of the noise itself, so that it is canceled before it can be emitted (Suter, 1994).

Reducing or controlling stresses associated with the interface between human and machine often involves redesigning equipment to conform to ergonomic principles. Strategies for reducing the occurrence of low-back pain among workers include the use of mechanical aids to lift heavy weights, appropriate workplace layout (to help workers avoid unnecessary twisting and reaching), seat design that allows adjustment and lumbar support, and appropriate packaging of products (to match package weights to human capabilities). Designs are available for safe stairs, floors, and work surfaces. Although back supports or belts and wrist orthotics are increasingly used to avoid workplace injuries, occupational health professionals in general recommend that such devices be limited to those prescribed by health-care providers for treatment of specific disorders. There is insufficient evidence that these devices prevent cumulative trauma disorders or low-back-pain episodes. Once again, those responsible for reducing injuries in the workplace should focus on job design (AIHA, 1994).

The General Outlook

Although many people work in conventional indoor settings, occupational health specialists must keep in mind that a host of people work outdoors. And many of them are exposed to occupational health hazards. Consider individuals who do abrasive blasting to remove surface coatings, scale, and rust in preparing large metal structures for finishing operations. This procedure is routinely performed during construction of bridges, buildings, and ships. Other examples are workers at hazardous waste sites, who may be exposed to a variety of toxic chemicals; those working at airports, who are exposed to air pollution and noise; building and highway construction workers, who are exposed to the hazards of lifting cranes and earth-moving machinery; and those on farms, who are exposed not only to a wide range of toxic chemicals and pesticides but also to higher rates of injury from accidents. Ironically, one of the interesting aspects of the Occupational Safety and Health Act of 1970 is that it exempted farmers from the enforcement efforts of the Occupational Safety and Health Administration. Experience to date certainly supports a reevaluation of this exemption (Kelsey, 1994).

Commercial fishermen and professional forest fire fighters also suffer high accident rates. Mention has already been made of the shift in employment in this country from manufacturing to service industries. Other significant changes are taking place as well—witness the increasing tendency of people to work at home. A key factor supporting this trend is the growing number of women workers, two-thirds of whom have children. Additional developments in electronic communications and robotics will further decentralize our workforce.

Also relevant is the fact that the U.S. workforce includes an increasing number of minorities, is growing older as a whole, and faces ever-stronger competition from overseas production. All of these factors may lessen resolve to maintain high occupational health standards. Furthermore, an increasingly complex array of materials, processes, and equipment is being used in industrial operations. Today many workplace hazards are less obvious and less clearly related to the job: effects on the reproductive system, for instance, and a host of subtle injuries, diseases, and disabilities resulting from low-level, on-the-job psychological stress. Assessment of the effects of these changes on occupational health will require input from a wide range of specialists, including social and behavioral scientists, public health research workers, medical-care specialists, and many others who may currently view the problems of occupational health as being outside their profession (Walker, 1988).

As a result of these changes, the success or failure of occupational health programs depends heavily on the awareness of corporate leaders that changes are occurring in workers' attitudes and needs; that provision of a safe workplace is increasingly important; and that occupational health program decisions have long-term economic impacts. In fact, improvements in the health of the workforce often depend more on changes in the values of corporate management than on technical considerations. Fortunately, many corporate leaders are beginning to emphasize the role of physical activity, nutrition programs, stress management, and other positive lifestyle behaviors in the health and productivity of the workforce.

The U.S. Department of Health and Human Services (HHS, 1990) has recognized the importance of a safe and healthy occupational environment in its *National Health Goals for the Year 2000.* Among the areas targeted for improvement are (1) increasing to 70 percent the number of smaller work sites (fewer than 50 workers) with health and safety programs, (2) increasing to 80 percent the number of larger work sites (more than 750 workers) with employer-sponsored opportunities for activities related to physical fitness, and (3) achieving an across-the-board reduction in repetitive-motion disorders and work-related injuries. That such efforts are needed is illustrated by the fact that repetitive stress injuries alone are estimated to cost U.S. industries over $10 billion annually. Experience has shown that proactive programs in areas such as improved physical fitness, avoidance of alcohol abuse, and injury prevention yield as much as a four-to-one economic benefit to employers through reduced employee turnover, illness, and absenteeism.

Another development that is influencing worker protection in the United States is the signing of the North American Free Trade Agreement (NAFTA). The migration of workers and the transfer of jobs between this country and Mexico and Canada may lead to pressures to reduce worker protection in order to ensure that U.S. industries maintain a reasonable share of the alliance's manufacturing capabilities and markets. A far better approach would be to establish uniform international standards, so that workers in all NAFTA countries are provided high-quality protection.

5

AIR IN THE HOME AND COMMUNITY

PEOPLE HAVE been aware for centuries of the effects of airborne pollutants on human health. Problems stemming from air pollution were noted during the Roman Empire, a time during which some 80,000–100,000 metric tons of lead, 15,000 tons of copper, 10,000 tons of zinc, and over 2 tons of mercury were used annually in industrial operations. Uncontrolled smelting of large quantities of related ores resulted in substantial emission of these materials into the atmosphere (Nriagu, 1996). In the 1300s, authorities in England banned silver and armor smithing because they realized that it contributed to air pollution. In 1895, Pittsburgh passed air pollution ordinances to reduce the amounts of contaminants being released by steel mills. An ordinance passed in Boston in 1911 was the first to acknowledge that air pollution has regional and national as well as local effects.

As is frequently the case, several major, acute episodes were required to demonstrate conclusively to policymakers and the public that air pollution could have significant effects on health. In 1930, for example, in Belgium's Meuse River valley, high concentrations of air pollutants held close to the ground by a thermal atmospheric inversion during a period of cold, damp weather led to the deaths of 60 people. The principal sources of pollution were industrial operations, including a zinc smelter, a sulfuric acid plant, and glass factories. Most of the deaths occurred among older people with a history of heart and lung disease. In 1948, in another river valley in Donora, Pennsylvania, about 20 people died as a result of air pollution from iron and steel plants, zinc smelting, and an acid plant. Again, cold, damp weather was accompanied by a thermal atmospheric inversion. In London in 1952, 4,000 people died as a result of domestic coal burning during similar meteorological conditions. Most of those admitted to hospitals were elderly or already seriously ill, affected by shortness of breath and coughing. Similar episodes occurred in 1959 and 1962, and analyses of death records have shown that additional episodes took place in 1873, 1880, 1882, 1891, and 1892 (Goldsmith, 1968).

Concern is mounting over the effects of decades of environmentally blind industrial development in eastern Europe and the former Soviet Union, which appears to have produced widespread threats to health and life from air pollution. Although specific data are lacking, reports indicate that tens of thousands of people there have developed respiratory and cardiovascular ailments as a result of various airborne contaminants; air pollution was so severe that drivers had to use their headlights in the middle of the day; and in many industrial areas 75 percent of children have respiratory disease (Munson, 1990). Outrage over environmental pollution is even said to have been a catalyst in the 1990 revolution against Communist rule in Poland (French, 1991).

Today the effects of air pollution on human health and on the global environment are widely recognized. Most industrialized nations have taken steps to prevent the occurrence of acute episodes and to limit the long-term, or chronic, health effects of airborne releases. All the same, estimates suggest that up to 8 percent of Americans suffer from chronic bronchitis, emphysema, or asthma either caused or aggravated by air pollution. Newer epidemiologic data suggest that tens of thousands of people in the United States may be dying annually, even though airborne concentrations are within federal limits (Graham and Dockery, 1994). The costs to society are enormous: a lower quality of life for the affected individuals, shorter life spans, and less productivity and time at work.

The Body's Responses to Air Pollution

The intake of pollutants into the lungs and retention at potential sites of injury depend on the physical and chemical properties of the pollutant as well as the extent of activity of the subject exposed. Gases that are highly water soluble, such as sulfur dioxide and formaldehyde, are almost completely removed in the upper airways. Less-soluble gases, such as nitrogen dioxide and ozone, penetrate to the small airways and alveoli.

The ease of entry and the sites for deposition of particulates are heavily influenced by their aerodynamic size and the anatomy of the space through which they are moving (Samet, 1991). Relatively large particles are suscep-

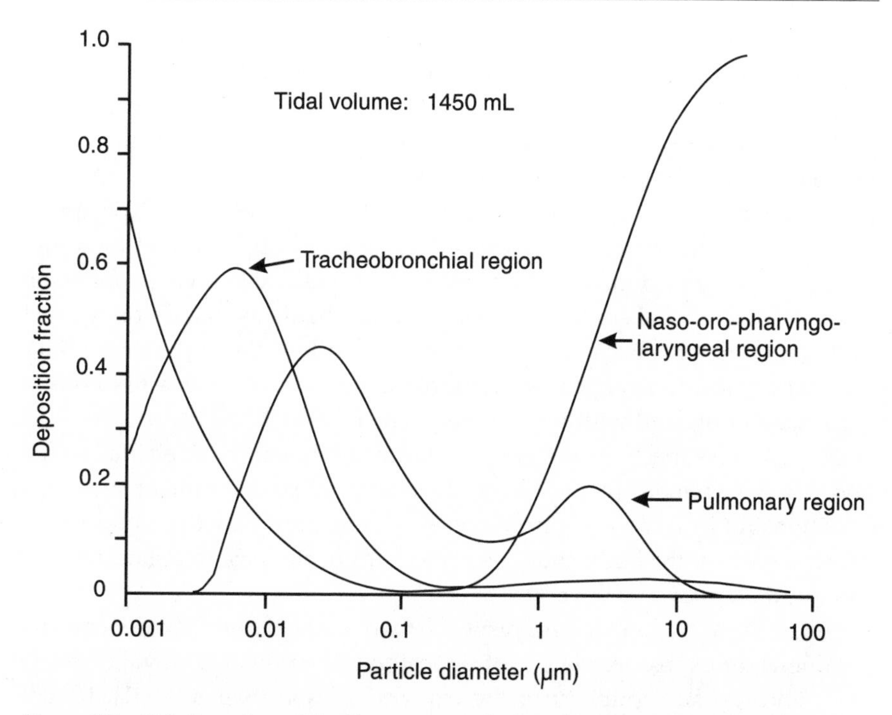

Figure 5.1 Relation of particle diameter to calculated regional deposition in the lungs for spherical particles of density 1 gram per cubic centimeter (assumed intake of about 1.5 liters per breath)

tible to inertial impaction in the airways, where the flow rate is high and the passageways change direction frequently. Particles that penetrate to the small bronchiolar and alveolar region can rapidly deposit in the lungs through settling and diffusion. Fractional depositions in various regions of the respiratory tract of inhaled particles within a range of sizes are shown in Figure 5.1. As may be noted, the total collection efficiency is lowest for particle sizes of about 0.5 micrometer (NCRP, 1995). The reason for this is that such particles do not settle rapidly and are too large to diffuse effectively. Another factor that influences particle delivery and deposition is the aerodynamics of respiration. Total deposition is higher and is more uniformly distributed with slow, deep breathing, as contrasted to rapid, shallow breathing (Valberg, 1990).

As with all kinds of environmental insults, the human respiratory system has a variety of protective mechanisms against airborne pollutants. Particles ranging in size from 5 to 10 micrometers and more are effectively removed by the nose, which acts as a prefilter. Particles that are inhaled and deposited in the upper respiratory tract can be removed by mucociliary action. Those that are deposited in the lower parts of the lungs can be engulfed and destroyed by cells called macrophages. Usually the cilia sweep the macrophages, along with dirt and bacteria-laden mucus, upward to the posterior pharynx, where they are expectorated or swallowed. Exposure to airborne gaseous irritants may cause sneezing or coughing and thus prevent their entry into the deeper parts of the lungs. Even if gases are taken into the lungs and absorbed, the body has mechanisms that detoxify most of them. (A notable exception is carbon monoxide.) Where the detoxification takes place depends on how soluble the gas is in various tissues and organs, and how and with what it reacts chemically.

Despite these mechanisms, some pollutants will still be deposited in the body. If they remain in the lungs, they may cause constant or recurrent irritation and lead to long-term illnesses. If transported by the bloodstream to other parts of the body, they can cause chronic damage to organs such as the spleen, kidneys, or liver.

The likelihood of an adverse response to an inhaled pollutant depends on the degree of exposure, the site of deposition, the rate of removal or clearance, and the susceptibility of the exposed person (Samet, 1991). Recent epidemiologic data suggest that the health effects of particulate matter 10 micrometers or less in a size (the size range on which current exposure standards are based) may be more serious than heretofore believed. Although more data and research are needed (Reichhardt, 1995), it has been estimated that nationwide such exposures may be causing 50,000–60,000 excess deaths annually.

Although the reasons for these suspected effects are yet to be determined, one major contributor could be the relatively high deposition in all regions of the lungs of particles in a size range less than about 0.1 micrometer (Figure 5.1). Another source of the increased biological effects may be the larger amounts of toxic and carcinogenic compounds that smaller particles can adsorb per unit mass (Graham and Dockery, 1994). Although the mass of one 10-micrometer particle is equivalent to that of one thousand 1-micrometer particles, the total surface area of the thousand smaller particles will be at least one hundred times that of the larger particle.

Standards for Air Pollution

Air pollution has been defined as the presence in the air of substances in concentrations sufficient to interfere with health, comfort, safety, or the full use and enjoyment of property. Substances released into the air therefore are considered potential pollutants in terms of their effects not only on human health but also on agricultural products and on buildings, statues, and other public landmarks. In fact, concentrations of some air pollutants considered acceptable for avoiding damage to agricultural products (so-called secondary standards) are far lower than those considered acceptable for humans (so-called primary standards). That standards are needed to protect property is confirmed by many instances of damage, one of the most recent being extensive discoloration of the marble of the Taj Mahal in India.

Under the Clean Air Act Amendments of 1970, 1977, and 1990 (see Chapter 13), the administrator of the U.S. Environmental Protection Agency (EPA) has established national ambient air quality standards (NAAQS) for six key pollutants: carbon monoxide, lead, nitrogen dioxide, ozone, particulate matter, and sulfur dioxide. The selection of these six pollutants is based on their health effects on humans as well as on the environment (EPA, 1994).

Carbon monoxide enters the bloodstream and reduces oxygen delivery to the body's organs and tissues. The health threat is most serious for those who suffer from cardiovascular disease, particularly people with angina or peripheral vascular disease. Exposures to elevated carbon monoxide concentrations are associated with impairment of visual perception, work capacity, manual dexterity, learning ability, and performance of complex tasks.

Lead accumulates in the blood, bone, and soft tissues. Because it is not readily excreted, it also affects the kidneys, liver, nervous system, and blood-forming organs. Excess exposure may cause neurological impairments such as seizures, mental retardation, and/or behavioral disorders.

Nitrogen dioxide can irritate the lungs and lower resistance to respiratory infections such as influenza. Although the effects of short-term exposure are not yet clear, continued or frequent exposure to high concentrations causes increased incidence of acute respiratory disease in children. Nitrogen oxides are also an important precursor of both ozone and acidic precipitation, and may affect both terrestrial and aquatic ecosystems.

Ozone damages lung tissue, reduces lung function, and sensitizes the lungs to other irritants. Scientific evidence indicates that ambient levels of ozone not only affect people with impaired respiratory systems, such as asthmatics, but healthy adults and children as well. In addition, ozone is responsible for several billion dollars of agricultural crop loss in the United States each year.

Airborne particulates can lead to respiratory symptoms, aggravate existing respiratory and cardiovascular disease, alter the defenses of the body against foreign materials, damage lung tissue, and produce latent cancers and premature mortality. The subgroups of the population that are likely to be most sensitive to the effects of particulate matter include individuals with chronic obstructive pulmonary or cardiovascular disease.

Sulfur dioxide, at high concentrations, affects breathing and produces respiratory illness, alterations in the defenses of the lungs, and aggravation of existing respiratory and cardiovascular disease. Subgroups of the population that are most sensitive to this pollutant include asthmatics and individuals with cardiovascular disease or chronic lung disease, as well as children and the elderly. Sulfur dioxide can also produce foliar damage on trees and agricultural crops. It, like nitrogen oxides, is a major precursor of acidic deposition.

While the benefits of the NAAQS are evident, it is important to recognize that the regulated pollutants serve primarily as surrogates for many other more toxic materials known to be present in the air. Such materials include carcinogens, mutagens, and reproductive toxins. Specific examples not covered by the NAAQS are acid aerosols, polynuclear aromatic hydrocarbons, chlorinated volatile organic vapors, and many toxic metals. Fortunately, some of these substances are being regulated through standards established by the EPA to control discharges of specific pollutants into the atmosphere—mercury, asbestos, beryllium, vinyl chloride, benzene, arsenic, and radioactive materials. This list has been expanded through the Clean Air Act Amendments of 1990, which require that similar attention be directed to a list of almost 200 additional air pollutants. For each such pollutant, the EPA has been required to identify and quantify its major sources and to specify the technology that must be applied in controlling related airborne emissions (Spengler, 1992).

The 1990 amendments also require that operating permits be established for stationary sources of air pollution (Lee, 1991). This change follows policies that have proved successful in the control of sources of water pollution. Another concept established by the law is a system of tradable emissions credits: the operator of an industrial facility who reduces emissions below the standard or ahead of the timetable set by the law earns credits that can be applied to future emissions or sold to an operator of another facility. Specific goals of the amendments include still-lower emission limits for cars and light trucks; additional pollution controls for oil refineries, chemical plants, gasoline stations, and dry cleaners; new controls for coal-burning electric power plants to reduce acidic deposition; and phased-out production of chemicals that contribute to depletion of the stratospheric ozone layer.

All the laws and regulations discussed above pertain to air pollution in the ambient (outdoor) environment. Unfortunately, similar attention has not been directed to problems of air pollution inside buildings. In those cases where individual indoor air pollutants have been addressed, the responsibility has been assigned to several federal agencies; in most cases, however, the responsible federal agency has not been specified.

One source of the problem is the general antipathy of the public toward any type of intervention inside the home. Another is the lack of resolution of public policy and public health questions relative to the proper role of government in safeguarding air quality inside public and private spaces (Sexton, 1993). With advances in computers and systems of electronic communications, increasing numbers of people in the more industrialized countries are using their homes as a secondary, or even a primary, place of work. For these and other reasons, the problem of indoor air pollution is likely to increase in the future.

Monitoring Air Pollution

In the past, epidemiologic studies designed to investigate the relationship between air pollution and adverse health effects have relied on crude descriptions of outdoor air quality, such as daily or annual averages for total suspended particulates, or for a specific chemical pollutant such as sulfur dioxide. Today it is recognized that these types of information are not adequate. What is needed is the concentration of the pollutant in the air being breathed and the length of time a person is exposed. Also important is the amount of the contaminant being taken into the body. Obtaining accurate data on the first two factors requires following the exposed individual throughout the day and making real-time measurements of the types and concentrations of pollutants in the air being breathed and the duration of exposure to the given types and concentrations. The integral of these two factors constitutes the "exposure" (Ryan, 1991). In contrast, the "dose" that a person receives is dependent on how much of the pollutant is taken into the body. A variety of factors come into play, including the physiological state of the exposed individual, whether breathing was taking place at a normal or rapid rate, and whether it was through the mouth or nose. The dose may also be influenced, for instance, by whether the exposed person was wearing a protective device such as a respirator.

Other factors that must be taken into account in estimating the dose include the composition and size of the contaminant, if it is an airborne particle, and variations in the concentrations of the pollutants with time. To assure that monitoring programs account for these elements, increasing attention has been directed in recent years to the development of instruments having the capabilities of making both continuous and discrete-period measurements of individual contaminants in the air, as well as of integrating these data to reflect cumulative exposures and to indicate the extent and frequency of periodic exposure peaks. Another advancement in assessing the doses to individuals has been the development of monitors to assure that the samples being collected and analyzed are representative of what the person is actually breathing. Small, inexpensive, lightweight units can be worn by individuals and are sufficiently sensitive to measure ambient concentrations of the more important individual airborne contaminants. Supplementing these developments are new techniques for determining the chemical composition of airborne particles (Spengler, 1992).

Outdoor Air Pollution

Air pollution in the ambient environment can be produced by both stationary and mobile sources and can lead not only to acute and chronic health problems but also to long-term global effects. For example, scientists are concerned that the discharge of carbon dioxide into the atmosphere could lead to a warming of the earth known as the greenhouse effect, and that continued releases of chlorofluorocarbons could destroy the stratospheric ozone layer; they already know that acidic deposition is produced by releases of nitrogen and sulfur oxides into the atmosphere. These macroscopic phenomena are discussed further in Chapter 19.

Outdoor air pollutants come in a wide range of particulates (for example, soot and metallic oxides) and gases (carbon monoxide, nitrogen and sulfur oxides, and hydrocarbons), each with its own chemical and physical properties. Sunlight can produce a series of chemical reactions in compounds such as nitrogen oxides and hydrocarbons and convert them into secondary pollutants, such as photochemical oxidants.

Ozone, the most abundant photochemical oxidant, is one of the most interesting atmospheric pollutants. At ground level it contributes to urban smog; when inhaled it can exacerbate asthma and reduce the ability of the lungs to remove infectious agents and toxins. In the stratosphere, however, ozone provides an essential shield against excess ultraviolet light reaching the earth from the sun. Reduced stratospheric ozone can lead to increased levels of ultraviolet light and accompanying skin cancers in humans, as well as damage to forests, crops, and wildlife.

SOURCES

Table 5.1 summarizes the relative contributions of various sources to five airborne pollutants in the United States in 1993: particulate matter, sulfur oxides, carbon monoxide, nitrogen oxides, and volatile organics (hydrocarbons). Lead is not included because the quantities emitted, although significant in terms of potential health impact, are far less than for the pollutants shown. In 1993, for example, the total quantity of lead emitted in the United States was less than 5,000 tons, with almost half being released as a result of industrial processes. Largely because of its removal from gasoline, airborne concentrations of lead decreased by almost 90 percent during the ten-year period 1984–1993 (EPA, 1994).

The principal industrial source of particulates and sulfur dioxide is the burning of fossil fuels in electricity-generating power plants. Particulate matter is also produced by industrial and natural processes and by agricultural practices. Other sources of sulfur dioxide are paper and pulp mills, smelters, and food-processing plants. Automobiles are a major source of carbon monoxide, nitrogen oxides, and volatile organics, but hydrocarbons are also released by petroleum refineries, solvent manufacturers, and distributors and users of their products, such as gasoline stations and dry cleaners. Various industrial operations (smelters, mills, refineries, factories) also produce a wide range of particulates and vapors, including compounds containing arsenic, asbestos, beryllium, cadmium, and mercury. The relative quantities of these pollutants vary geographically. For example, emissions and airborne concentrations of carbon monoxide, nitrogen oxides, and hydrocarbons are higher in urban areas because of the larger number of motor vehicles and the more prevalent use of dry-cleaning fluids

Table 5.1 Sources of pollutant emissions, United States, 1993

	Particulate matter[a]		Sulfur oxides		Carbon monoxide		Nitrogen oxides		Volatile organics	
Source	10^6 tons	%	10^6 tons	%	10^6 tons	%	10^6 tons	%	10^6 tons	%
Transportation	0.6	22	0.7	3	75.3	77	10.4	45	8.3	36
Fuel combustion	1.2	46	19.3	88	5.4	6	11.7	50	0.6	3
Industrial processes	0.6	21	1.9	9	5.2	5	0.9	4	3.1	13
Solid waste	0.3	11	0.0	0	1.8	2	0.1	0	10.4	44
Miscellaneous	—	—	0.0	0	9.5	10	0.3	1	0.9	4
Total	2.7		21.9		97.2		23.4		23.3	

Note: Because of rounding, the totals may not represent the sum of the individual source contributions; similarly, the percentages may not add to 100.

a. Since 1987, particulate matter smaller than 10 micrometers has been reported instead of total suspended particulates, to represent more closely airborne particles in the respirable range. Natural and miscellaneous nontraditional sources (such as agricultural tilling, mining and quarrying, and wind erosion) contribute 10 to 15 times the quantities or airborne particulates tabulated here.

and industrial solvents; motor vehicles alone can account for 50 percent or more of urban releases of nitrogen oxides.

CONTROL

The best way to control air pollution is to prevent it in the first place, by altering the processes that produce it or by substituting nonpolluting substances for those that generate contaminants. Some controls can be implemented on a generic basis. For example, installation of emission-control devices in automobiles is required as part of the manufacturing process. And for major industrial operations (such as power stations, solid-waste incinerators, and metallurgical plants) that have uniform characteristics, emission controls can be specified nationwide by federal regulations.

Other controls must be tailored to a wide range of characteristics, such as the size of the operation, the processes used, and the age and condition of the facility. Many of these are similar to the approaches used to control airborne contaminants in the workplace. Approaches to controlling the release of air pollutants from industrial and commercial operations include the following:

Atmospheric dilution. This minimal form of control uses the dilution capacity of the local atmosphere to reduce the concentrations of a pollutant to an acceptable level. In many cases, however, it simply spreads the risk over a larger area. The approach is applicable only if the amount of pollution and the number of sources in the area are limited and if regulations permit.

Substitution or limitation. This approach either eliminates the pollutant by substituting materials or methods that do not produce it, or limits the amounts of key chemical elements available for pollutant production. Examples are using substitutes for lead to improve the octane rating of gasoline, and limiting the permissible sulfur content in coal and oil burned in electric power plants.

Reduction in quantity produced. Relevant procedures include improving the combustion efficiency of furnaces, adding exhaust and emission controls to motor vehicles, and keeping the engines of motor vehicles properly tuned to minimize their emissions.

Process or equipment change. Typical approaches include the use of fully enclosed systems in processes that generate vapors, floating covers on tanks that store volatile fluids, and electric motors instead of gasoline engines.

Air-cleaning technologies. Common examples are filters, electrostatic precipitators, scrubbers, adsorbers, or some combination of these, to remove pollutants from airborne exhaust systems.

These control approaches are primarily technical. Air pollution can also be reduced through changes in lifestyle: using economic incentives to promote mass transportation and carpooling; managing land use to restrict certain areas to residential, commercial, or industrial use; and promoting the use of products that are recyclable.

Stimulated by the more restrictive controls on air pollution required by the Clean Air Act Amendments of 1990, state and local agencies have been reviewing and evaluating a wide range of control strategies. A substantial portion of this effort has been directed to motor vehicles which, in many of the critical areas, are the most important source of such pollution. A number of factors, however, make the development of an effective motor vehicle control strategy complicated. (1) Even though emissions of volatile organic compounds (VOC) from new cars are about 95 percent below those from cars sold in the late 1960s, reductions in average emission rates have not been comparable. One reason is that emission control systems tend to fail as cars grow older. (2) Past regulations have focused on emission rates only, ignoring the number of vehicle miles traveled. Since 1970 the annual number of miles traveled by vehicles in the United States has increased 69 percent, the annual number of trips by 22 percent. The latter change is especially significant because emissions are higher during cold starts and during initial operation following startup. (3) Past regulations have been directed primarily to the control of tailpipe emissions; they did not address fuel evaporation, which accounts for upward of 10–15 percent of total VOC emissions. For this reason the 1990 amendments include specific requirements related to the use of reformulated gasoline in areas with severe ozone problems. In the case of certain high-mileage commercial and government fleets, the amendments require the introduction of alternative-fuel vehicles, such as cars that operate on methanol or compressed natural gas. Finally, the amendments require enhanced vehicle inspection and maintenance programs in problem areas (Harrington and Walls, 1994).

Other approaches being considered include incentives for accelerating the retirement of older vehicles, increasing the taxes on gasoline, assessing vehicle registration fees based on emission rates, and imposing fees on motorists during peak commuting periods. Although accelerated retirement programs could be accomplished at about the same cost per ton of emissions reduced as enhanced inspection and maintenance, the cars removed from the road under such programs have at most only a few years of life remaining.

In addition to the development and application of strategies such as these, achieving the goals of the Clean Air Act Amendments of 1990 will require the use in power plants of alternative fuels or coal with a lower sulfur content than at present, coupled with the installation of emission controls such as scrubbers, or the application of innovative technologies such as fluidized bed plants and integrated gasification combined-cycle plants (Abelson, 1990). As the principal sources of air pollution are brought under control, other previously neglected sources will assume importance. Items currently receiving attention are lawn mowers and weed whackers, both of which use small gasoline engines, are used in the millions, and release contaminants into the air. Other possible approaches include expanded uses of solar and nuclear power.

All such strategies or approaches involve trade-offs. Efforts to reduce one type of pollutant may lead to other types of environmental and public health problems. For example, using methanol as a fuel for cars would significantly reduce hydrocarbon emission but increase formaldehyde emission, and using methanol as a fuel may be less cost-effective than other approaches to reducing hydrocarbons (Walls and Krupnick, 1990). Similarly, the use of electric-powered automobiles would reduce airborne emissions in metropolitan areas, but certain aspects of their operation—particularly the use of lead-acid batteries, at present the most cost-effective type for powering such vehicles—could represent an additional source of environmental pollution (Lave, Hendrickson, and McMichael, 1995). Use of electric-powered vehicles could also lead to increased pollution from electricity-generating stations, and the installation of scrubbers to remove sulfur from power-plant emissions will produce large quantities of solid waste. Any proposed solution to an existing environmental problem must be considered in terms of the total system.

TRENDS

Shown in Figure 5.2 are the average ambient concentrations of sulfur dioxide, carbon monoxide, and airborne particulates in the United States for the years 1988–1995 (CEQ, 1996). These graphs and the data in Table 5.2 reflect trends nationwide in the ambient concentrations and releases of six key pollutants in the United States. Volatile organic compounds, listed in this table and Table 5.1, are among the precursors of atmospheric ozone.

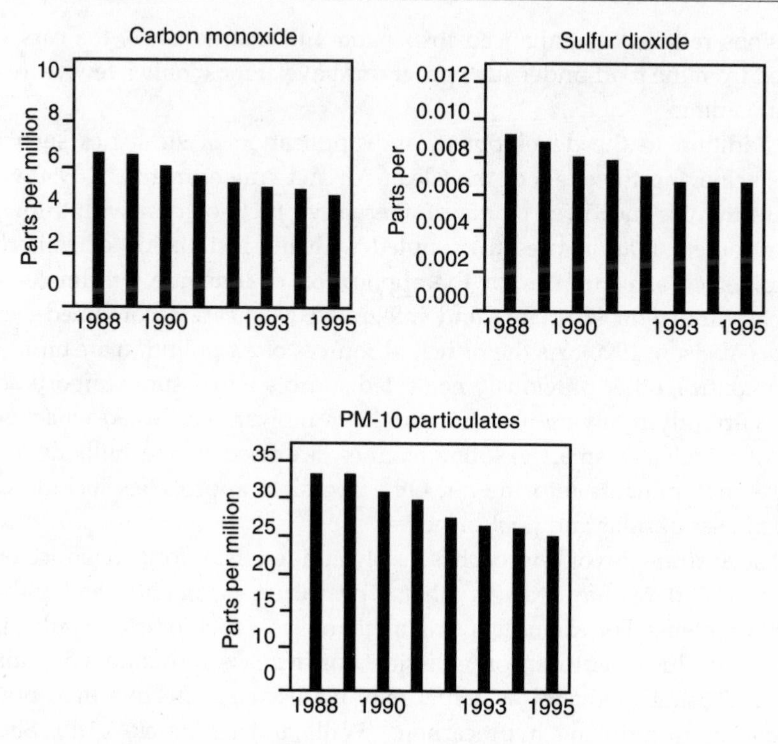

Figure 5.2 Ambient concentrations of carbon monoxide, sulfur dioxide, and airborne particulates (PM-10), 1988–1995, in the United States.

Significant progress has been made over the past decade in reducing both the quantities emitted and the ambient atmospheric concentrations for many of the major air pollutants. Although a 20 percent reduction in the concentrations of airborne particulates was noted, this may not be as meaningful as implied; airborne particles, especially those in the smaller-size fractions, may have more damaging health effects on humans than heretofore estimated. The least progress appears to have been made in controlling the emissions of nitrogen oxides. Emissions of this pollutant showed a 1 percent increase over the past decade; even more troubling is the 2 percent increase for the one-year period 1992–1993 (EPA, 1994). Also of significance is the fact that nitrogen oxides serve as a secondary precursor in the formation of ozone.

Because of substantial variations in the concentrations of airborne pollutants in different areas of the country, many people continue to be ex-

Table 5.2 Sources, annual emissions, removal mechanisms, and trends in concentrations of key atmospheric pollutants, United States

Pollutant	Major source		Estimated annual emission, 1993 (10^6 tons/year)		Removal mechanism	Estimated time in atmosphere	% change, 1984—1993[a]	
	Natural	Man-made	Natural	Man-made			Source[b] emission	Airborne concentration
Particulate matter	Volcanoes, wind, erosion, agricultural crops, forest fires	Industrial processes, combustion, transportation	8–40	2.7	Larger particles settle to earth; smaller particles brought down with precipitation	Minutes to a few days	−10	−20 (arithmetic mean for 799 sites)
Sulfur dioxide	Volcanoes, reactions of biogenic sulfur emissions	Fossil fuel combustion, industrial processes	2	21.9	Oxidation of sulfate by photochemical reactions and incorporation into precipitation	1–4 days	−6	−30 (arithmetic mean for 474 sites)
Carbon monoxide	Forest fires, photochemical reactions	Transportation, general combustion	9.5	97.2	Photochemical reactions with CH_4 and OH	1–3 months	−15	−37 (second-highest 8-hour average for 314 sites)

Table 5.2 (continued)

Pollutant	Major source		Estimated annual emission, 1993 (10^6 tons/year)		Removal mechanism	Estimated time in atmosphere	% change, 1984—1993[a]	
	Natural	Man-made	Natural	Man-made			Source[b] emission	Airborne concentration
Nitrogen oxides	Lightning, biogenic processes in soil	Combustion	18[c]	23.4	Oxidation to nitrate and incorporation into precipitation	2–5 days	+1	−9 (annual mean for NO_2 at 201 sites)
Volatile organic compounds	Biogenic processes in soil and vegetation	Transportation, industrial processes, combustion	0.2–1	23.3	Photochemical reactions with NO and ozone	Hours to a few days	−9	−12[d] (second-highest daily 1-hour maximum for ozone at 532 sites)
Lead	Wind-blown soil	Transportation, fuel combustion, lead smelting and refining, storage-battery manufacturing	0	0.005	Larger particles settle to earth; smaller particles brought down with precipitation	Minutes to a few days	−88	−89 (maximum quarterly average at 204 sites)

a. Data for particulates are for 1988–1993.
b. Based on man-made emissions only.
c. As NO_2.
d. Based on ozone concentrations.

posed in excess of current standards. The concentrations of fine particles in the ambient air in the eastern half of the United States are increasing; the same is true of parts of the Southwest and urban areas on the West Coast (Spengler, 1994). According to EPA estimates, 59 million people in the United States in 1993 (almost one-quarter of the total population) resided in counties where ambient concentrations of at least one pollutant exceeded the NAAQS.

Ground-level ozone appears to be the most pervasive problem: the figure of 51 million people living in counties that exceeded the standard for this pollutant in 1993 was almost double the total population of all the areas in which the standards for the other five pollutants were exceeded. Compounding the problem is that these numbers are based on data for only a single year, 1993, and that only about two-thirds of the population lives in counties that have ozone monitoring equipment. If data on carbon monoxide and particulates are also considered, approximately 200 major metropolitan areas in this country are classified by the EPA as nonattainment areas, that is, they represent areas in which the concentrations of one or more contaminants exceed the applicable health-based ambient air quality standards (Hofeldt, 1996).

Indoor Air Pollution

As noted above, environmental health professionals are directing increased attention to the assessment and control of airborne contaminants in residential and office buildings. There are two major reasons for this broadened emphasis. First, the average person in this country spends between 80 and 90 percent of his or her time in the home or some other indoor environment, 5–7 percent in transit, and usually less than 5 percent of the time outdoors (UNSCEAR, 1993; Brown, 1984). Urban populations and some of the most vulnerable people (the young, the infirm, and the elderly), typically spend more than 95 percent of their time indoors. Second, indoor air pollution has a host of potentially significant sources. In many cities indoor concentrations of nitrogen oxides, carbon monoxide, airborne particulates, and other volatile organic compounds exceed measured outdoor pollutant concentrations. Even if indoor concentrations proved to be low, the longer duration of indoor exposures could render them significant in terms of total exposure of the general population.

As a result, the concentrations of airborne contaminants in these indoor environments are important contributors to exposure, discomfort, irritation, and health effects. Although the effects of pollutants (such as ozone, acidic particles, and polychlorinated biphenyls) of outdoor origin need to be assessed, a variety of pollutants (such as nitrogen dioxide, tobacco smoke, lead, asbestos, volatile organic compounds, formaldehyde, and radon) of indoor origin may have a much greater impact on human health.

Consideration of indoor environments, however, should not be restricted to homes and office buildings. The concentrations of volatile components due to the combustion of gasoline can be six to ten times higher inside a passenger car during rush-hour traffic than at standard urban outdoor monitoring sites. Relatively high concentrations of nitrogen dioxide have been observed in homes with unvented gas cooking stoves, and in the air at hockey rinks because of the use of gasoline- or propane-powered vehicles to resurface the ice (Brauer and Spengler, 1994). Similarly high concentrations of airborne pollutants have been observed in the passenger cabins of airplanes where, because of fuel costs and the desire to restrict operating expenses, up to 50 percent of the air is recirculated. Even with the ban on cigarette smoking on most flights, recirculation results in a buildup of pollutants within the cabin. This, combined with the low relative humidity of the air, often leads to complaints of eye irritation and respiratory problems (Fotos, 1991).

SOURCES

Indoor air pollution has received attention only in the last few decades, yet soot found on ceilings of prehistoric caves provides evidence of millennia of such exposure. Relatively high concentrations of airborne pollutants are common where people cook over open fires fueled by charcoal, wood, coal, kerosene, or oil. In technologically advanced nations the use of synthetic materials in construction, the design of buildings to be energy efficient, and the increased use of wood-burning and coal-burning stoves and kerosene heaters (all of which emit toxic and carcinogenic particles and gases) have increased the quantity and complexity of airborne exposures despite code-specified ventilation requirements. Mobile homes, which account for 20 percent of new housing stock, combined with prefabricated housing units, contribute significantly to these problems. Such homes and units have smaller volumes and lower air-exchange rates, are constructed of more materials containing volatile organic resins, and are more likely to use propane as a cooking fuel than conventionally built dwellings (Spengler and Sexton, 1983).

There are six major sources of indoor air pollution.

Combustion by-products—These products include carbon monoxide, carbon dioxide, sulfur dioxide, formaldehyde, hydrocarbons, nitrogen oxides, and a variety of airborne particles. Depending on the source and type of fuel and the air-exchange rate, average long-term indoor nitrogen dioxide concentrations can exceed the National Ambient Air Quality Standards. (Peak hourly nitrogen dioxide concentrations from conventional gas cooking may range from two to seven times the NAAQS.) Airborne nitrogen dioxide and carbon monoxide are commonly present in homes and schools that use unvented kerosene and gas heaters.

Microorganisms and allergens—Sources of these contaminants include detergents, humidifiers, air-cooling towers, household pets, and insects that live in dust and ventilation ducts, which can release pollens, molds, mites, chemical additives, animal dander, fungi, algae, and insect parts. Air-cooling equipment, cool-mist vaporizers, humidifiers, nebulizers, flush toilets, ice machines, and carpeting can incubate and distribute airborne bacteria indoors. Related bacterial infections include Legionnaires' disease and humidifier fever. Ultrasonic humidifiers using tap water can dispense minerals into the air in the form of aerosolized droplets (Highsmith, Rodes, and Hardy, 1988). These droplets evaporate into particles small enough to enter the lungs, causing respiratory irritation and associated health problems. Tightly sealed buildings in humid climates enhance the growth of molds and fungi.

Formaldehyde and other organic compounds—Formaldehyde is a common ingredient in many building materials (such as plywood and particleboard), in furnishings (draperies and carpets), and in some types of foam insulation. Excess formaldehyde in such products can continue to be released over several years. Other sources include unvented gas combustion units and tobacco smoke. Low concentrations of formaldehyde can cause eye discomfort; higher concentrations can cause lower respiratory irritation and pulmonary edema and may also affect the central nervous system, producing subtle changes in short-term memory, increased anxiety, and slight changes in a person's ability to adapt to darkness. Other sources of organic contaminants in indoor environments include the combustion of wood, kerosene, and tobacco products, which yield polycyclic aromatic hydrocarbons; and personal-care products, cleaning materials, paints, lacquers, and varnishes, which generate chlorinated compounds, acetone, ammonia, toluene, and benzene.

Asbestos fibers—Until 1980 asbestos was used in many building materials, including ceiling and floor tiles, pipe insulation, spackling compounds, concrete, and acoustical and thermal insulation. Workers exposed to asbestos have shown increases in lung cancer, pleural and peritoneal mesotheliomas, and gastrointestinal tract cancers. People living or working in buildings containing asbestos need be concerned only if the asbestos is friable (shedding). In most cases exposures are minimal. Throughout the 1980s EPA programs promoted the removal of asbestos from buildings. Experience has shown, however, that it may be safer to leave the material in place, especially if it is well contained.

Tobacco smoke—More than 2,000 compounds have been identified in cigarette smoke, many of them known carcinogens and irritants. Specific airborne contaminants present in cigarette smoke include respirable particles, nicotine, polycyclic aromatic hydrocarbons, carbon monoxide, acrolein, and nitrogen dioxide. Concentrations in dwellings and in public places vary widely, depending on air filtration rates, the frequency and amount of smoking, the number of air-cleaning devices, and the type of ventilation system. Airborne particle concentrations in a home with several heavy smokers can exceed ambient air quality standards (Spengler and Sexton, 1983). Epidemiologic studies reveal a direct correlation between the extent of maternal smoking and various reported illnesses in children (Ferris et al., 1986).

Radon—Elevated concentrations of the naturally occurring radioactive gas radon (Rn-222) are present in many houses in the United States and are a significant source of radiation exposure; in fact, radon and its airborne decay products account for 55 percent of the radiation dose to the U.S. population (NCRP, 1987). As many as 5–10 percent of homes in the United States may have radon concentrations higher than the level at which remedial action is recommended (Figure 5.3). Based on epidemiologic studies of uranium miners who developed lung cancer as a result of occupational exposures to airborne radon decay products, the presence of this contaminant in U.S. houses may cause 5,000–20,000 lung cancer deaths annually (AMA, 1987).

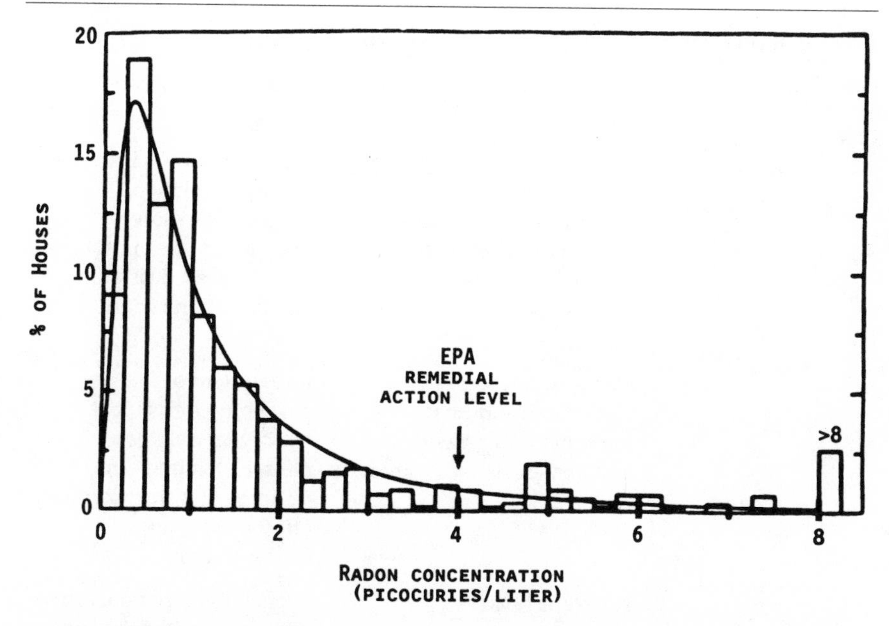

Figure 5.3 Frequency distribution of radon-222 concentrations in 552 U.S. houses, 1986. Adapted with permission from A. V. Nero, M. B. Schwehr, W. W. Nazaroff, and K. L. Revzan, "Distribution of Airborne Radon-222 Concentrations in U.S. Homes," *Science* 234, no. 4779 (21 November 1968), fig. 2, p. 995. Copyright 1986 the American Association for the Advancement of Science.

In addition to the previously discussed chronic health effects, airborne pollutants frequently also cause acute effects—which may result from allergic reactions, nonallergic reactions, or both. Allergic reactions may involve various antibody mechanisms or cell-mediated responses; nonallergic reactions may involve physical irritation, infection, cytotoxic effects, or combinations of these. Table 5.3 summarizes the major sources and acute effects of several categories of common indoor contaminants (several of which are not listed in Tables 5.1 and 5.2).

CONTROL

Problems of indoor air pollution have been highlighted by the so-called sick building syndrome, in which people living or working in certain buildings have reported various types of chronic ailments. One of the origins of this problem was the energy crisis of the 1970s, which led to the application of a number of conservation measures, including reduction in the

Table 5.3 Indoor air pollutants, sources, and effects

Pollutant	Source	Acute effects/symptoms
Gases		
NO_2	Gas stoves, malfunctioning gas or oil furnaces/hot water heaters, fireplaces, wood stoves, unvented kerosene heaters, tobacco products, vehicle exhausts (garages)	Respiratory tract irritation and inflammation; increased air-flow resistance in respiratory tract; increased risk of respiratory infection
CO	Garages, transfer of outdoor air indoors, malfunctioning gas stoves and heaters, tobacco smoke	Impairment of psychomotor faculties; headache, weakness, nausea, dizziness, and dimness of vision; coronary effects with high concentrations
SO_2	Kerosene heaters	Bronchoconstriction, often associated with wheezing and respiratory distress; impairment of lung function; increased asthmatic attacks
Volatile organic compounds		
Formaldehyde	Tobacco smoke, glues, resins	Irritation of eyes and respiratory tract; headaches, nausea, dizziness; bronchial asthma at high doses; allergic contact dermatitis and skin irritation (occupational)
Reactive chemicals		
Isocyanates	Paints, foams, structural supports	Upper and lower respiratory tract irritation; bronchoconstriction; contact dermatitis; pulmonary sensitization

Table 5.3 (continued)

Pollutant	Source	Acute effects/symptoms
Trimellitic anhydride	Plastics, epoxy resins, paints	Bronchial asthma, asthmatic bronchitis, rhinitis; contact dermatitis
Environmental particulates		
Biologic allergens: dust mites, cockroaches, animal dander, protozoa, insects (dusts, fragments), algae, pollen	Pets, insects, plants	Hypersensitivity pneumonitis, causing cough, dyspnea, and fatigue; allergic rhinitis; asthma
Toxins: fungi (including molds) and bacteria (endotoxins)	Fungi and bacteria (especially in high-humidity environments)	Hypersensitivity pneumonitis, causing cough, dyspnea, and fatigue, allergic rhinitis; humidifier fever, causing flulike illness with fever, chills, myalgia, and malaise
Airborne infectious agents		
Legionella pneumophila	Bacteria (in contaminated water sources such as humidifiers and cooling systems)	Pneumonia, Pontiac fever (flu-like symptoms including fever, chills, myalgia, and headache)
Complex mixtures		
Tobacco smoke	Indoor smoking	Eye, nose, and throat irritation; nasal congestion, rhinorrhea; inflammation of lower respiratory tract

amount of outdoor air that was brought into buildings for purposes of ventilation. Prior to the energy crisis, the accepted rate of bringing in outside air was 20 cubic feet per minute per person, a long-standing quantity that was based on the control of odors. Stimulated by the need to conserve energy and reduce associated costs, this rate was reduced to 5 cubic feet per minute. The accompanying development of the "sick" (or "tight") building syndrome led to the rate's being subsequently restored to

20 cubic feet per minute, where it remains today (Mendell and Fine, 1994). Although this change was partially due to increasing recognition of the importance of ventilation in controlling indoor air pollutants, it was made possible through the development of electronic controls that provide better and more efficient operation of ventilation systems; high-efficiency, low-pressure drop filtration units; and more efficient heating and cooling units that make maximum use of heat recovery. As a result, it is possible today to have buildings that are both healthy and energy efficient (Shahan, 1994).

Effective control of indoor air pollution depends on an understanding of several factors. First, the characteristics of the contaminant (concentration, reactivity, physical state, and particle size) must be assessed. All such characteristics affect its removal. Second, the nature of the emissions must be ascertained. Are they continuous or intermittent, from single or multiple sources, primarily inside or outside? Third, the quantitative relationship between the exposure and the resulting health effects must be considered. Are individuals to be protected primarily from long-term chronic exposures to low concentrations, or from periodic short-term exposures at peak concentrations? Finally, the nature of the facility may be important: some controls are more readily applied in residential buildings, others in commercial or office buildings. Also influential are the age and condition of the building.

Control measures for some of the more important indoor air contaminants are summarized in Table 5.4. They closely parallel the steps previously described for the control of ambient air pollution. As is frequently the case, however, different pollutants require different control measures. Adding to the complexity is the fact that indoor air pollution often arises through the interaction of a host of factors that are constantly changing: the temperature and humidity of the air, as well as any contaminants it may contain. Various environmental factors may impact on the building occupants: improper lighting, noise, vibration, and overcrowding. Ergonomic factors and job-related psychosocial problems (such as job stress) also may be important. Each of these alone or in combination can readily produce symptoms that are similar to those associated with poor air quality.

Given our present state of knowledge, it is difficult to relate complaints of specific health effects to exposures to specific indoor pollutant concentrations, since the exposures may involve mixtures of pollutants. Even small problems can have disruptive and potentially costly consequences if the building occupants become frustrated and mistrustful. Effective communi-

Table 5.4 Control measures for indoor air pollutants

Control measure	Pollutant	Example
Ventilation: dilution of indoor air with fresh outdoor air or recirculated filtered air, using mechanical or natural methods	Radon and radon progeny, combustion by-products, tobacco smoke, biological agents (particles)	Exhaust canopy to vent gas-stove emissions; air-to-air heat exchangers for ventilation with energy conservation
Source removal or substitution: removal of indoor emission sources or substitution of less hazardous materials	Organic substances, asbestiform minerals, tobacco smoke	Restrictions on smoking in public places; removal of asbestos
Source modification: reduction of emission rates through changes in design or processes; containment of emissions by barriers or sealants	Radon and radon progeny, organic substances, asbestiform minerals, combustion by-products	Plastic barriers to reduce radon levels; containment of asbestos; catalytic oxidation of CO to CO_2 in unvented kerosene burners
Air cleaning: purification of indoor air by gas adsorbers, air filters, ion generators, and electrostatic precipitators	Particulate matter, combustion by-products, biological agents (particles), airborne radon decay products	Residential air cleaners to control tobacco smoke or wood smoke; formaldehyde-sorbent filters; fan-ion generators to remove airborne radon decay products
Behavioral adjustment: reduction in human exposure through modification of behavior patterns; facilitated by consumer education, product labeling, building design, warning devices, and legal liability	Organic substances, combustion by-products, tobacco smoke	Smoke-free zones; architectural design of interior space; certification of formaldehyde and radon concentrations for home purchases

Source: Reprinted with permission from J.D. Spengler and K. Sexton, "Indoor Air Pollution: A Public Health Perspective," *Science* 221, no. 4605 (1 July 1983), table 2, p. 13. Copyright 1983 American Association for the Advancement of Science.

cation among facility managers, staff, contractors, and building occupants is the key to cooperative problem solving. Another key is recognition that the expense and effort required to prevent most indoor air quality problems is much less than that required to resolve problems after they develop. Many indoor air problems can be prevented by following commonsense recommendations: maintaining proper sanitation, providing adequate ventilation, and isolating pollutant sources. When problems do arise, successful mitigation often requires the application of a combination of techniques, one of the most important being proper maintenance and operation of the heating, ventilating, and air conditioning (HVAC) system (EPA/NIOSH, 1991).

The General Outlook

Despite significant progress in reducing the concentrations of certain airborne contaminants in the ambient environment, much work remains to be done, especially in connection with long-term problems of acidic deposition, global warming, and depletion of the ozone layer. Even if the outdoor air were pollution free, indoor air pollutants would remain a cause for concern. Legislation to control indoor releases is unlikely to be enforceable, both because of the sheer number of housing units and because of the long-established principle of the sanctity of private dwellings. Only intensive public education is apt to provide reductions in pollution concentrations in existing dwellings. The best long-term approach will be to incorporate control measures in new building design, construction, and appliances, following the long-established basic principles of public health.

That air pollution represents a serious environmental and public health problem is exemplified by other developments, such as the determination by the National Institute for Occupational Safety and Health that environmental (so-called secondhand) tobacco smoke is potentially carcinogenic, and recent revelations that fine airborne particles may be more harmful than heretofore suspected. Air pollution also poses a danger to water, fish stocks, forests, natural vegetation, and agricultural crops. In addition, it can lead to deterioration of buildings, statues, and other types of physical structures. Compounding the problem is that air pollution readily crosses national boundaries to affect areas far distant from the emission sources (Spengler, 1992).

6

FOOD

GIVEN the central importance of food in our personal environment, one would expect it to be an aspect of our lives that we control. Yet this is far from the case. In recent years both the number of individuals affected and the number of food poisoning outbreaks have been increasing. *Salmonella enteritidis,* for instance, showed a fourfold increase in outbreaks from 1973 to 1987 (Bean and Griffin, 1990). Some of the increase may be a result of better reporting; nevertheless, food poisoning in the United States is grossly underreported despite the best surveillance efforts. The reported data could well be the tip of the iceberg; most foodborne diseases occur as isolated or sporadic events rather than as part of large dramatic outbreaks that attract the attention and investigative resources of public health authorities. Exacerbating the situation is the fact that many victims do not seek medical care. Current estimates are that the consumption of contaminated food causes some 6.5 million cases of preventable disease and some 9,000 deaths annually in the United States (Walker, 1993).

As is well known, food is essential to life and many foods are known to have ingredients that are beneficial to health. In fact, dietary factors are implicated in the cause and prevention of a variety of ailments, including cancer, coronary heart disease, birth defects, and cataracts. Strong evidence indicates that the consumption of vegetables and fruits protects against these diseases; however, the active constituents have not been completely identified. Whether fat per se is a major cause of disease is a question being debated. However, many nutritionists believe that the consumption of saturated fats probably increases the risk of coronary heart disease. One conclusion from existing epidemiologic evidence is that many individuals in the

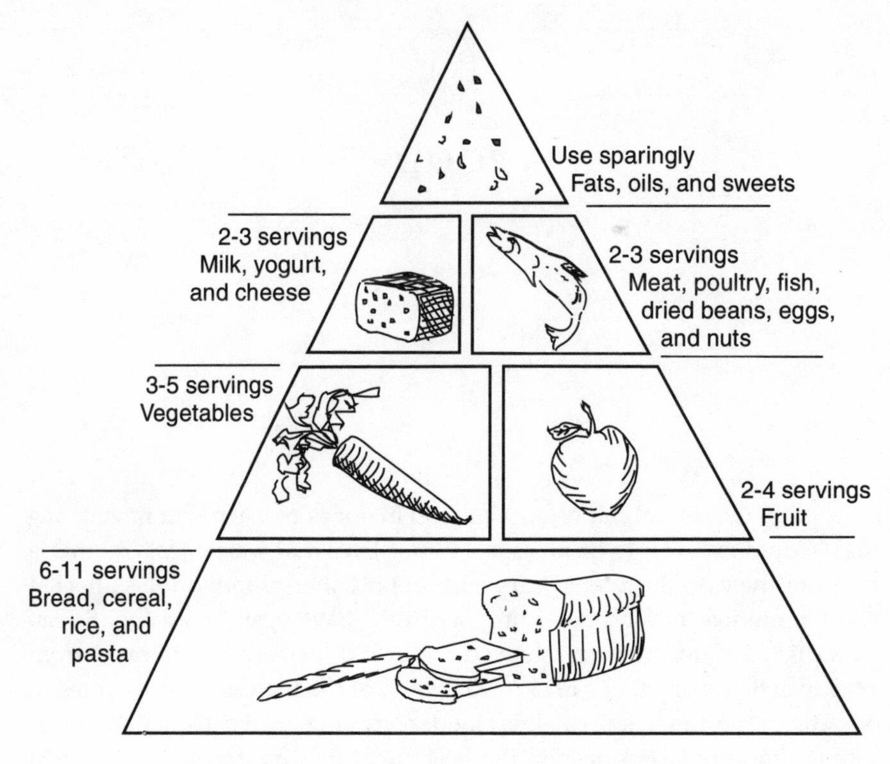

Figure 6.1 The food pyramid—guides for a healthy diet

United States have suboptimal diets and that the potential for disease prevention via improved nutrition is substantial (Willett, 1994).

In recognition of the important role that diet can play in the health of the public, the Department of Agriculture has developed a "food pyramid" (Figure 6.1), which provides advice on what people should eat. Foods at the bottom of the pyramid—grains, vegetables, and fruit—should be consumed in larger quantities than those at the top—meat and dairy products, fats, oils, and sweets. The highest listed numbers of daily servings are for tall, active men; the lowest are for short, inactive women.

The focus in this chapter will be on contaminants that are commonly found in food, their effects on health, and the steps that must be taken in the preservation and handling of food to assure its safety. Aside from objectionable materials, such as rust, dirt, hair, machine parts, nails, and bolts, the contaminants found in food fall into two broad categories: (1) biological agents, such as bacteria, viruses, molds, antibiotics, parasites, and their

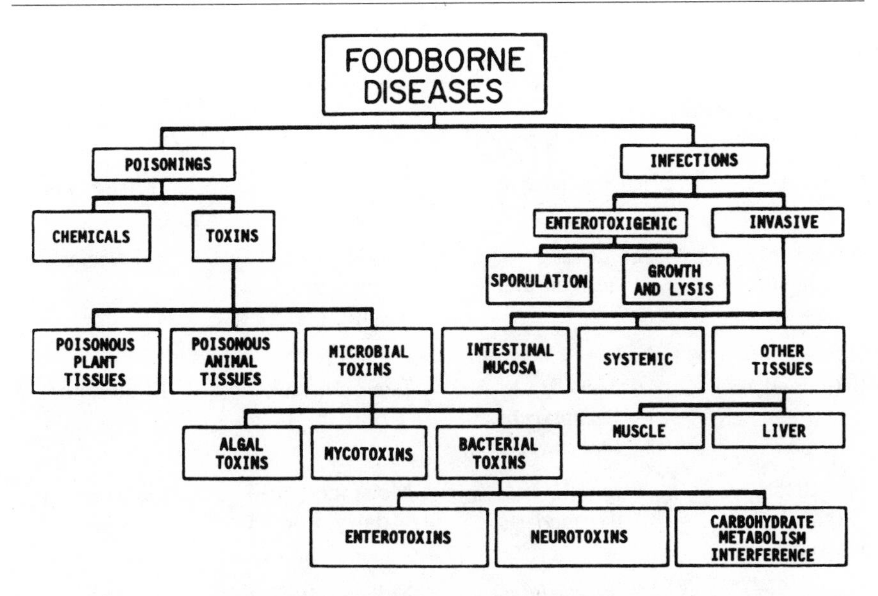

Figure 6.2 Classification of foodborne diseases

toxins, which can cause a wide range of illnesses; and (2) chemicals, such as lead, cadmium, mercury, sodium, phosphates, nitrites, nitrates, and organic compounds, which can have both acute and chronic health effects (Figure 6.2). Such contaminants can gain access to the food chain at any of a multitude of stages during growing, processing, preparation, or storage.

Foodborne Illnesses and Their Causes

Table 6.1 summarizes the major foodborne illnesses, with their causes, the food usually involved, and the incubation period. These illnesses may be caused by foodborne parasites, bacterial infections, viral infections, or toxins that either occur naturally in food or are produced by bacteria or viruses.

PARASITES

Typical parasitic foodborne illnesses include trichinellosis (caused by *Trichinella spiralis*) and giardiasis (caused by *Giardia lamblia*).

Trichinellosis usually results from the consumption of inadequately cooked pork infested with *Trichinella spiralis,* a nematode or worm. It can also be contracted through beef or horsemeat or through cross-contamina-

Table 6.1 Major foodborne illnesses

Illness	Causative agent	Food usually involved	Incubation period
Foodborne parasites			
Trichinellosis	*Trichinella spiralis*	Pork	8–15 days
Giardiasis	*Giardia lamblia*	Raw salads and vegetables	7–10 days
Amebiasis (amebic dysentery)	*Entamoeba histolytica*	Food contaminated with fecal matter	2–4 weeks
Bacterial infections			
Salmonellosis	*Salmonella typhimurium* and *enteritidis*	Eggs, chicken, pork, beef	12–36 hours
Shigellosis (bacillic dysentery)	*Shigella dysenteriae*	Moist food, milk, dairy products	1–3 weeks
Streptococcal sore throat and scarlet fever	*Beta-hemolytic streptococcus*	Milk products, egg salad	1–3 days
Viral infections			
Viral hepatitis	Hepatitis A	Sandwiches, salads	28–30 days
Foodborne toxins			
Paralytic shellfish poisoning	Dinoflagellates (neurotoxins)	Shellfish	Up to 24 hours
Cancer	*Aspergillus flavus* (mycotoxins)	Peanuts, corn, cereal grains	Years
Staphylococcal food poisoning	*Staphylococcus aureus*	Meat, poultry, custards, salad dressings, sandwiches	2–4 hours
Botulism	*Clostridium botulinum*	Home-canned vegetables, fruit	12–36 hours
Gastroenteritis	*Clostridium perfringens*	Meat, poultry, vegetables, spices	10–12 hours
Diarrhea	*Escherichia coli*	Meat, poultry, shellfish, watercress	24–72 hours

tion of food. This serious and painful illness manifests itself in abdominal pain, vomiting, and malaise at the beginning, and in muscular pain, fever, and fatigue over a longer period. The severity of the illness appears to be related to the number of larvae ingested. Prevention includes the thorough cooking of any garbage fed to pigs and of any pork products consumed by humans.

Giardiasis is a protozoan infection principally of the upper small intestine. *Giardia* cysts are present in human stools worldwide, ranging in prevalence from 1 percent to 30 percent, depending on the community and age group surveyed. Children are infected more frequently than adults. The asymptomatic carrier rate is high. The disease shows itself in diarrhea, fever, or both—and in flatulence, nausea, malaise, or abdominal cramps. Often the organism enters the body through the consumption of contaminated water or person-to-person contact. Localized outbreaks may also occur from ingestion of food that has been fecally contaminated by handlers. *Giardia* outbreaks are on the increase, one source being the open salad bars that have become common in many restaurants. Specific sources that have been identified include home-prepared salmon and fruit salads contaminated by *Giardia* cysts (Porter et al., 1990).

BACTERIAL INFECTIONS

Certain bacteria can gain access to foods and be ingested and transported to the digestive tract, where they can multiply and cause illnesses. Common examples are *Salmonella, Shigella,* and *Streptococci.*

Salmonella exist in the intestines of chickens, dogs, and rodents. These bacteria can also live in the ambient environment and can survive conditions that many other organisms cannot. This resistance accounts for their transmission through drinking water as well as food. Once ingested, *Salmonella* multiply in the intestines and cause fever and diarrhea within 12–36 hours.

In 1986, 37 percent of the chicken, 12 percent of the pork, and 5 percent of the beef inspected by the U.S. Department of Agriculture was contaminated with *Salmonella* (Rubel, 1987). Recent data confirm that comparable rates of contamination continue to be experienced. Eggs and poultry are primary sources of infection. Formerly, transmission of *Salmonella* through eggs was thought to involve contamination on the outside of the shells. Today the organisms are often present inside the eggs, as a result of infections in the ovaries of chickens. All protein foods requiring a large amount of handling are subject to contamination. Low-acid foods, such as meat pies, custard-filled bakery products, and improperly cooked sausages, are common sources of *Salmonella* outbreaks. Transmission can be prevented by pasteurization of frozen eggs and thorough cooking.

Bacillic dysentery is caused by *Shigella dysenteriae*, an infectious agent common wherever sanitation is a problem. Two-thirds of all cases, and most deaths, occur in children under 10 years of age. Illness in infants less than 6 months old is unusual. Secondary attack rates in households can be as high as 40 percent. *Shigella* are commonly present in human feces, and transmission is favored by crowded conditions where personal contact is unavoidable. Food handlers can readily spread the infection through contamination of food. Flies can also transfer the organisms to nonrefrigerated food, where they can multiply. Ingestion of a relatively large number of organisms is required, and onset of the disease is delayed for 1–3 weeks while the bacteria multiply in the body. Personal cleanliness, particularly in handling food, is an important factor in the control of this disease (Benenson, 1995).

Beta-hemolytic streptococcus causes streptococcal sore throat and scarlet fever. The sore-throat patients frequently exhibit fever, sore throat, tonsillitis, and tender lymph nodes. Scarlet fever is characterized by a skin rash. The two ailments are common in temperate zones, well recognized and diagnosed in semitropical areas, and less frequently recognized in tropical climates.

Explosive outbreaks of streptococcal sore throat may follow ingestion of contaminated food. Milk and milk products have been associated most frequently with foodborne outbreaks; egg salad and deviled hard-boiled eggs have recently been implicated with increasing frequency (Benenson, 1995). Although scarlet fever is more often conveyed by direct person-to-person contact, transmission has also been traced to consumption of raw milk, which is also a principal carrier in cases of streptococcal sore throat. To prevent the growth of the organism, milk should be held at a temperature of 50° F (10° C). Milk products and protein mixtures such as egg salad may also carry the organism if they are not made from pasteurized ingredients.

VIRAL INFECTIONS

Infectious hepatitis (hepatitis A) is a highly contagious disease caused by a virus whose symptoms are fever and general discomfort. The disorder occurs most frequently among school-age children and young adults, with the infectious agent commonly being present in feces. Common-source outbreaks have been related to food, such as sandwiches and salads, that is not cooked or is handled after cooking by infected food handlers. Raw or undercooked mollusks harvested from contaminated waters may also be sources of infection. In the United States an estimated 1,000 cases each year are attributed to suspected foodborne or waterborne sources. An outbreak in 1988 was traced to commercially distributed lettuce. Certain commercially processed foods, such as frozen raspberries, appear to be associated with transmission of the virus in the United Kingdom (Rosenblum et al., 1990). One of the latest concerns is bovine spongiform encephalopathy, or "mad cow disease," which was observed in large numbers of beef and dairy cattle in England in 1996. It was feared that, without adequate controls, the disease would spread to herds in other countries and that consumption of infected meat could lead to similar disorders in humans (O'Brien, 1996). To control the problem, areas with infected cows were quarantined and the diseased animals were destroyed.

TOXINS

Foodborne illnesses may also be caused by the ingestion of toxins naturally present in plants or animals, or subsequently produced in food as a result of contamination by bacteria, viruses, and fungi.

Naturally occurring toxins. Laboratory studies of natural foodstuffs and cooked food show that they contain a surprising array of toxins that would normally not be permitted as regulated additives. Carrots, for example, contain *carotatoxin,* a fairly potent nerve poison; *myristicin,* a hallucinogen; and *isoflavones,* which have an estrogenic effect similar to female hormones. Peanut butter contains *aflatoxins,* some of which (aflatoxin B, for instance) are acutely poisonous and carcinogenic to laboratory animals. The common assumption that "natural" is safe, and "man-made" is suspect, is contrary to current scientific knowledge. Data show that a typical diet contains far more natural carcinogens than synthetic (NRC, 1996).

Of the many toxins that occur naturally in plants and animals, only a few have been specifically associated with human illness. The most dramatic example is paralytic shellfish poisoning, caused by a highly poisonous neurotoxin that is a metabolite of certain marine dinoflagellates. Poisonous concentrations accumulate in shellfish (mussels, clams, and occasionally scallops and oysters) that feed in areas where the dinoflagellates are undergoing rapid growth (that is, blooming). Because the blooms impart a characteristic red color to the water, the phenomenon is commonly referred to as red tide.

The shellfish show no disturbance as a result of the toxin; when the contaminated seafood is consumed by humans, however, symptoms of toxicity usually develop within one to three hours—numbness of the lips and fingertips, ascending paralysis, and finally, in cases of severe poisoning, death from respiratory paralysis. Should the person survive the first 24 hours, recovery is generally uneventful. Prevention depends on avoiding the consumption of shellfish during periods of red tide (Newberne, 1994).

The problems of toxins in seafood are not restricted to shellfish. Toxins can also accumulate in certain types of fish. Consumers of tropical reef fish, such as snapper, grouper, and barracuda, may be at risk for ciguatera intoxication, a food poisoning that can cause serious and persistent neurological symptoms. With the increasing consumption of seafood by people seeking to reduce their intake of animal fat, public health officials are directing increased attention to these types of problems (Ahmed, 1991).

Toxins can also be produced in plants by naturally occurring fungi. Since many toxigenic fungi can thrive in a wide variety of environmental situations, it is not surprising that numerous recorded episodes of human and livestock illnesses and deaths result from the ingestion of food or feed contaminated with fungal toxins. The toxic metabolites produced by fungi are referred to as *mycotoxins*, and diseases resulting from exposure to mycotoxins are called *mycotoxicoses*.

The best-characterized mycotoxicoses are those in which mycotoxins were consumed in amounts high enough to result in immediate signs of illness or death. In most of these cases, the level of fungal invasion of the responsible food or feed was substantial, and mold damage was highly visible. Humans consumed the infected food either because they were ignorant of the possible consequences or because no other food was available. Mycotoxicoses have been far more common in animals than in humans, simply because livestock are more likely to be fed mold-damaged feed. The discovery in the 1960s of a specific group of mycotoxins known as the aflatoxins added new dimensions to this problem.

Mycotoxicoses can no longer be considered as occurring only after severe mold invasion of food, as limited mainly to agriculturally primitive regions, and as preventable simply by handling and storing agricultural commodities in ways that avert serious mold growth. Aflatoxins can readily contaminate certain primary food and feed, such as peanuts, corn, and other cereal grains, and tree nuts. Studies are needed to determine precisely whether contamination takes place in the field or during harvest, transport, or storage. Significant evidence has accumulated to suggest that contamination by mold growth occurs in the field and progresses further during storage. Some aflatoxins have immediately acute effects. Others take their toll through chronic exposure; in one case the result is primary liver cancer (Nelson and Whittenberger, 1977).

Toxins from improper food handling. In contrast to the bacterial infections described above, which are caused directly by the organisms, some foodborne illnesses are caused by the toxins produced by bacteria that are not in themselves harmful. Four principal types of bacteria produce harmful toxins.

(1) If present in food under the proper conditions, *Staphylococcus aureus* will produce one or more enterotoxins that can lead to foodborne illness. This organism is one of the principal sources of acute foodborne illnesses in the United States. Ingestion of contaminated food causes acute intestinal disturbance, usually within two to four hours. *Staphylococcus aureus* can readily be transmitted to food from infected cuts, boils, sores, postnasal drip, or sprays expelled during coughing or sneezing. Meat (especially ham), meat products, poultry, poultry products, and poultry dressing are common sources. Custards used for pastry fillings have likewise been involved in outbreaks of food poisoning. *Staphylococci* are present in air, water, milk, and sewage. They grow rapidly, especially in food held at room temperature for several hours before being eaten. Refrigeration after cooking provides effective control.

(2) Botulism is caused by the ingestion of food containing the toxins produced by *Clostridium botulinum.* This organism will grow in food and it thrives under conditions of reduced oxygen. The spores of the bacillus may be present in all types of food, especially spoiled vegetable or animal matter. Underprocessed food and preserved food provide ideal environments. The toxins can exist for long periods, are very resistant to destruction by heat, and are extremely potent; a small quantity can cause death. In fact, these toxins are generally considered to be the most potent produced by any biological organism.

Five types of *Clostridia* have been identified as causing botulism, each varying in the degree of toxicity to its host. Specific antitoxins are available for each type. Heating food to 212° F (100° C) will destroy the organisms; destruction of the spores, however, requires prolonged heating. Swelling or bulging canned goods do not necessarily indicate the presence of spore-forming bacteria, but goods in such containers should not be consumed.

(3) *Clostridium perfringens* (gastroenteritis) is a mild illness, and those afflicted seldom seek medical treatment. *Clostridium* is widely distributed and can usually be found in soil, dust, human feces, and animal manure. Meat

and poultry are frequently contaminated, as are vegetables and spices. Although *Clostridium* rarely produces spores while growing in food, it does produce spores in the intestinal tract. They are very resistant to heat; some can withstand boiling temperatures for up to six hours, so even cooked food is frequently contaminated. Once in the human intestine, the organism multiplies and generates an enterotoxin, which in turn produces diarrhea and abdominal cramps. The illness is self-limiting, so treatment is supportive, not curative. Prevention requires inhibiting germination of the spores and proliferation of the vegetative cells. It can be accomplished by holding cooked food at a temperature either high enough (above 133° F, 56° C) or low enough (below 64° F, 18° C) that *Clostridium perfringens* cannot multiply (Bryan, 1980).

(4) *Escherichia coli,* a common cause of traveler's diarrhea, are present in the lower intestinal tract of most warm-blooded animals and are the most prevalent oxygen-tolerant bacteria in the large intestine of humans. They are transferred from feces and intestinal contents to carcasses and meat during processing. Shellfish and watercress can be contaminated if grown in sewage-contaminated waters. The enterotoxigenic strains cause illness 8 to 44 hours after infection. Diarrhea usually ceases within 30 hours. However, a virulent strain of *Escherichia coli,* specifically *E. coli* 0157:H7, caused three deaths, and more than 350 people became ill in the northwestern United States in 1993. The source of the contamination was found to be undercooked hamburgers sold by a fast-food chain.

Contamination with *Escherichia coli* can be minimized or prevented by careful personal hygiene during food handling. Ill effects can also be prevented by heating food long enough and at temperatures high enough to destroy the bacteria, then cooling the food at temperatures sufficiently low (below 40° F, 4° C) to prevent their proliferation. Also essential to preventing the spread of these organisms are a safe drinking-water supply and proper methods of sewage disposal (Bryan, 1980).

As a result of the 1993 episode, the Department of Agriculture now requires all raw meat and poultry to be labeled with instructions regarding proper cooking and handling (Associated Press, 1993).

Chemical Contaminants

Many compounds in food plants have been chemically characterized, and a number have been found to be toxic. Although it is generally assumed that the natural components of food, even those known to be toxic, do not constitute an acute health hazard, very little information exists about the toxic effects from ingestion of these compounds over long periods. That certain foods have been consumed for centuries without obvious adverse effects is not sufficient proof that they are safe.

Abnormal and toxic metabolites are frequently produced when plants are subjected to stress. These metabolites include protease inhibitors (which are found in many legumes and can inhibit the proteolytic activity of certain enzymes), hemagglutinins (which are found in castor beans, soybeans, black beans, and yellow wax beans and can agglutinate red blood cells), goitrogens (which are found in cabbage, turnips, rutabagas, mustard greens, horseradish, and white mustard and can cause hypothyroidism), and allergens (which are present in peanuts, certain fruits, and grains and cause allergies of the skin and respiratory and gastrointestinal tracts). Some plants, such as the bracken fern, are poisonous to animals, and the toxins may be present in milk from cows that have eaten the plants.

Some chemicals may be added to foods during production and processing. Nitrates and nitrites are common ingredients of nitrogen fertilizers and thus can be taken up by vegetables. The concentrations in vegetables vary widely, depending primarily on the species, the concentration of nitrate and molybdenum in the soil, light intensity, and drought conditions. High concentrations of nitrate in baby food, much of which is converted into nitrites, can be a cause of methemoglobinemia in infants. Nitrite modifies hemoglobin compounds in the blood so that they cannot transport oxygen from the lungs to the tissues. Methemoglobinemia can be detected in the blood of infants who consume formulas made from water containing high nitrate concentrations. The recommended limit for the nitrate ion in drinking water is 45 milligrams per liter, which corresponds to about 10 parts per million of nitrate nitrogen. A concentration of about 2 milligrams of nitrate ion per kilogram of body weight will produce methemoglobinemia in infants.

Sodium is the seventh most plentiful element in the crust of the earth. In its ionized form, sodium enters the food chain during processing or domestic cooking, or at the table. Numerous studies of the effects of sodium chloride intake on human health have yielded contradictory and confusing results, but the data do indicate that excessive salt intake is related to hypertension and gastric injury in some individuals. The estimated average current daily intake of sodium chloride by adults in the United States is 6.8 grams, approximately double the recommended daily intake of 3–4 grams. The recommended daily intake for children and adolescents is 2–3 grams.

Phosphate is being increasingly used in poultry processing and in the production of soft drinks and modified starches. An excessive daily intake of phosphate can lead to premature cessation of bone growth in children and may contribute to osteoporosis in the elderly. How the human body processes phosphate is influenced by its intake of other chemicals, such as calcium. Good health depends on maintaining a proper ratio in the intake of these two chemicals. The recommended daily intake for adults is about 1,200 milligrams of phosphate and 1,200–1,800 milligrams of calcium; that is, the ratio of calcium to phosphate should be about 1.5:1. The recommended daily intake of phosphate for infants is 150 milligrams.

Metals such as mercury, lead, and cadmium can have severe effects on human health. Mercury discharged into rivers, lakes, and oceans in the form of inorganic salt or as the metallic element (which is not harmful to humans) can be converted by microbes to alkyl mercury, primarily methyl mercury. In this form mercury can be a significant health hazard. Large-scale poisonings by such compounds have caused deaths and cases of permanent damage to the central nervous system. In a classic episode in Japan in the early 1950s, industrial wastes containing mercury were discharged into Minamata Bay. More than 100 people who ate contaminated fish were poisoned, and 46 died. In the early 1970s tuna and swordfish were recalled from stores in the United States because of mercury concentrations in excess of 0.5 parts per million, the limit set by the Food and Drug Administration. In 1969 a family in New Mexico was severely poisoned after eating pork that contained a methyl mercury fungicide. Similarly, in Iraq in the 1950s people were poisoned as a result of using cooking oil made from seeds that had been treated with fungicides containing mercury.

The hazards of lead have been known for centuries. Its most salient adverse effects are on the nervous system, the hematopoetic system, and the kidneys. Lead can be ingested indirectly in the form of lead-based paint, drinking water contaminated by the lead pipes that distribute it, metal vessels, pottery glazes, and fungicides that enter the food chain. Before the use of lead was banned from gasoline in the United States, vegetables and other crops from farms located near major roadways were often contaminated with lead by means of atmospheric deposition. Shellfish may also concentrate lead from contaminated water.

Cadmium has effects similar to those of lead and moves through the environment and into the food chain by similar pathways. A major source of environmental contamination with cadmium is discarded batteries such as those used in pocket calculators, cameras, radios, and flashlights. Cadmium affects the kidneys and may be related to hypertension and renal cancer.

Organic Contaminants, Antibiotics, and Hormones

Organic contaminants in food include the chlorinated hydrocarbons, polychlorinated biphenyls, chlorinated dibenzo-*p*-dioxins, and chlorinated dibenzofurans. Their presence in the environment, and subsequently in food, stems primarily from the use of pesticides and herbicides and of polychlorinated biphenyls in heat exchangers in electrical transformers, which sometimes leak or explode. Organic compounds can produce pathological changes in the body, including the stimulation of certain metabolizing enzymes. Such stimulation may have serious implications for human health, particularly if the affected enzymes activate or deactivate certain environmental chemicals. Currently, very little is known about the mechanisms or significance of these responses.

Organic contaminants can also be produced in foods, especially meat, through the cooking process. Tests have shown, for example, that the browned or burned portions of meats that have been charbroiled, whether fried or smoked, contain heterocyclic aeromatic amines—many of which have been shown to be highly mutagenic. Although some of these compounds originate from the smoke deposited on the meat, others are produced by the breakdown of meat protein during the cooking process. Benzo-*a*-pyrene, ubiquitously produced when organic matter is burned, is one of these components. Not only is it mutagenic and carcinogenic, but under laboratory conditions it can promote the carcinogenic action of other chemicals.

Additional chemicals that have been identified as being produced by the cooking of meat include the polycyclic aromatic hydrocarbons, as well as numerous breakdown products of common dietary amino acids. Even the roasting of beef, which takes place at much lower temperatures, has been shown to produce substances thought to be mutagenic. Measures that have been suggested to avoid the production of these compounds include using alternative processes such as stewing, poaching, or boiling, to cook meat, and employing a microwave oven to cook fish and poultry (Newberne, 1994).

Today nearly half of all antibiotics produced in the United States are fed to farm animals to prevent disease and to promote growth. Most are broad-spectrum antibiotics such as penicillin or tetracycline (Wright, 1990). The transmission of antibiotics through milk and dairy products could affect people who have adverse reactions to certain drugs and, of course, infants and small children. Such practices could also lead to the development of microorganisms that are more resistant to antibiotics. Already resistance is increasing in *Salmonella* strains (Cohen and Tauxe, 1986). This resistance can lead in turn to the ineffectiveness of antibiotics currently used in medical treatment and a need for newer, often costlier, antibiotics. Although the addition of antibiotics to animal feed in the United States is continuing, no action has been taken to ban this practice. As a minimum, however, some experts have called for restricting such use to narrow-spectrum antibiotics that will not result in the development of resistant strains (Wright, 1990).

Hormones are also being used to promote growth in various farm animals. As long as the hormones are used under prescribed conditions, there is no evidence of any associated risks to human health. Although the Codex Alimentarius Commission has endorsed the use of three natural and two synthetic hormones for such purposes, it has been careful to stipulate that approval is contingent upon the hormones being used according to "good veterinary practice," supported by analyses to confirm that hormone residues in meat remain below prescribed limits (Kaiser, 1995).

Care in Food Preservation and Handling

A variety of safe methods are available for preserving wholesome food, preventing contamination, and destroying organisms or toxins that may have gained access to or been produced within the food. The effective application of these methods requires an understanding of the factors that affect bacterial growth. These will be briefly reviewed prior to discussing the methods themselves.

Acidity and alkalinity. Bacteria grow best in a neutral medium; highly acid or alkaline media inhibit growth. Most bacteria that contaminate food require oxygen for growth.

Moisture. Most bacteria will grow on moist surfaces or in water. Each kind has an upper and a lower limit of growth activity in solution, depending on whether salt, sugar, or other materials are present.

Temperature. Each kind of bacteria has maximum, optimum, and minimum temperatures at which growth proceeds. Most disease organisms grow best at the normal temperature of the human body (98.6° F, 37° C). Temperatures above 160° F (71° C) will kill most organisms; temperatures below 40° F (5° C) will retard their growth.

In short, most bacteria will grow rapidly under warm and moist conditions, contaminating any food in which they are present. However, certain bacteria are very useful, particularly in the fermentation of foods such as bread, yogurt, and wine. In these cases, the growth of the bacteria and the changes they produce are essential to their beneficial effects.

Now for a review of the methods that have proved effective for preserving food.

Cooking—Cooking renders food digestible and palatable. Although it also tends to kill many bacteria, this process alone will not preserve food. In fact, partial cooking may render protein foods (meat, eggs, milk, milk products) more susceptible to bacterial growth, permitting active increases in the number of harmful organisms or the toxins they may produce. Unless food is heated thoroughly and to a sufficiently high temperature, any microorganisms present will not be killed. Even if they are killed, the cooked food must be eaten promptly or protected from subsequent spoilage.

Canning—The process of canning consists in heating food enough to kill any microorganisms present and then sealing it in a container to keep it sterile. The combination of time and temperature required to preserve food by canning varies with the product and its likely contaminants. Acid foods—tomatoes and some fruits—need to be heated to the boiling point for only a few minutes. Nonacid foods—corn and beans—must be heated to higher temperatures (under pressure) for a longer time to prevent undesirable changes in appearance and flavor, as well as to destroy the anaerobic microorganisms that produce the botulism toxin.

Drying and dehydration—Air drying, one of the most economical and effective ways of preserving food, has been practiced for centuries. Today food can be dried in the sun or by artificial heating processes. Other methods include spray drying, freeze drying, vacuum drying, and hot-air drying. Once the food is reconstituted by the addition of water, bacterial activity resumes and sanitary control is essential.

Preservatives—Certain preservatives can be used to inhibit the growth of microorganisms or to kill them. Salt, sugar, sodium nitrate, and nitrites are commonly used for curing and pickling meats and vegetables; often other agents such as salicylic acid and sodium benzoate are added too. Ordinarily, meats can be preserved by a combination of salting, curing, and smoking. Corned beef, however, must subsequently be refrigerated. Smoking often improves flavor and helps inhibit microbial growth. Propionates and sorbic acid are commonly used to prevent mold formation in breads.

Refrigeration—Storing food at temperatures lower than 40° F (5° C) will retard the growth of pathogenic organisms and the more important spoilage organisms, but it does not prevent all changes. The level of humidity is also important: too little results in moisture loss; too much promotes the growth of spoilage organisms. Proper air circulation and regular cleaning and sanitizing of chill spaces are mandatory.

Freezing—Bacteria that cause food spoilage do not multiply at freezing temperatures; but once thawing begins, frozen food becomes vulnerable to bacteria and the associated toxins they may produce. Refreezing will not make the food safe. Nor will freezing improve the original quality of the product. Thus, the selection of appropriate products for freezing is essential. One variation is "dehydrofreezing," in which the food is partially dehydrated (but still perishable) and then frozen. This process provides the space and weight savings of dehydration without depriving the food of its fresh color, flavor, and palatability.

Pasteurization—Pasteurization is an excellent method of preserving food for a short time. Combined with refrigeration, it extends the useful shelf life of dairy products. Milk is generally heated to 145° F (63° C) for 30 minutes—or to 161° F (72° C) for 15 seconds—to kill the pathogenic organisms. Although some heat-resistant organisms will survive, subsequent refrigeration will preserve the milk for up to several weeks.

Irradiation—This process subjects food to ionizing radiation at sufficiently high doses to kill a large fraction of any microorganisms it contains. At the doses proposed, irradiation would not sterilize most foods—meats and poultry, for example, would still require refrigeration and proper handling. No radioactive material (frequently the source of the radiation) is permitted to come in contact with the food, and the final product is not radioactive. However, in some foods the process produces unwanted changes in taste and palatability, which has given it something of a negative reputation. Fears have been expressed about other changes that take place during irradiation, especially the formation of radiolytic compounds. However, the types and quantities of these compounds found in irradiated foods are no different than those in foods processed by other methods of preservation. In 1983 the Codex Alimentarius Commission, a joint body of the Food and Agriculture Organization (FAO) and the World Health Organization (WHO), representing more than 130 countries, adopted worldwide standards for irradiating foods. The process has been approved by the Food and Drug Administration (FDA) for the preservation of pork, chicken, herbs and spices, fresh fruits and vegetables, and grains (FDA, 1990; Fumento, 1994). Irradiation is commercially applied in the United States, as well as in other countries of the world.

In addition to care in processing, prevention of foodborne illnesses requires an effective sanitation program. An essential part of such a program is the collection of data on the nature and sources of major outbreaks. Experience has shown that the principal sources of foodborne disease outbreaks are improper storage or holding temperature, inadequate cooking, and poor personal hygiene (Bean and Griffin, 1990).

Also essential to a sound food sanitation program are a safe water supply, adequate garbage and refuse disposal, proper wastewater and sewage disposal, and effective insect and rodent control. Other factors involve equipment and facilities, personnel training and habits, standards and regulations, and enforcement and monitoring.

Equipment and facilities—Equipment used in the preparation or processing of food should be designed to facilitate cleaning. Cutting boards should be made of nonporous materials. Vehicles used to transport food products must be clean and should not carry other products. Refrigerated vehicles must be available for the transport of perishable foods. Facilities should be designed so that all foods, particularly vegetables, can be stored above the floor, where they will remain dry and will not come in contact with powders and sprays used to control insects and rodents.

Personnel training and habits—Personal hygiene is indispensable in the proper handling and preparation of food products. Antimicro-

Table 6.2 Ten rules for safe food preparation and consumption

1. Choose food processed for safety.
2. Cook food thoroughly.
3. Eat cooked food immediately.
4. Store cooked food immediately.
5. Reheat cooked foods thoroughly.
6. Avoid contact between raw and cooked foods.
7. Wash hands repeatedly.
8. Keep all kitchen surfaces meticulously clean.
9. Protect foods from insects, rodents, and other animals.
10. Use pure water.

bial cleaners should be used on the surfaces on which foods are prepared, and cleaning rags and sponges should be disinfected regularly or replaced. Food handlers must wash their hands after toilet use and before and after work; must avoid contact between open wounds and foodstuffs; must wear clean outer garments, including a cap over the hair; and must avoid using tobacco products while working. The use of tobacco can lead to contamination of the fingers and hands with saliva and may promote spitting, which can transfer disease organisms to the food or to the surfaces with which it comes in contact. Food handlers should also be trained in appropriate methods of food storage, garbage disposal, and insect and rodent control. Summarized in Table 6.2 are the essential rules for safe food preparation and consumption.

Standards and regulations—The basic requirements are for national standards and regulations governing the proper methods of processing, preparing, and selling food products; for limitations on the types and quantities of chemicals that can be added to foods; for restrictions on the quantities, types, and manner in which pesticides can be used on agricultural food crops; and for commercial food products to be properly labeled. Listed in Table 6.3 are the principal federal agencies that have responsibilities related to food safety, along with a brief description of their duties. Summarized in Table 6.4 are the safety standards or limits currently being applied for the control of a range of contaminants and ingredients that may be present in foods. Excellent guidance on the proper preparation of foods to minimize the risks of foodborne illnesses, and to ensure the safety of foods served in restaurants, grocery stores, and institutions such as nursing homes, is given in the *Food Code* published by the Food and Drug Administration (FDA, 1995). Providing guidance on food safety at the international level is the previously mentioned Codex Alimentarius Commission. This organization serves as a vehicle for the development of standards and codes involving basic principles, technical specifications for products, and satisfactory manufacturing practices (Institute of Food Technologists, 1992).

Enforcement and monitoring—Within the United States, agencies at the state and local levels have primary responsibility for the inspection

Table 6.3 Federal agencies responsible for safety of the U.S. food supply

Department of Health and Human Services:

- Food and Drug Administration, which is responsible for the regulation of food labeling, safety of food and food additives, inspection of food processing plants, control of food contaminants, and establishment of food standards
- Centers for Disease Control and Prevention, which analyze and report incidents of foodborne diseases
- National Institutes of Health, which conduct research related to diet and health.

Department of Agriculture:

- Food Safety and Inspection Service, which is responsible for inspection and labeling of meat, poultry, and egg products, as well as grading of foods
- Animal and Plant Health Inspection Service, which inspects food and animal products imported into the United States
- Human Nutrition Information Service, which establishes food consumption standard tables for the nutritive value of food and provides educational materials related to food

Other agencies:

- Environmental Protection Agency, which develops standards for the use of pesticides on food crops
- National Marine Fisheries Service, within the Department of Commerce, which conducts inspections and establishes standards relative to the quality of seafood
- Bureau of Alcohol, Tobacco, and Firearms, which regulates alcoholic beverages, and the Customs Service, which inspects food products imported into the United States (both within the Department of the Treasury)
- Federal Trade Commission, which regulates food advertising

Table 6.4 Federal regulation of food ingredients

Ingredient	Definition	Safety standard or limit
Unavoidable contaminants	Inherent food substances that cannot be avoided	Adulterated if substance "may render food injurious to health"
GRAS substances	Substances "generally recognized as safe" by the scientific community	Must be "generally recognized as safe"
Food additives	Substances added for specific intended effects, including GRAS substances, color additives, new animal drugs, and pesticides	"Reasonable certainty of no harm"
Substances previously sanctioned	Substances explicitly approved for use by FDA or USDA prior to 1958	Adulterated if substance "may render food injurious to health"
Pesticides	Substances intended for preventing, destroying, repelling, or mitigating any pest, or intended for use as plant regulator	Tolerance based on whether substance is "safe for use," considering its benefits
New animal drugs	Substances intended for food-producing animals, excluding antibiotics	"Reasonable certainty of no harm"
Color additives	Dyes, pigments, or other substances capable of imparting color, excluding substances that also have other intended functional effects	"Reasonable certainty of no harm"
Prohibited substances	Substances prohibited from use because they present a potential risk to public health or because the data are inadequate to demonstrate their safety in food	Must not be present in detectable amounts

of restaurants, retail food establishments, dairies, grain mills, and other food establishments. Their goals are to assure the safe handling, proper labeling, and fair marketing of food products. Methods used to meet these responsibilities include inspection at the point of production or processing, examination of products at the retail or wholesale level of distribution, and licensing of establishments that manufacture or handle foods. Because it is impossible to inspect every food at every site of production, processing, and distribution, the incentives to comply with regulations depend heavily on the probability of detection and the penalties for noncompliance (which can include fines and legal proceedings). In addition, compelling economic and business factors encourage food handlers, processors, and distributors to want to comply with the regulations. No food processor wants to purchase a product that will have to be discarded because of safety concerns or quality defects. Similarly, none of them wants to suffer the loss of customer confidence that can accompany a highly publicized foodborne disease outbreak (Institute of Food Technologists, 1992).

Biotechnology and Genetic Engineering

Although still in their infancy, biotechnology and genetic engineering appear destined to have a tremendous impact on how food is produced and the associated quality and health implications. Through the science of biotechnology, farmers are reaping the benefits of crops that resist disease, drought, and frost; more innovations are on the way. New product varieties will provide higher yields with less input. Upcoming products will include fruits and vegetables that can be picked and delivered at the peak of ripeness and flavor, cooking oils with improved fat profiles (that is, less saturated fat), and leaner meats. Studies show no evidence that the ingredients in such products pose any unique or unforeseen environmental or health hazards (Huttner, 1996).

In 1993 the FDA approved the first biotechnology product, bovine somatotropin (BST), for commercial use in animals in the United States. Administration of this product to cows results in a 10–15 percent increase in milk yield, without affecting the composition of the milk. Tests have also shown that meat derived from BST-treated animals is safe for human consumption (Etherton, 1994). As with any such new development, expressions of concern have been voiced. One possible negative aspect is that BST may cause an increase in the prevalence of mastitis in the cows to which it is administered (Stone, 1994).

Another product that has been approved by the FDA for public consumption is a tomato that has been genetically engineered to stay fresh up to 10 days longer than conventionally grown varieties. Thus farmers can allow the fruit to ripen on the vine, rather than pick and ship it while still green (Moffat, 1994a). Being considered for approval is porcine somatotropin (PST), another biotechnology product, which when administered to growing pigs produces a 10–14 percent increase in growth rate accompanied by as much as an 80 percent reduction in carcass fat. The resulting leaner pork will permit consumers to reduce their intake of total and saturated fat (Etherton, 1994).

In other developments, human genes are being introduced into livestock to produce human hormones and certain other drugs. Efforts are also under way to replace the cow's milk-producing protein genes with genes from humans. Proponents claim that the milk produced by such animals will approximate human breast milk and should be substantially more healthful for human infants than conventional milk from cows or infant formulas (Woods, 1993).

On the basis of a policy issued by the White House Office of Science and Technology, existing U.S. laws have been deemed sufficient for regulating the products of genetic engineering (Institute of Food Technologists, 1992). Approval of new applications, however, is much slower here than in many other countries. As a result, a number of genetically altered crops, such as potatoes, cotton, rice, and tomatoes, as well as papayas, are being field tested in the developing countries of the world, well ahead of similar activities in the United States. Field tests of potatoes developed to be resistant to specific viruses have been pursued in Mexico for several years. In China, vegetables such as tomatoes that have been genetically engineered to resist viruses have not only been tested but are now being grown and marketed. Such plants are still awaiting approval in the United States (Moffat, 1994a). A host of other genetically engineered food products are in the developmental stages. These include oils and salad dressings with less saturated fat, potatoes that absorb less fat when fried, and grains with higher concentrations of protein (Whelan, 1994). Furthermore, the future may bring the development of coffee plants that are genetically engineered to produce beans that are naturally decaffeinated (Moffat, 1994b).

Although these benefits alone are substantial, the advantages of using such products do not end here. There are substantial benefits relative to environmental pollution. Reductions in the amount of feed required to produce a unit amount of milk or meat decreases the need for fertilizers, pesticides, and other inputs associated with growing, harvesting, and storing animal feed. Applications of biotechnology also reduce the production of animal wastes, including methane.

The General Outlook

Modern technology has transformed the production, storage, and preparation of food. What once was a relatively simple system of local farming and home preparation is today a complex involving producers, processors, distributors, and retailers. As a result, at many points in the food chain food safety can be compromised (Walker, 1993). At the same time, applications of newer technologies have tended to separate consumers from their food sources and make them less familiar with what they eat. Consumers are apt to become more anxious about food safety, particularly if they do not trust the system responsible for producing, processing, and safeguarding their food supply (Institute of Food Technologists, 1992).

To cope with these challenges, public health officials have identified a number of steps that need to be taken to improve the production, management, and safety of food products (Hoffman, 1993). These range from restructuring the supporting federal laws and the responsible regulatory agencies, to educating the public on how to recognize and control foodborne illnesses, to achieving better health through proper choices of the foods they eat. Individuals need to recognize that a safe food supply requires water of adequate quality and quantity, proper disposal of human excreta, and education of food handlers in sanitary procedures. All persons must assume responsibility for ensuring that their food is as clean as reasonably possible, that their overall diet is balanced in nutrients, and that they handle and store food safely.

7

DRINKING WATER

RECORDS show that the quest for pure water began in prehistoric times. Information on methods for treating water has been found in Sanskrit medical lore, and pictures of apparatus to clarify water have been discovered on Egyptian walls dating back to the fifteenth century B.C. Treatment methods, such as boiling, filtration through porous vessels, and even filtration through sand and gravel, have been prescribed for thousands of years. In his writings on public hygiene, Hippocrates (460–354 B.C.) directed attention to the importance of water in the maintenance of health (NRC, 1977). The Romans demonstrated a similar awareness of the merits of pure water; witness the extensive aqueduct systems they developed, as well as their use of settling reservoirs to purify water, their rulings that unwholesome water should be used only for irrigation, and the passage of laws prohibiting the malicious polluting of waters (Frontinus, A.D. 97).

The first positive evidence that public water supplies could be a source of infection for humans was based on epidemiological studies of cholera in the city of London by John Snow in 1854. His study is particularly impressive when it is realized that, at the time he was working, the germ theory of disease had not yet been established. A similar study by Robert Koch in Germany in 1892 provided evidence of the importance of filtration as a mechanism for the removal from water of the bacteria that caused cholera. Subsequent experiments in the United States relative to the control of typhoid fever confirmed his observations and revealed the further benefit of the addition of chemicals to coagulate the water prior to filtration (NRC, 1977).

The most important technological development in the treatment of water was the introduction in 1908 of chlorination, which provided a cheap, reproducible method of ensuring the bacteriological quality of water. Although scientists now realize that chlorination can produce unwanted byproducts, this process remains one of the major means of guaranteeing the safety of our drinking-water supplies.

Approximately 70 percent of the earth is covered by water, but only 2.5 percent of the total volume on earth is not salty and therefore potentially available for consumption by plants and animals. Of this 2.5 percent, about two-thirds is unavailable, frozen in the polar icecaps and glaciers. The remaining 0.8 percent is held in aquifers, soil pores, lakes, swamps, rivers, plant life, and the atmosphere (Postel, Daily, and Ehrlich, 1996). Much of the water below the earth's surface is not readily available for human use. Only about 0.3 percent of the total is available as surface sources. However, even this small percentage represents a tremendous quantity. The earth's freshwater lakes are estimated to contain nearly 30,000 cubic miles of water, of which more than 300 cubic miles are in the world's rivers and streams at any one time (U.S. Geological Survey, 1986). The volume of groundwater in the upper half-mile of the earth's crust in the contiguous 48 states is estimated to be 50,000 cubic miles (CEQ, 1989).

The basic source of all water on earth is precipitation—rain, snow, and sleet. Of the precipitation reaching the earth's surface, only about 30 percent falls on land areas—and that is not evenly distributed. Average annual precipitation in the United States is about 30 inches per year; however, the range varies from a few tenths of an inch in the desert areas of the Southwest to 400 inches or more in some parts of Hawaii. About 40 percent of the contiguous United States (primarily areas east of the Mississippi River) receives over 75 percent of the rainfall (CEQ, 1989). Worldwide, the distribution of rainfall is still more uneven. In any given year, departures from average conditions can be extreme.

About 70 percent of the precipitation that reaches land areas is evaporated or transpired (through vegetation) directly back into the atmosphere; 10 percent soaks in and becomes groundwater, and 20 percent runs off into lakes, streams, and rivers. Most of this surface and groundwater ultimately flows into the oceans. The overall movement of water from precipitation through various pathways on earth and back into the atmosphere is called the hydrologic cycle (Figure 7.1).

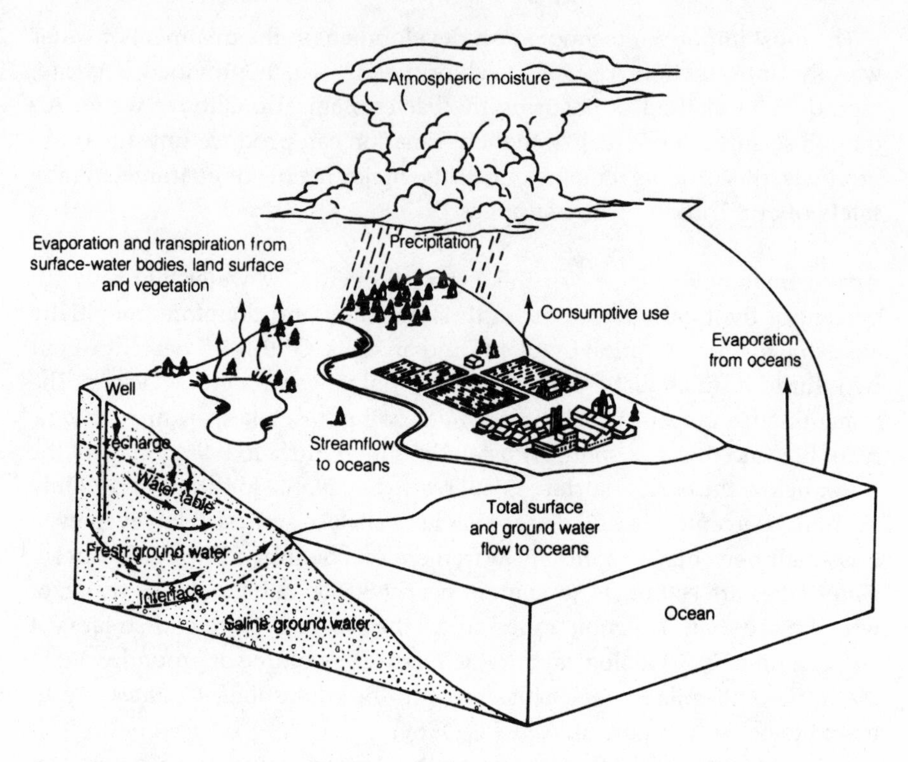

Figure 7.1 The hydrologic cycle

Human Uses of Water

Water is absolutely essential to life. From 50 to 65 percent of the human body is composed of water, and variations of as little as 1–2 percent will cause thirst or pain. The loss of 5 percent of body water can cause hallucinations; a loss as large as 10–15 percent can be fatal. Although humans can live several months without food, under hot, dry conditions they can survive only a day or two without drinking water.

People use water for a wide variety of purposes, many of them indirect or almost unnoticed, and most of them wasteful. In 1993 almost 350 billion gallons of water per day were withdrawn from aquifers and streams in the United States for public water supply, rural use, irrigation, and industry. For purposes of comparison, it might be noted that the average streamflow in the contiguous United States is about 1,200 billion gallons per day—only slightly more than three times the current rate of use (USGS, 1986). Of the amount of water being utilized, 250 billion gallons are eventually discharged into rivers and streams, and about 90 billion gallons are consumed and incorporated into manufactured products, agricultural crops, and animal tissue and hence are no longer available for immediate use. Although direct human consumption accounts for only about 230 million gallons per day, water that meets drinking-water standards is routinely used for irrigating lawns, fighting fires, washing cars, cleaning streets, and recreational and aesthetic purposes. With the increasing shortages of water in many areas of the world, this approach is rapidly changing. Dual water systems have been constructed in many arid areas, whereby separate plumbing systems deliver high-quality water for human consumption and less-pure water for uses such as irrigation and waste disposal. In many locations, reclaimed water (for example, treated sewage) is being used to irrigate greenbelts, parks, and other recreation areas. Uses of water can be broadly grouped as follows.

Personal use: Personal use includes drinking, cooking, bathing, laundering, and excreta disposal. On a daily basis, flushing the toilet consumes some 15–25 gallons (60–90 liters); bathing consumes another 15–20 gallons. Total personal (domestic) water usage depends, of course, on whether a home contains a washing machine and dishwasher, whether it has a swimming pool, the extent to which water is employed to irrigate lawns, and other factors. On average, each person in the United States uses a total of about 50 gallons of water each day. Of this, the amount actually consumed (used for cooking and drinking) averages only about 2 quarts (2 liters) per day.

Industrial use: The four largest industrial users of water are the manufacture of paper, petroleum, chemicals, and primary metals. Daily industrial consumption totals more than 36 billion gallons. Eight billion gallons and more are consumed by commercial users, including military bases, college campuses, office buildings, and restaurants (Parfit, 1993).

Waste disposal: The water-carriage method of excreta disposal, an outgrowth of the development of the flush toilet, uses 250 gallons of water, purified at enormous economic cost, to transport a single pound of fecal material to a sewage treatment plant for disposal.

Recreational and aesthetic use: Boating, sailing, water skiing, spray fountains, and the like fall in this category. Except for discharges of oil and gasoline from powerboats into lakes and rivers, few of these uses result in significant pollution.

Irrigation: About 140 billion gallons of water are used daily in the United States for irrigation. About half comes from surface water and half from groundwater sources. In the years 1940–1985 the use of water for irrigation in this country doubled, and over the past three decades the removal of groundwater for this purpose has increased even more rapidly (CEQ, 1989).

Other: Other uses of water include providing supplies for farm animals and aquaculture, transportation (waterways and canals), and generation of power. The last application includes the use of water in hydroelectric plants, and as a coolant in fossil-fueled and nuclear-fueled electric generating stations.

Few people in industrialized nations are aware of the many ways in which water is used to support their accepted standard of living. Consider the following requirements: more than 50 glasses of water to grow the oranges to provide one glass of orange juice (Anonymous, 1991); 8 gallons to grow a single tomato; 120 gallons to produce one chicken egg (including the water required to grow a chicken to the age at which she can lay eggs, and to feed her subsequently); 3,500 gallons for a steak; and 60,000 gallons for one ton of steel, roughly the amount in an automobile (Canby, 1980).

Health Risks

Historically, considerations of the effects on health of water contamination have focused on waterborne diseases, such as typhoid fever and cholera. Although today the most frequently reported waterborne disease in the United States is acute gastrointestinal illness (or gastroenteritis), this concern has expanded to include giardiasis and cryptosporidiosis. The latest data show that almost one million people in this country become sick each year from drinking contaminated water. Careful examination, however, shows that water can have effects on the health of a population far beyond those that result from its ingestion. Water can be a source of disease in people through at least four avenues (Bradley, 1977).

Water-borne diseases. These result from the *ingestion* of water that contains the causative organisms, as in enteric diseases such as typhoid, cholera, and infective hepatitis. Prevention depends on avoiding the contamination of raw water sources by human and animal wastes or purifying them prior to consumption.

Water-contact diseases. These can be transmitted through direct contact with organisms in the water. One example is Guinea worm disease; another is schistosomiasis, which can be transmitted to people who swim or wade in water containing snails infected with the organism. The larvae, which leave the snail and enter the water, can readily penetrate the skin of humans. Prevention requires the proper disposal of human excreta and deterring people from contact with infested waters.

Water-insect related diseases. Examples are malaria and yellow fever, where water serves as a habitat for the disease transmitter, in this case the mosquito. Prevention requires elimination of the insect transmitter or its contact with people.

Water-wash diseases. These result from lack of sufficient water for personal hygiene and washing. Shigellosis, trachoma, and conjunctivitis are among the diseases that may ensue.

Although the transmission of infectious diseases by water has been virtually eliminated in the developed nations, concern is increasing about the health implications of a host of other contaminants, particularly toxic chemicals, that can be present in drinking-water supplies. Evaluation of the health effects of these newly recognized contaminants is complicated by the problems faced in essentially all other fields of environmental and occupational health, namely, the latency (time delay) in the appearance of effects and the lack of definitive data on dose-response relationships (see Chapter 2). In addition, the federal government has been slow to acknowledge the importance of drinking water to health; the first federal law directly addressing the subject, the Safe Drinking Water Act, was not passed by the U.S. Congress until December 1974.

An early scientist who studied the role of drinking water in health was H. A. Schroeder of Dartmouth College. One of the results of his studies was what appeared to be a clear correlation between heart disease and the "hardness," or mineral content, of water (Schroeder, 1974). He found that people who drank "soft" water (containing few minerals) had a higher incidence of heart disease—apparently because soft water, being corrosive, dissolves toxic substances (such as lead, cadmium, and other harmful chemicals) from plumbing systems before being ingested. Schroeder's studies have been extended and expanded by other scientists under the aegis of organizations such as the National Institute of Environmental Health Sciences and the Environmental Protection Agency. An excellent reference that provides detailed information on the health effects of various contaminants in drinking water is the National Research Council's nine-volume *Drinking Water and Health* (NRC, 1977–89).

Sources of Drinking Water

The primary sources of drinking water are groundwater and surface water. In addition, precipitation (rain and snow) can be collected and used. Groundwater comes from wells that tap into underground reservoirs or aquifers. Dug wells (excavated by hand or with the use of power equipment) are shallow, extending 5–20 feet beneath the surface of the ground—far enough to reach the water table; driven wells may extend somewhat deeper, and drilled wells (constructed using either percussion or rotary-drilling machines) may penetrate to depths of 1,500–2,000 feet (Figure 7.2). Springs represent another source of groundwater, being outcrops where the underground aquifer intersects the surface of the earth (Symons, 1992). Sources of surface water include lakes, reservoirs, and rivers. Surface water may also come from protected watersheds, and private systems can be installed to collect rainwater from the roofs of houses. Each of these sources has its advantages and disadvantages.

GROUNDWATER

Groundwater serves as a source of drinking water for almost half of the U.S. population. Its widespread use stems not only from its general availability but also from economic and public health considerations. Groundwater is commonly available at the point of need at relatively little cost, and reservoirs and long pipelines are not necessary. Groundwater is also normally free of suspended solids, bacteria, and other disease-causing organisms. As long as it does not contain contaminants introduced by human activities, it does not require extensive treatment prior to use.

Unfortunately, more than half of the nation's land area possesses geologi-

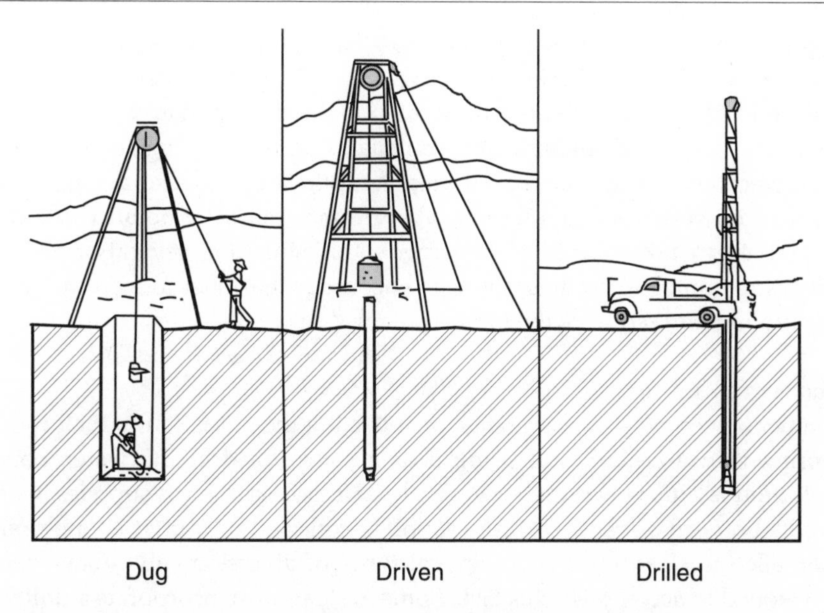

Figure 7.2 The three principal types of wells—dug, driven, and drilled

cal features that have permitted groundwater to be contaminated by agricultural runoff or through land surface and subsurface disposal of liquid wastes. As of 1992, it was estimated that more than 10 percent of the community water supply wells and almost 5 percent of the rural domestic wells in the United States contained detectable concentrations of one or more contaminants, primarily agricultural pesticides, with about 1 percent containing one or more contaminants in excess of health-based limits. Elevated concentrations of nitrates, for example, have been detected in many such supplies. Because groundwater provides base flow to streams, the potential for adverse impact on surface-water quality also exists, especially under conditions where dilution is minimal (CEQ, 1993).

Accessible groundwater sources are limited in volume and, once depleted, are essentially irreplaceable. Withdrawal at too rapid a rate can significantly lower the water table. Declines of over 40 feet have occurred in aquifers in more than 30 states. In some cases the water table has been lowered so much that it is below the pump at the bottom of the well. At this point the well ceases to be a source of water. Excessive withdrawals in some areas in Texas and California have caused the land to subside as much as

3–6 feet; in Mexico City some land areas have sunk as much as 30 feet. In many cases, the resulting low-lying areas have become vulnerable to flooding. In Florida, where 90 percent of the population depends on groundwater as its source of drinking water, some land areas overlying aquifers have collapsed, and in certain coastal areas withdrawals have so depleted the volume of fresh water underlying the ground that salty ocean water has moved in to take its place. To remedy this situation, several states are diverting rainfall to holding ponds to encourage the water to seep back into the ground and recharge the local aquifers.

PROTECTED RUNOFF

Many homeowners have systems for collecting the rainfall from their roofs, storing it in a cistern, and using it as a source of drinking water. Some pollution is almost certain, however, to gain access to such supplies. Contamination can be reduced if the runoff is collected only after enough rain has fallen to cleanse the roof. Several types of diversion valves have been developed to accomplish this task. Some systems also incorporate a unit for filtering the water prior to use. Cisterns should be of watertight construction, and manhole or other covers should be designed to prevent the subsequent entrance of contaminants.

Rainfall and accompanying runoff can also be collected on a wider scale to supply large municipalities. Cities employing this approach include New York, Boston, and Lisbon, where foresighted planners set aside large land areas for collecting precipitation and runoff in man-made lakes. The collected water serves as the municipal drinking-water source. For many years the only treatment of such supplies was disinfection. The situation is changing now primarily because of the realization that disinfection may not provide adequate protection against *Giardia* and *Cryptosporidium,* two disease organisms being detected even in supposedly well protected surface supplies. Fortunately, filtration is an effective method for removing *Giardia* and *Cryptosporidium* organisms. In fact, records show that waterborne disease outbreaks are eightfold higher and waterborne disease cases are sixfold higher within populations served by nonfiltered supplies from protected watersheds (an example is New York City), as contrasted to those served by filtered supplies (Okun, 1993). Because tests show that *Cryptosporidium* is present in 65 percent or more of the surface waters in the United States, the EPA now requires that all large drinking-water systems be monitored for *Cryptosporidium* and that all surface waters, including those obtained from protected watersheds, be filtered prior to being distributed for human consumption.

SURFACE SUPPLIES

Lakes, streams, and rivers are sources of drinking water for people in many areas. Such water, however, usually requires extensive treatment before use, and industrial pollution has increased the costs of the associated purification processes. The adequacy of such supplies is questioned in many parts of the world, especially in light of other demands for the water, such as irrigation, fisheries, and habitats for wildlife. Heated debates have ensued—for example, in the West—on how the limited surface-water supplies should be managed and allocated. In the final analysis, it will be necessary to move toward the principle that all users must accept limitations and share responsibility for these resources, and to recognize that water resources must be managed as interconnected, essential parts of large aquatic ecosystems (CEQ, 1993).

Drinking-Water Standards

The basic federal law pertaining to drinking water is the previously cited 1974 Safe Drinking Water Act, which was expanded and strengthened by amendments passed in 1977, 1986, and 1996. Guided by this act, the EPA has developed a series of primary standards, designed to protect human health, and a series of secondary standards, designed to assure that drinking water is pleasing in terms of temperature, color, taste, and odor. As established, the primary standards include maximum contaminant levels (MCLs) for selected inorganic contaminants (such as arsenic, barium, cadmium, chromium, lead, and mercury), volatile organic chemicals (including pesticides such as endrin, lindane, toxaphene, and methoxychlor, and certain of the chlorinated hydrocarbons such as the trihalomethanes), and radioactive materials (such as radium), as well as limits for turbidity and the presence of coliform organisms (EPA, 1975). The secondary standards include limits for iron, which can discolor clothes during laundering; sulfates and dissolved solids, which can have the same effect as a laxative; and minerals that can, for example, interfere with the taste of beverages (EPA, 1979).

Limits have also been specified for the amount of suspended solids (turbidity) in drinking water both for aesthetic reasons and because the efficacy of disinfection is related to the clarity of the water being treated. Disinfection of very turbid water can vary from difficult to impractical. Fortunately, clear (nonturbid) water is readily accomplished through application of normal water treatment processes.

To assure that the limits specified in the primary standards are within the capabilities of current technologies, the EPA has also identified treatment processes that it considers adequate for reducing concentrations of individual regulated contaminants to acceptable levels. Although apparently no single treatment technique is effective for the removal of all inorganic chemicals, conventional coagulation, sedimentation and filtration, or lime softening treatment (discussed below) has proved effective for removing many of them. One of the problems with contaminants such as pesticides and organic compounds, is that water purification plant operators must anticipate which contaminant will be present and be ready to remove it. In light of these uncertainties, the use of activated carbon beds as chemical adsorbers appears to offer the best barrier against such contamination (EPA, 1977).

Although a number of disease-causing organisms, ranging from bacteria and viruses to various forms of parasites, may exist in drinking-water supplies, detection and identification of these contaminants on an individual basis is difficult and tedious. In the past, the efficacy of the disinfection process was not measured by tests for the absence of individual pathogens, but rather by measurement of the concentrations of the coliform group of bacteria. In reality, coliform bacteria served as indicators or surrogates for the disease organisms. Because coliforms originate primarily in the intestinal tracts of warm-blooded animals, including humans, they are ready indicators of possible fecal contamination.

Certain pathogenic micro-organisms, particularly cysts, are more resistant to disinfection than are coliform bacteria. Concerns over these types of contaminants helped prompt passage of the 1996 amendments to the Safe Drinking Water Act. Under these amendments, analyses of drinking-water supplies for contaminants, such as *Cryptosporidium,* are mandatory in many cases. It is hoped that the development of new, sophisticated methods may soon make such analyses readily possible. Molecular probes, for example, can be used not only to detect the presence of human feces, but also to determine whether an organism, such as *Salmonella,* is present in drinking water (Mitchell, 1996). Standard methods for the sampling and analysis of a wide range of physical, chemical, and bacteriological contaminants in drinking water have been jointly developed by the American Public Health

Association, the American Water Works Association, and the Water Environmental Federation (Greenberg et al., 1992).

Water Purification

Preparing water for human consumption is a major industry. Approximately 60,000 municipal water purification and distribution systems exist in the United States alone. Most are small. In fact, well over half provide water in towns with 500 people or fewer; only 250 are in towns with populations of 100,000 or more (Symons, 1992). About 11,000 of these systems obtain their water from surface supplies, providing drinking water to over 100 million people. The remainder use groundwater sources. An additional 13 million people (5 percent of the population) obtain their water from private wells; in fact, about 95 percent of rural households depend on groundwater as a source of supply. In addition, some 165,000 "noncommunity" drinking-water suppliers—motels, remote restaurants, and similar establishments—serve the traveling public.

The combined output of these systems is 40–50 billion gallons per day, or 160–200 gallons for each person in the country. Fifty-seven percent of this water is used domestically, 32 percent by commercial groups and industry, 9 percent for public services (fighting fires, cleaning streets, watering parks), and 2 percent is lost through leaks in the distribution system (USGS, 1988). In addition, many industries have their own purification systems. Most water used for irrigation is taken directly from rivers, lakes, or the ground and used without prior treatment.

The capital investment in municipal water treatment facilities totals about $250 billion, and the annual cost of operating them is roughly $5 billion. Some $2 billion is spent annually in capital improvement of the facilities. Nonetheless, the cost of drinking water remains low, still well under a dollar a ton, or about one cent for more than 150 eight-ounce glasses. In actuality, this does not reflect the true cost of producing the water or the fact that many sources currently being used, particularly those that are underground, will be depleted within the next few decades. As more and more contaminants gain access to surface and groundwater, the cost of purification and distribution can be expected to increase substantially.

The primary purposes of a water purification or treatment system are to collect water from a source of supply, purify it for drinking if necessary, and distribute it to consumers. About half of the groundwater supplies are distributed untreated. In the future, except in special cases, the minimum allowable treatment for such water will be disinfection (Symons, 1992). The section that follows focuses on the treatment of drinking water obtained from surface-water supplies, and the two principal methods of purifying such supplies, namely, slow sand filtration and rapid sand filtration.

SLOW SAND FILTRATION

In this relatively simple process, the raw water supply is passed slowly through a sand bed 2–3 feet (60–90 centimeters) deep. Soon after a bed becomes operative, a biological growth develops on top of and within the sand, removing and retaining particles from the raw water. This process removes most bacteria and disease organisms, including the cysts of *Giardia lamblia*, the organism that causes giardiasis. Because excess turbidity in the raw water supply will rapidly plug the filter bed, preliminary settling is recommended. A filter bed area of 2,000 square feet (185 square meters) will provide approximately 100,000 gallons of treated water per day. With proper care, slow sand filter beds can be operated 30–200 days before the top layers of sand have to be scraped, cleaned, or replaced. In other respects, such systems require minimal maintenance (Allen et al., 1988; Leland and Damewood, 1990).

RAPID SAND FILTRATION

Figure 7.3 shows the principal steps in this purification process. First, water is pumped or diverted from a river or stream into a raw water storage basin. Such storage provides a carryover or reserve in case the raw water supply becomes unfit for use for several days—for instance, through accidental release of a contaminant upstream of the supply. Storage also removes color and reduces turbidity and bacteria.

The initial step in the process is to add chemicals to the water to create a coagulant. The chemical most commonly used in the United States is $Al_2(SO_4)_3 \cdot 14H_20$, commonly called alum. A less frequently used chemical is ferric chloride ($FeCl_3$). The basic reactions are almost identical:

$$Al^{+++} + 3HCO_3^{-} \rightarrow Al(OH)_3 + 3CO_2;$$

$$Fe^{+++} + 3HCO_3^{-} \rightarrow Fe(OH)_3 + 3CO_2.$$

The highly positively charged Al^{+++} and Fe^{+++} ions also attract the negatively charged colloidal suspended matter in the water and together with the $Al(OH)_3$ or $Fe(OH)_3$ form a gelatinous mass called floc. Rapid mixing is

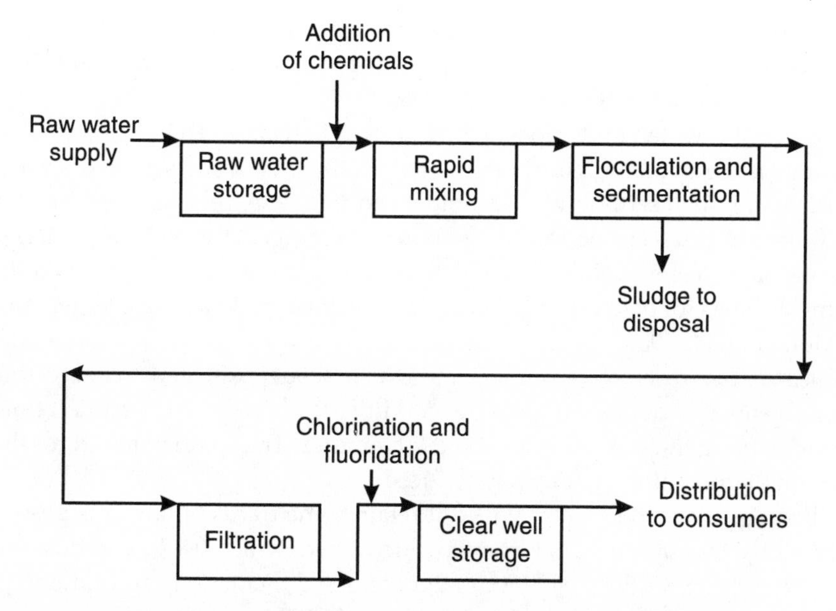

Figure 7.3 Principal steps in the water purification process, using rapid sand filtration

essential to provide maximum interaction between the positive metallic ions and the negative colloidal suspended matter.

Although theoretically a knowledgeable chemist could calculate the proper amount of alum or ferric chloride to add to the raw water for optimum coagulation and flocculation, the required quantity is most commonly determined on the basis of a jar test. A known amount of water is placed in a series of jars or beakers, a different concentration of coagulant is added to each, then one observes which concentration (dosage) coagulates and flocculates best. This is the dosage used for the full-scale operation. Since the characteristics of the water supply are subject to change, jar tests are generally conducted once a day.

Once the water has been rapidly mixed to assure proper coagulation, it is slowly and gently stirred to enable the finely divided floc to agglomerate into larger particles that will rapidly settle. This process, called flocculation, is accomplished by moving large paddles slowly and gently through the water. Since water treatment is performed as a continuous flow-through process, flocculation often takes place as the water enters one end of a large tank, with settling of the floc (sedimentation) occurring at the other end. During flocculation, relatively large particles in the water (including bacteria) are enmeshed in the floc, and ionic, colloidal, and suspended particles are adsorbed on its surface. This process does not remove contaminants that may be dissolved in the water (Chanlett, 1979).

Next, the water undergoes a period of quiescence. The settled floc or sludge is removed from the bottom of the settling tank and sent to disposal. Originally, such settling was accomplished in a large rectangular tank and required a quiescent period of 2–4 hours. More recently, high-rate settling tanks have been developed in which the water is passed through small-diameter tubes (or between parallel plates) set at an angle within a larger tank (Montgomery, 1985). Because the solids in the water travel a shorter distance before reaching a surface on which to deposit, and because this arrangement provides unique flow conditions, the required detention time for clarifying the water is only about 20 minutes. The space required for the settling tank is also significantly reduced.

Because the settled water will still retain some traces of floc, it is passed through a bed of sand 2–3 feet deep. Such a bed, which uses a combination of adsorption, additional flocculation and sedimentation, and straining, removes even the smallest particles (Chanlett, 1979).

Sand filters become loaded with floc and must be cleaned by backwashing with purified drinking water every 12–72 hours. About 2 percent of the water produced in a water purification facility is required for backwashing. In many cases, the wastewater is sent to a sewer. The accompanying solids, however, present a formidable disposal problem. A typical water treatment plant will produce about 250 cubic feet of sludge (three large truck loads) per million gallons of water processed.

Once the clarified water is disinfected and fluoride has been added, it is sent to storage and is ready for distribution to consumers. Although some groups continue to oppose the addition of fluoride to drinking water, this chemical has been used in the United States for more than 50 years, and epidemiological studies have clearly demonstrated its benefits in the reduction of dental caries. In fact, data show that people in areas with optimal fluoride in their drinking water have only 30–40 percent of the dental caries experienced by people living in communities without fluoride. Epidemiological studies have also demonstrated that people who consume fluoridated water are not being subjected to increased risk of cancer, kidney failure, or bone disease (Hicks, 1993). For this reason, the EPA has continued to affirm its endorsement of this additive (EPA, 1993). Today more than 60 percent of the people living in the United States have access to water supplies that are fluoridated.

Although sedimentation and filtration remove a significant portion of the microorganisms present in water, these processes alone do not provide adequate protection. Some form of disinfection is required. The most common agent in the United States is chlorine, which is usually added in sufficient quantity that a small residual will accompany the water entering the distribution system. Consumers are thereby protected in case bacterial contaminants later gain access to the supply. Unfortunately, the addition of chlorine to water that contains organic contaminants will produce chlorinated hydrocarbons, which are known to be carcinogenic. Because certain of the chlorinated hydrocarbons are volatile, they are readily released into the air and can be inhaled when water is used in dishwashers and clothes washers, in showers, and in flushing a toilet. In fact, this path of intake appears to be more important than ingestion (McKone, 1987).

One of the steps that can be taken to avoid the production of chlorinated hydrocarbons is to use another type of disinfectant. Two acceptable substitutes are ozone and chlorine dioxide, each of which has for some time been relegated to a secondary status owing to certain disadvantages. Chlorine dioxide, for example, cannot be transported because of its potential explosiveness and therefore must be generated at the point of use. This compound also yields by-products and end products in the disinfected water, which may be cause for concern. Although ozone produces no unwanted by-products, it too must be generated on site, since it is highly reactive and cannot readily be transported. It is expensive and cannot be added in a form that will maintain a disinfecting residual in the treated water.

In spite of these drawbacks, ozone, used on a regular basis in many European countries, is gaining favor in the United States (Dimitriou, 1994). At the present time it is used in several dozen water purification facilities in this country. Because of stricter regulations being imposed by the EPA, particularly with respect to the by-products of chlorination, ozone is likely to see increasing application in the future. This trend is confirmed by the full-scale water purification facility, having a capacity in excess of 50 million gallons per day and employing ozone as the disinfectant, that began operation in Portland, Maine, in 1994 (Anonymous, 1994).

Ionizing radiation, heat, silver, and ultraviolet light are also employed as water disinfectants, but on a very limited basis (EPA, 1977).

Additional steps that can be taken in the purification of drinking water include procedures to remove hardness, iron and manganese, organic compounds, and tastes and odors.

Hardness. Hardness in drinking water is caused by the presence of calcium and magnesium, two chemicals that are otherwise harmless to humans. These chemicals make the water "hard" in the sense that they make it difficult to develop a lather when a person attempts to use soap when taking a bath or washing dishes or clothes. These are also the chemicals that leave a scum or ring in the bathtub (Symons, 1992). Although hardness is not generally a problem where supplies are derived from surface-water sources, it frequently is where supplies are derived from groundwater sources. The relative amounts of hardness in groundwater supplies in various portions of the conterminous United States are indicated in Figure 7.4 (AAVIM, 1973).

Removal of calcium and magnesium from small volumes of water, as in systems serving an individual household, can be accomplished through the ion-exchange process. When large volumes are to be treated, the lime-soda process is apt to be used. In it, calcium hydroxide (lime) and sodium car-

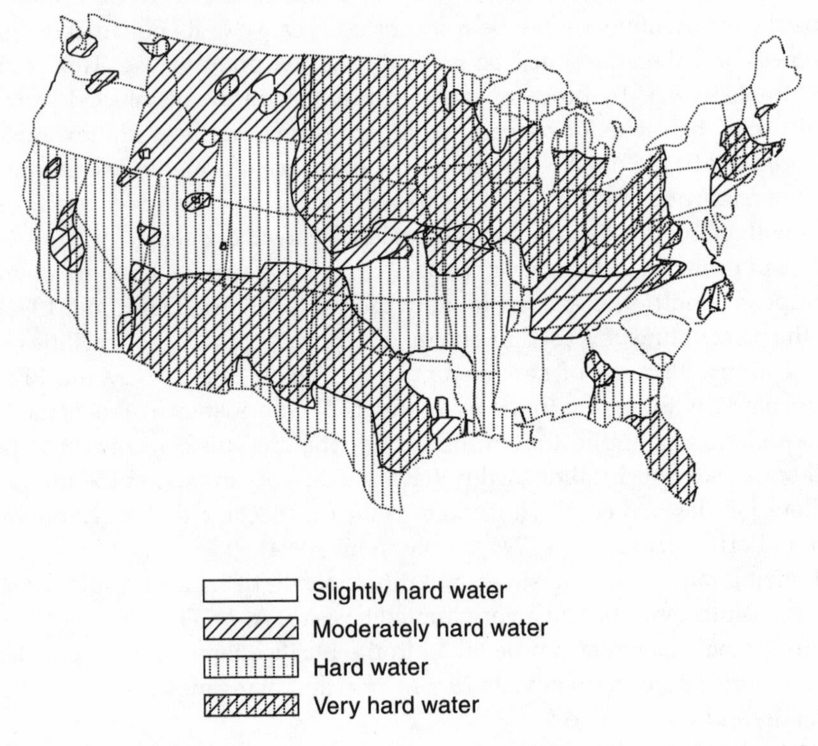

Figure 7.4 Variations in the hardness of groundwaters in different sections of the conterminous United States

bonate are added to the water and interact with dissolved calcium to form insoluble calcium carbonate, which precipitates and thus reduces the concentration of calcium.

Unfortunately, the ion-exchange process adds significant quantities of sodium to the treated water. Since sodium is believed to cause high blood pressure in some people, care must be exercised in consuming water treated by this process. An approach that can be used for an individual household is to connect the softener only to the hot-water line, thus restricting use of the treated water primarily to taking baths and washing clothes and dishes. The cold-water supply, normally used for drinking and cooking, is not softened and therefore does not contain the added concentrations of sodium.

Another approach is to install a water purification system, such as a reverse osmosis unit, to purify that portion of the water supply used for cooking and drinking. Although the output of such units is small, units that are adequate to meet household needs for cooking and drinking are readily available at reasonable cost (Symons, 1992).

Iron and manganese. These two chemical elements are soluble in water only in the reduced chemical state. If oxidized, they immediately become insoluble and precipitate. Thus, they are readily removed by aeration.

Organic compounds, and tastes, and odors. The most common sources of bad tastes or odors in drinking water are hydrogen sulfide and algae growths, and byproducts resulting from reactions of chlorine with various chemicals in the water (Symons, 1992). In general, though, none of the contaminants in drinking water that affect your health can be tasted or smelled.

One of the best methods for avoiding bad tastes and odors is to remove organic compounds and similar chemicals from the water prior to chlorination. This procedure will also significantly reduce the production of chlorinated hydrocarbons. As noted above, one of the best methods is to adsorb the organic compounds on activated carbon (charcoal).

The effects of various steps in the water purification process on specific characteristics of the raw water supply are summarized in Table 7.1.

Because of fears of harm from various contaminants, many homeowners have installed treatment devices to further purify their drinking water

Table 7.1 Effects of purification processes on specific characteristics of water

Process	Characteristic						
	Bacterial content	Color	Turbidity	Taste and odor	Hardness (calcium and magnesium)	Corrosiveness	Iron and manganese
Raw water storage	+	+	+	±	+	0	+
Aeration	0	0	0	+	0	+	+
Coagulation and sedimentation	+	+	+	0	0	−	+
Lime-soda softening	+	0	+	0	+	+	+
Sand filtration	+	+	+	0	0	0	+
Chlorination or ozonation	+	+	0	±	0	0	+
Carbon adsorption	−	+	+	+	0	0	0

Note: 0 = no effect; + = beneficial effect (aids in alleviating the problem); − = negative effect (adds to the problem); ± = sometimes beneficial, sometimes negative effect.

supplies. Although not needed to make drinking water safe (assuming that the public supply meets federal, state, and local drinking-water standards), certain of these devices can improve the quality (particularly the taste) of drinking water. These include the following (Symons, 1992).

Particulate filters that remove suspended materials from water.

Chemical adsorbers (such as the previously cited activated carbon or charcoal) that remove chlorine, tastes and odors, and some organic compounds such as pesticides. Unfortunately, microbes frequently grow in these units and can degrade the quality of the water.

Reverse osmosis units that remove hardness, chemicals such as nitrates, and some organic chemicals. They do, however, allow some organic chemicals to pass into the treated water.

Water softeners (such as the previously cited ion-exchange units) that remove hardness. To avoid the problem of excess sodium in the treated water, resins charged with hydrogen (as opposed to sodium) ions can be employed.

Distillation units that remove some organic and inorganic chemicals (such as calcium, magnesium, and nitrates). Some organic chemicals, however, will be volatilized and carried over with the steam into the water.

All of the above units require maintenance, should be purchased only from a reputable dealer, and should be approved by the National Sanitation Foundation (Symons, 1992).

Still another approach being used by an increasing number of people is the purchase of bottled water. In California it is estimated that one-third of the residents do so (Allen and Darby, 1994). Even though bottled water generally costs about a thousand times as much as the public supply, purchases of such products have doubled in the last six years. Consumption of bottled water may comfort the consumer, but this option is not necessary if the local public water supply meets federal, state, and local standards. One exception might be for individuals on a highly restricted diet with respect to sodium intake. They may want to seek a brand of bottled water that contains less sodium than the normal supply (Symons, 1992). Applicable regulations have been issued by the Food and Drug Administration. These require that bottled water, like all other foods, be processed, packaged, shipped, and stored in a safe and sanitary manner and be truthfully and accurately labeled. Such products must also meet specific quality standards for contaminants (FDA, 1996).

The General Outlook

The provision of water supplies of high quality (almost totally through the use of chlorine as a disinfectant) and in sufficient quantity, and the development of processes for treating and disposing of municipal sewage have promoted an ever-higher level of public health, particularly in the developed countries. The availability of high-quality water has also enhanced industrial growth in the United States over the past century. But many problems remain. These include the presence of an increasing number of toxic chemicals (albeit at low concentrations) such as lead in drinking-water supplies, the need to dispose safely of sludges and other wastes produced in the purification of drinking-water supplies, the rapid depletion of many groundwater supplies, and the contamination of many of those supplies with toxic chemicals.

As in essentially all aspects of public health, experience has demonstrated that an adequate defense against waterborne disease requires a systems approach: watershed management to protect high-quality surface and groundwater sources from human and animal waste, altering the processes that generate wastes to minimize their production, and assuring that those wastes that are produced are adequately treated prior to release. In addition, water supplies intended for human consumption must be properly treated, including the removal of any harmful contaminants. The final requisites are disinfecting the water to guarantee that all disease organisms are killed, adding a residual to protect against any organisms that may subsequently gain access to the water, and ensuring that the distribution system (including the plumbing within buildings and homes) does not add harmful contaminants. The importance of clean water is well illustrated by the fact that waterborne diseases account for an estimated 80 percent of all illnesses in the developing world.

Also to be faced on a worldwide basis are the ever-increasing shortages of water, particularly in arid and semiarid regions. Already more than two dozen countries do not have enough fresh water to meet the needs of their populations; more are expected to join this list in the near future. In Africa, it is estimated that 300 million people, one-third of the population of that continent, will be living in drought-stricken areas by the end of the decade. In Poland, three-fourths of the river water is too contaminated even for

Table 7.2 Potential water savings from using water-efficient instead of conventional household systems

System	Water consumption		Savings (%)
	Gallons	Liters	
Toilets[a]			
Conventional	5	19	
Common low-flush	3.5	13	32
Washdown	1	4	79
Air assisted	0.5	2	89
Clothes washers[a]			
Conventional	37	140	
Wash recycle	26	100	29
Front loading	21	80	43
Showerheads[b]			
Conventional	5	19	
Common low-flow	3	11	42
Flow limiting	2	7	63
Air assisted	0.5	2	89
Faucets[b]			
Conventional	3	12	
Common low-flow	2.5	10	17
Flow limiting	1.5	6	50

a. Consumption per use.
b. Consumption per minute.

industrial use. With the passage of time, conflicts and the potential for conflicts will undoubtedly develop as demands on our limited water resources increase (Frederick, 1996). Yet all over the world rivers are being used as sewers—for human excrement, industrial and agricultural wastes, and urban runoff. Many water shortages stem from the widespread failure to value water at anything close to its true worth. People have yet to learn that the water resources of the world are finite. Grossly underpricing water perpetuates the illusion that it is plentiful (Postel, 1992).

Efforts are under way worldwide to reduce the quantity of water used for various daily activities. Current federal law in the United States requires manufacturers to produce only low-flow toilets and showerheads. Treated sewage is being recycled for use in industrial processes, and is utilized (instead of drinking water) to irrigate recreational areas such as golf courses. Since the food processing industry is a major consumer of water, membrane filtration and separation systems are being explored as methods for reducing both its use of fresh water and its discharge of wastewater (Moore, 1994).

Many opportunities for conserving water are available to individual homeowners through the use of water-efficient household fixtures (Table 7.2). Congress is considering bills that would mandate limits on the water consumption of various household devices such as dishwashers and clothes washers. These requirements would parallel those imposed on the energy efficiency of other household products such as hot-water heaters, refrigerators, and furnaces. Even though some present-day household appliances increase our uses of water, this is not always the case. An average dishwasher, for example, uses 50 percent less water than that used when washing and rinsing dishes by hand (Symons, 1992).

Higher prices and stricter water pollution laws have further stimulated industries to recycle water. In fact, while the total U.S. consumption of water has quadrupled since 1950, industrial use has decreased by 19 percent (Parfit, 1993).

8

LIQUID WASTE

ONE OF the most common types of liquid waste is human sewage. Basic guidance on its disposal can be found in verses 12 and 13 of the twenty-third chapter of Deuteronomy, where God provided the following instructions to Moses: "You shall have a place outside the camp and you shall go out to it; and you shall have a stick with your weapons and when you sit down outside, you shall dig a hole with it, and turn back and cover up your excrement."

An early and simple method for disposing of human excreta follows this guidance almost to the letter: the pit privy, a hole in the ground with a small closed shelter and toilet built above it. Generally the hole is 3–4 feet in diameter and about 6 feet deep. Privy designs range from those in which excreta are deposited on the surface of the ground to those in which they are collected in a bucket or tank for later removal and disposal elsewhere. Double-vault privies are used by many people in the developing countries. Alternating the pits each year provides sufficient retention and decomposition to assure the destruction of most pathogenic organisms in the wastes. An essential part of later improved versions is a screen-covered vent pipe (Figure 8.1), which provides a natural pathway for removing odors and for trapping flies and other insects.

With the development of the water closet or flush toilet, sewage treatment and disposal entered a new era. Although the Minoans in Crete in 2800 B.C. had toilets that could be flushed either with rainwater or with water stored in cisterns, it was not until 1596 that the modern flush toilet was invented by Sir John Harrington (Chanlett, 1979). Although Queen Elizabeth I admired the concept and had one built for her own use, the idea

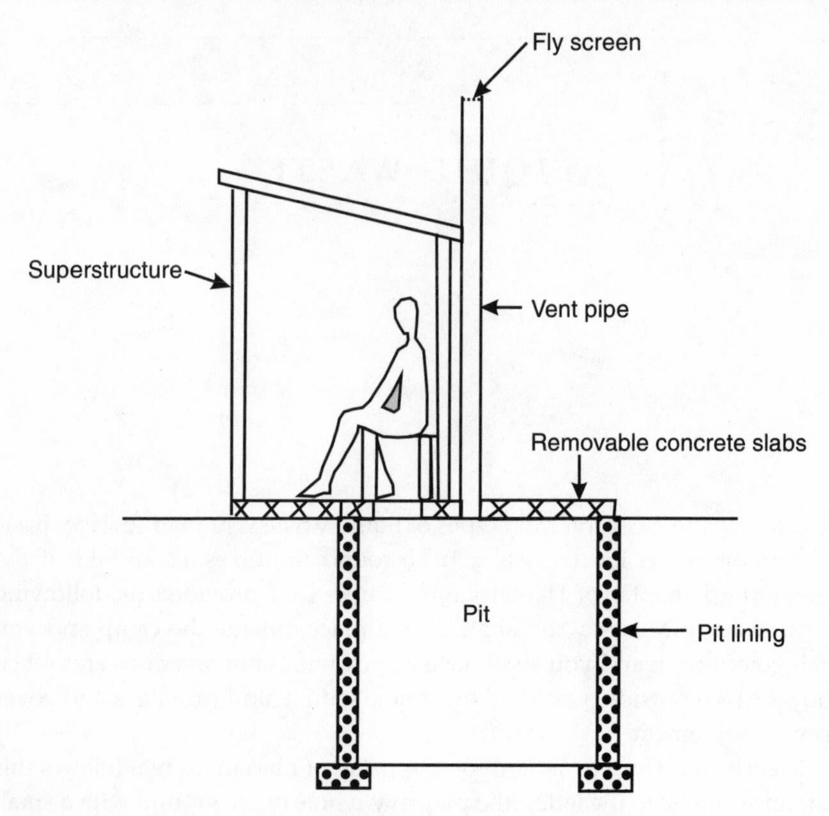

Figure 8.1 Pit privy with ventilation pipe

had to wait until the eighteenth century for further development. In 1775 Alexander Cummings patented a water closet with a valve for initiating the flush and a trap to seal off odors. However, the valve controlling the inflow of water allowed considerable leakage. (This problem was solved by Sir Thomas Crapper, who in 1872 invented the first valveless water waste preventer. The principles of his design continue in use today.) Thomas Jefferson had a water closet installed in the White House in 1800. Within the next several decades, most wealthy people had at least one indoor water closet that discharged either onto the ground or into a cesspool, an underground pit. In 1855 George Vanderbilt had the first bathroom (consisting of a lavatory, porcelain tub, and flush toilet) built inside an American house. As late as the 1880s, however, only one of every six people in U.S. cities had access to modern bathroom facilities.

Individual Household Disposal Systems

The advent of the flush toilet necessitated methods for disposing of the discharged wastes. Most municipalities constructed systems to transport the effluent to a sewer and then to some form of municipal sewage treatment plant. But even today some 30–35 percent of the U.S. population—75–90 million people—are not served by sewers. They depend instead on some form of on-site subsurface sewage disposal system. The most common of these is the septic tank.

A septic tank is usually constructed of concrete, with an inlet for sewage to enter and an outlet for it to leave (Figure 8.2). As sewage passes through the tank, solids settle to the bottom and are digested through the action of anaerobic bacteria that naturally develop and grow in the tank. Although some groups advocate the addition to the tank of special types of organisms to enhance digestion, the majority of experts agree that such augmentation is not necessary. Most septic tanks have a divider in the bottom and a baffle at the top near the outlet end to help prevent carryover of settled solids and floating material in the effluent. Under proper operating conditions, the effluent is clear and is discharged into a drain field consisting of open-jointed or perforated pipe through which the liquid can seep into the surrounding soil. The purposes of the drain field are several. It acts to disperse the septic tank effluent over a wide surface area and thus promotes infiltration of the waste into the soil. Furthermore, natural bacterial populations in the soil continue the digestion of soluble organic materials in the

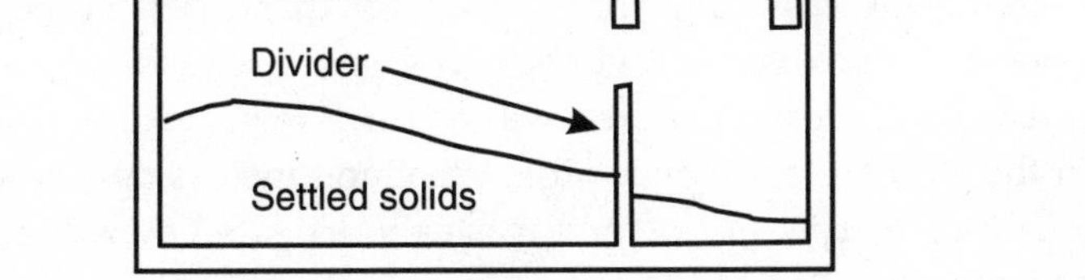

Figure 8.2 Cross section of a typical septic tank

waste that has passed through the septic tank. The soil also acts as a filter mechanism to adsorb pathogenic organisms remaining in the waste (Olson, 1992).

For proper performance, it is generally recommended that (1) the tank hold a volume of at least 500 gallons, (2) the soil in which the drain field is located be sufficiently porous to absorb the liquid effluent, (3) the land area be adequate for complete absorption of the effluent flow, and (4) the tank be cleaned (solids removed) every three to five years. The last recommendation is extremely important, because if solids are permitted to build up too long in the tank they will be carried out with the effluent and will seal up the drain field.

Only a third of the soils in the United States are suitable for absorbing septic tank effluents. About 25 percent of the tanks malfunction either periodically or continually because they have been installed improperly or in unsuitable soils. Under these conditions the effluent cannot be absorbed and will break through to the ground surface or find its way into a flowing groundwater source, resulting in bacterial and viral contamination of the surface soil or contamination of drinking-water supplies.

Because of the problems with septic tanks, a wide variety of alternative household treatment systems have been developed for the disposal of human wastes. These include aerobic systems wherein the sewage is collected in a tank, mixed by a pump to break up the solids, and aerated. An aerobic system is less prone than a septic tank to produce disagreeable odors, and it yields an effluent that contains dissolved oxygen, which helps prevent clogging of the drain field. Modifications of such systems are available in which the effluent can be recycled and used for flushing the toilet again.

Other approaches include biological toilets, composting toilets, incinerating toilets, oil-flushed toilets, and vacuum disposal systems. One of the most popular is the Clivus Multrum household excreta and garbage disposal system developed in Sweden. All such systems feature low water usage. The vacuum disposal systems require only a quart of water per flush.

Even with improved systems for toilet wastes, disposing of other domestic wastewater (from dish and clothes washers, bathtubs, and showers) remains a problem. However, these wastes carry fewer pathogenic organisms than the effluent from toilets. It is becoming increasingly common to use these and other types of treated wastes to irrigate lawns, recreational areas, and agricultural crops.

Liquid Wastes—A Broader Perspective

In today's society, the sources of liquid wastes extend far beyond individual households. Within a modern city such sources include commercial and office buildings, schools, restaurants, and hotels, as well as a wide range of industrial operations. The nature of industrial wastes is often different from municipal sewage; they frequently contain toxic chemicals and other hazardous substances, as well as heated water and various types of suspended materials. If discharged into rivers and lakes without treatment, liquid wastes can be a major source of pollution. If discharged onto the land, they can contaminate the soil and groundwater.

Many of the sources of liquid wastes, such as industrial and municipal discharges, are readily identifiable and are defined as point sources. Severe water pollution problems are also caused by less obvious and more widespread sources of pollution, the so-called non-point sources. In fact, these sources, many of which have not been previously controlled, may contribute more to water-quality degradation than point sources (CEQ, 1993). Data indicate that one such source, liquid runoff from agricultural lands, accounts for 40 percent of the pollution being discharged into U.S. rivers (Figure 8.3). In localized cases, it can account for up to 80 percent of the degradation of such waters. The quantities of this runoff can be particularly large during spring thaws.

Livestock farms, particularly the so-called hog factories, can be significant sources of liquid runoff. On such farms, the number of which has increased dramatically in recent years, the pigs are grown in pens within large buildings, where they are automatically fed and watered. The excreta from as many as 12,000 animals on a single farm are flushed into giant lagoons which, depending on the size of the farm, may hold in excess of 25 million gallons. Unfortunately, surface overflow and underground seepage from the lagoons have in many cases contaminated nearby surface and groundwater supplies. Odors and swarming flies often create noxious conditions for nearby residents. Excess spraying of wastes onto fields has led to runoff that pollutes surface waters. Exacerbating the problem is that pigs produce nearly twice as much waste as beef cattle, and about three and a half times as much waste as chickens (Satchell, 1996).

Another primary contributor to nonpoint source pollution is liquid runoff from urban areas. Where storm sewers are combined with sanitary sewers, designed to transport domestic and industrial wastes, the runoff

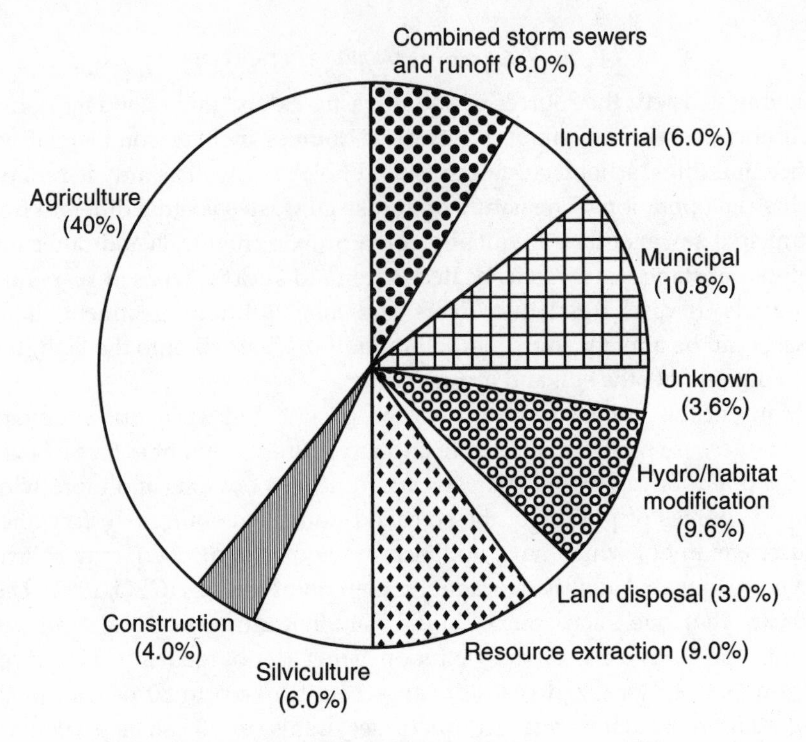

Figure 8.3 Sources of pollution in U.S. rivers, 1993

from rain enters the sewers. Although as it falls rainwater may be relatively clean, it quickly picks up a wide variety of contaminants as it flows over and essentially washes the paved areas within a city. Typical contaminants in urban runoff include suspended solids, bacteria, heavy metals, floatable materials and other types of debris, nutrients, oxygen-demanding organic compounds, and oil and grease. Furthermore, the increased flow during periods of rainfall overloads the municipal treatment plant and permits the waste to pass essentially untreated into nearby lakes or streams (Mitchell, 1996).

One of the most common methods used in the United States for disposing of treated liquid wastes is to discharge them into a large body of water, such as a lake or a river. If the wastes contain toxic chemicals and/or pathogenic organisms, and the receiving waters later serve as raw sources for drinking-water supplies, the resulting contaminants may have direct effects on the health of those who consume the water. If the wastes are applied to land areas for irrigation or other purposes, other types of contact with humans may occur. Any organic matter contained in wastes discharged to a receiving body of water will be decomposed by bacterial action. This process requires oxygen, which must be obtained from that which is dissolved in the water. At the same time, eddies and other turbulence, and the growth of green algae and various plants, will serve as sources for replacement of the oxygen removed from the water.

As long as the amount of organic matter discharged into a river or stream is small enough in relation to the volume of the diluting water (and if additional pollution is not discharged into the same receiving body), some dissolved oxygen will always be present. If oxygen is depleted faster than it is replaced, the concentration of dissolved oxygen (DO) may become too low to support fish and other forms of aquatic life. The better varieties of fish usually are affected first. As a result, one of the initial impacts of the discharge of an excessive amount of liquid waste into a stream will be that the more desirable varieties of fish will be replaced by pollution-resistant lower orders, such as carp. If all the dissolved oxygen is consumed, anaerobic conditions will result. Instead of releasing carbon dioxide, anaerobic decomposition gives off methane or hydrogen sulfide, and the stream or lake turns dark and malodorous. After reaching a minimum concentration of DO, the stream will in most cases ultimately recover. Figure 8.4 shows the "oxygen sag curve," a schematic plot of the DO concentration in a stream as a function of time or of distance downstream from the point of sewage discharge.

The harmful effects of liquid wastes on aquatic life in rivers, lakes, and streams are not restricted to oxygen-demanding pollutants. Discharges of suspended solids, toxic chemicals, heavy metals, and other hazardous substances can impair aquatic life in much the same manner as they can be injurious to humans. Both the receiving waters (even after treatment) and fish and shellfish harvested from such waters can become unsafe for human consumption (Findley and Farber, 1992). Discharges of heated water, so-called thermal pollutants, can also harm aquatic life. Higher temperatures accelerate biological and chemical processes, reducing the ability of the water to retain dissolved oxygen and other gases. The net result is that the growth of aquatic plants such as algae is hastened, and fish reproduction may be disrupted.

Additional effects may be caused, as described above, by the discharge into lakes of excess nutrients, such as detergents, fertilizers, and human and animal wastes, combined with soil runoff from agricultural and other lands.

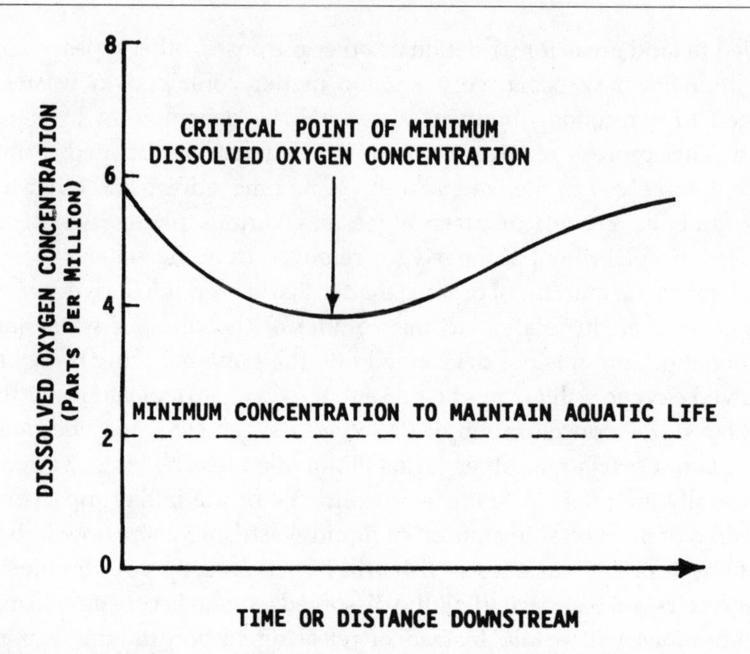

Figure 8.4 An oxygen sag curve, showing dissolved oxygen concentrations as a function of time, distance, or both in a stream into which sewage has been discharged

The National Research Council (1993) considers nutrients a high-priority pollutant, meaning that they pose significant risks to human health or ecosystems well beyond their points of discharge and are not under demonstrable control (Table 8.1). Releases of such pollutants lead to eutrophication or the "dying of lakes," as exemplified by several of the Great Lakes. Over time, lakes undergoing this process become shallower and biologically more productive, eventually evolving into swamps and finally into land areas. Although such a sequence normally requires thousands of years, human activities can accelerate the process (Findley and Farber, 1992).

That better control of these releases is needed is illustrated by several recent discoveries. One is that the enhanced growth of water plants as a result of excess discharges of nitrogen and phosphorus can block sunlight and destroy coral reefs through the inhibition of calcification (Burke, 1994). Another is that the increased growth due to excess discharges of nutrients may be responsible for recent increases in the frequency of toxic algal blooms—for example, the so-called red tide or paralytic shellfish poisoning that occurs in many of the world's oceans (Culotta, 1992).

Table 8.1 Anticipated national-level priorities for constituents of concern in liquid wastes

Priority	Pollutant group	Example
High	Nutrients	Nitrogen
	Pathogens	Enteric viruses
	Toxic organic chemicals	Polynuclear aromatic hydrocarbons
Intermediate	Selected trace metals	Lead
	Other hazardous materials	Oil, chlorine
	Plastics and floatables	Beach trash, oil, grease
Low	Organic matter	Municipal sewage
	Solids	Urban runoff

Water Pollution Regulations

The main items of federal legislation pertaining to the control of water pollution comprise the original Federal Water Pollution Control Act passed in 1956, amendments to this act passed in 1972 and 1977, and the Water Quality Act of 1987. The 1956 act directed primary attention to the establishment of water quality standards applicable to interstate and navigable waters. With the passage of the 1977 amendment, attention shifted to requirements for the treatment of point sources of industrial wastes and, through the National Pollution Discharge Elimination System (NPDES), the requirement that the EPA set pretreatment standards for pollutants destined for discharge into public waters or sewer systems. In order to comply, industries discharging such wastes were required to install the best available control technology, and all industries discharging such wastes into municipal sewer systems had to meet secondary sewage treatment standards. The focus of the regulations was on toxic pollutants that would not be adequately controlled by municipal treatment systems. Under the NPDES, the EPA has the authority to enter and inspect industrial sources of water pollution, to sample direct and indirect discharges, and to inspect the monitoring equipment and accompanying records (Findley and Farber, 1992).

With the passage of the Water Quality Act of 1987, primary attention was directed to the control of nonpoint sources of waste. The EPA was directed to promulgate regulations requiring municipal and industrial stormwater dischargers to obtain permits to release such wastes to U.S. waters. Included in the permits was the requirement that industries collect and analyze samples of runoff during initial portions of the rain (when contaminant concentrations are at a maximum), as well as estimate the quantities of individual contaminants released during the entire storm event. The permit application also had to describe a stormwater pollution prevention plan, outlining procedures for preventing releases of hazardous substances and oil onto the area within and surrounding the plant or municipality (Anonymous, 1993).

A variety of methods is available for measuring the quantities of contaminants in a liquid waste. One indicator is the concentration of suspended solids. Another is the amount of nutrients in the waste. Still another is the amount of chlorine required to oxidize the organic matter in the waste. The acidity or alkalinity of the waste may also be used as an indicator of its "strength," or polluting potential. In general, methods for treating liquid wastes, particularly domestic sewage, are designed primarily to stabilize or oxidize the organic matter therein, through biological processes. Measuring the amount of organic matter provides an indication of how effective biological treatment would be in stabilizing wastes, and how detrimental the effect would be if the wastes were directly discharged into the environment. Since there is a direct correlation between the amount of organic matter in liquid wastes and the amount of oxygen required to stabilize them, assessment of the oxygen demand of a waste is one of the more effective measures of its potential environmental impact. This test does not, however, indicate the quantity of nutrients or toxic chemicals. Since in many cases nutrients and toxic chemicals are high-priority pollutants (Table 8.1), tests to evaluate their potential contribution to the impact of a waste are also necessary.

As described above, fish cannot live in a stream that does not contain sufficient dissolved oxygen. The concentration of DO in a stream is expressed in milligrams of oxygen per liter of water. Since this unit represents one part of oxygen per million parts of water, oxygen concentrations in water are generally reported in this way. The same unit is used to express the oxygen demand of liquid wastes containing organic matter. The procedure most commonly used to make this determination is measurement of what is called the five-day, 20° C biochemical oxygen demand (BOD) of the waste. (A related chemical test has been developed to assess the oxygen demand of toxic wastes that inhibit bacterial growth and therefore do not permit use of a BOD test. The COD, or chemical oxygen demand test, is routinely used to assess the potential polluting effects of industrial wastes.)

The BOD test essentially measures the amount of dissolved oxygen needed to stabilize the decomposable organic matter in waste through aerobic bacterial action. A typical BOD test is conducted in the laboratory and measures depletion of the dissolved oxygen concentration in a diluted sample of the waste after an incubation period of five days at 68° F (20° C).

In general, the BOD after five days will be approximately 70 percent of that which would be exhibited if the sample were incubated until all the organic matter in it was stabilized. The selection of 68° F, or 20° C, is based on the fact that it is a representative outdoor temperature on a spring or fall day. At higher temperatures, the rate of depletion of oxygen in the sample will be higher; at lower temperatures, it will be lower.

Treatment of Municipal Sewage

Three stages of treatment can be applied to municipal sewage: primary, secondary, and tertiary. Primary treatment consists of holding the wastes undisturbed in a tank for a sufficient period of time to permit the solids within the waste to settle and be removed. Secondary treatment is the use of a biological process to oxidize the nutrients in the waste. Tertiary treatment involves a variety of processes tailored to the intended uses of the finished product. One of the more common tertiary or advanced methods for treating liquid wastes is very similar to the coagulation, settling, and filtration processes used in treating surface waters to make them acceptable for drinking.

Each of these processes represents a progressive level of purification, and the number of treatment stages applied depends on the degree of treatment required. On average, secondary treatment plants remove only about 30 percent of the phosphorus and a maximum of about 20 percent of the nitrogenous materials in municipal wastes. With modifications, however, higher removals are possible. As shown in Figure 8.5, all municipal sewage treatment processes must begin with the primary stage.

As indicated above, *primary treatment* consists in simply holding sewage in a large tank to permit the removal of solids by sedimentation. Before entering the settling tank, the sewage is commonly sent through a chamber or collector to remove sand, grit, and small rocks that might damage pumps or other equipment in the treatment plant. The settling tanks are operated on a flow-through basis and are large enough to hold the material for several hours. During that time approximately half the solids settle out, providing a BOD reduction of 30–50 percent.

Grease and light solids that float are removed from the settling tank by a scraper and are pumped along with the settled solids to a large closed tank

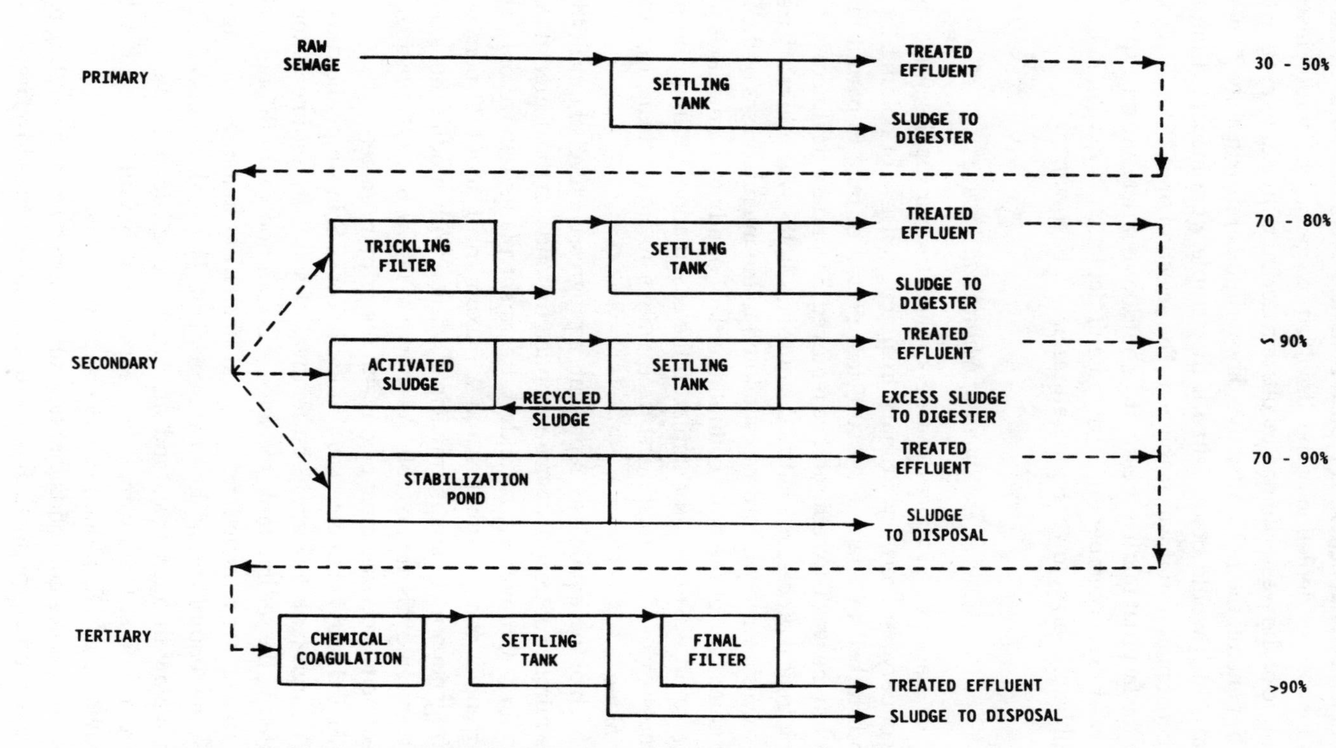

Figure 8.5 Primary, secondary, and tertiary stages in treatment of municipal sewage

called a digester, where they are held for anaerobic digestion. Digestion is most effective when the sludge is heated to 90° F (32° C) or more. At 95° F the sludge is digested in about 24 days; at 130° F, in about 12 days. The methane gas produced in the process provides fuel for heating the digester.

After primary treatment (settling), sewage can be subjected to *secondary* or biological *treatment*. In most cases this is accomplished by use of a trickling filter, the activated sludge process, or a waste stabilization pond. The first two methods are aerobic; the last combines aerobic and anaerobic systems. The overall objective is to make conditions ideal for the biological stabilization processes that normally occur in nature. No special organisms are added; those that are present develop and flourish naturally.

The trickling filter is one of the most common forms of secondary treatment. The term *filter* is a misnomer, since the system does not filter the sewage. Rather, a trickling filter consists of a large tank, roughly 6 feet deep, filled with stones 2–4-inches in diameter over which sewage is intermittently trickled or sprayed from a distributor. The stones rapidly become coated with a biological film or slime. The solids in the sewage percolating through the bed are incorporated into the bacterial growth, where the microorganisms convert the organic matter into cell protoplasm and inorganic matter.

When the bacterial growth on the stones becomes too thick and heavy, it sloughs off and is carried away in the liquid effluent leaving the bottom of the filter bed. The effluent is sent to a secondary settling tank, where the bacterial sloughings settle to the bottom as a sludge. The settled effluent represents the treated product. As in the case of primary treatment, the settled solids or sludge is sent to a digester for anaerobic decomposition. The total BOD removal from a trickling filter plant is 70–80 percent. For somewhat greater BOD removals, two trickling filters can be used in series, or a single unit can be used and a portion of the settled effluent recycled through the filter bed. Figure 8.6 shows the trickling-filter treatment process.

The activated sludge process is another form of aerobic secondary treatment for municipal sewage. Sewage is sent into a large open tank, where it is held for several hours and its oxygen content maintained by means of aerators (air diffusers) or mechanical agitators (paddles or brushes). Rather than growing on the surfaces of stones as in the trickling filter, the microorganisms float as suspended particles in the aerated sewage. The effluent is sent to a secondary settling tank, where the microorganisms settle out, and the settled sewage is the treated product. The accompanying BOD reduc-

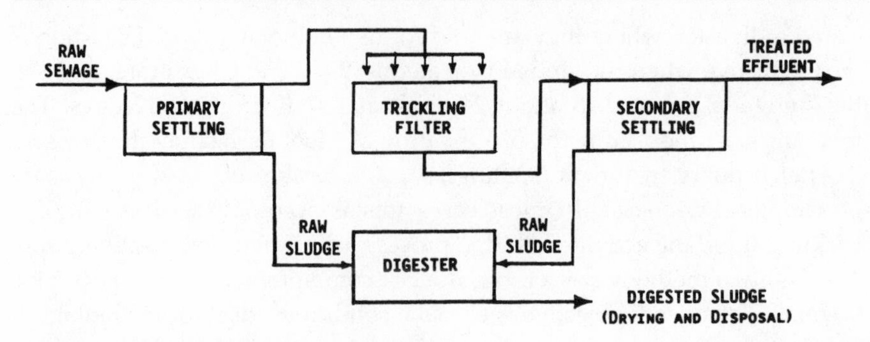

Figure 8.6 Trickling-filter sewage treatment

tions are about 90 percent. Some of the microorganisms that have settled out in the secondary tank are pumped back into the aerated tank to maintain an adequate population of microbial growth. The rest of the growth is treated as sludge and sent to a digester.

Used in other countries for many years, waste stabilization ponds were largely ignored in the United States until the 1950s (Gloyna, 1971). Since then, more and more have been built, particularly in the South. These simple ponds dug in the ground are typically 3–6 feet deep, 90 feet wide, and 300 feet long. Operated on a flow-through basis, they are usually designed to provide a retention time of 30–80 days. They are cheap to construct, easy to operate, and require minimal maintenance (EPA, 1992a). One pond can serve between 1,000 and 2,000 people. However, care should be taken not to locate the ponds in soils with fissures that allow the sewage to move through the ground without filtration, thereby contaminating groundwater supplies.

Waste stabilization ponds can be used singly or in series. They can be designed to receive either raw sewage or sewage that has undergone primary treatment. Most ponds operate biologically at two levels: the lower portion is anaerobic, the upper portion is aerobic. In the border area, facultative bacteria (which can live under either aerobic or anaerobic conditions) are active. In some cases, wind-driven mixers are used to increase the amount of air in the pond water. Algae growth at the surface also helps assure aerobic conditions. When the pond fills with sludge, it must be cleaned and the cycle begun anew.

Most methods used in the *tertiary treatment* of sewage are modeled on those used in the purification of drinking water: a coagulant is added, a floc is formed and settled, the liquid is passed through a sand filter, and a disinfectant is added. For wastes that contain unusual amounts of organic compounds, or heavy metals and viruses, additional steps may be required. To remove organic compounds, for example, the tertiary treatment process is supplemented by passing the treated waste through two granular carbon beds, each of which provides 30 minutes of contact time. Ozone is applied to disinfect the waste as it passes from the first carbon bed to the second. Heavy metals and viruses are removed by using lime as a coagulant. This process, however, creates large volumes of highly toxic sludge that must be handled and disposed of carefully.

Treatment of Industrial Wastes

As noted above, early wastewater treatment systems were based on the traditional methods for treating municipal sewage, that is, they were designed to stabilize the organic matter in the waste. Later it was recognized that nutrients such as nitrogen and phosphorus also had to be removed. Today treatment systems are designed to remove toxic chemicals as well. Application of one or more of these expanded approaches is essential for many types of industrial wastes. Such approaches include physical and chemical, as well as biological processes, and they can be applied either singly or in combination.

Physical processes include those designed to:

1. Remove suspended solids. These processes can range from simply holding the wastes undisturbed in a tank to permit the solids to settle, to passing the wastes through filters to remove the solids, to centrifuging to separate the solids from the liquid wastes.
2. Remove suspended oils, greases, and emulsified organics. Aeration of the wastes causes such materials to float to the surface, where they can be removed by skimming devices. Sometimes foaming agents are employed to enhance the process. Aeration can also be used to oxidize and precipitate chemicals such as iron and manganese.
3. Remove dissolved materials such as organic and inorganic chemicals. One way of removing organic chemicals is to pass the wastes through beds of activated carbon that physically adsorb and remove these materials. Another is to apply the process of reverse osmosis, which entails passing the wastes under pressure through a semiper-

meable membrane. The net result is that the clean water passes through the membrane and leaves the contaminants behind. This process is used, for example, to remove toxic ions in liquid wastes from industrial metal plating operations, to recover specific chemicals from pharmaceutical wastes, to remove organic compounds from vegetable and animal wastes, and to recover acids from pickling liquors.

4. Recover acids, such as nitric and hydrofluoric, from stainless steel pickling liquor, and pulping liquor from organic mixes. One such process is electrodialysis, which employs a membrane filter coupled with an electric charge (Chanlett, 1979).

Chemical processes include:

1. Addition of acids to neutralize wastes that are alkaline, or addition of bases to neutralize wastes that are acid. Frequently, this results in the production of precipitates that settle and can be removed from the wastes.
2. Addition of chemicals to liquid wastes to coagulate and precipitate suspended solids. This process is very similar to the tertiary or advanced systems applied to the effluents from municipal sewage treatment plants. If desired, specific chemicals can be added to react with and precipitate dissolved chemicals in the wastes.
3. Use of ion-exchange resins to exchange innocuous chemicals for the contaminants in the wastes.
4. Use of oxidants such as chlorine, ozone, hydrogen peroxide, and ultraviolet light to convert volatile and nonvolatile organic contaminants into nontoxic compounds.

Biological processes include:

1. Predigestion of brewery, winery, and meat packing wastes in a tank under *anaerobic* conditions as an initial step in their treatment. Frequently, this is done at elevated temperatures to accelerate the process. The approach is especially beneficial in the stabilization of wastes with high concentrations of organic solids.
2. Oxidation of certain types of industrial wastes under *aerobic* conditions similar to those applied in the treatment of domestic sewage.

Trickling filters and oxidation ponds are used, as well as the activated sludge process.

Under normal circumstances, industrial wastes containing compounds such as formaldehyde, phenols, and pickle factory residues would be disruptive to the biological organisms present in municipal (sewage) treatment systems. If, however, wastes of this type are fed for a time to smaller-scale biological systems set up especially for the purpose, biological organisms suitable for treating the wastes will selectively develop and colonize the systems. The approach may be augmented with organisms known to have specific capabilities for treating given types of toxic chemicals.

Another method used for the disposal of certain types of liquid chemical wastes is deep well injection. This technique, often applied without prior treatment of the wastes, involves injecting them into a deep underground formation using a specially designed well. On the average, injection zones are one-quarter to one mile below the ground surface and are specifically selected so as to be separated from sources of groundwater by impermeable overlying rock formations. According to EPA regulations, operators of injection wells must be able to show that hazardous concentrations of the waste will not migrate from the injection zone for a period of 10,000 years (O'Mara, 1991). Operators must also be able to demonstrate that the wastes do not contain certain hazardous materials, for example, solvents and dioxins, and that they contain less than specified concentrations of certain materials such as arsenic, cadmium, lead, mercury, and nickel, or halogenated organic compounds and polychlorinated biphenyls (EPA, 1992c).

Because of the intermittent flow rate of nonpoint sources and the difficulties in designing facilities to treat them, primary efforts to control such releases are being directed to reducing the volumes and improving the quality of such wastes, whether agricultural or human in origin.

One practice that has proved especially useful in reducing the presence of pollutants in agricultural runoff is the optimization of pesticide application rates and timing. Other controls include improving the way irrigation water is used; applying different conservation techniques such as reduced tillage, crop rotation, and winter cover crops; establishing buffer zones, such as vegetative cover along streambanks; and planting strategically placed grass strips and artificial wetlands to intercept or immobilize pollutants.

The presence of contaminants in urban runoff can be reduced by providing convenient disposal sites for used oil and household hazardous waste; collecting leaves and yard trimmings on a frequent basis; and using vacuum equipment for street cleaning (Richards, 1991). Public education can also play a pivotal role in changing behaviors that lead to minimization of local land disposal of pollutants. Other techniques, such as structural controls, are available to slow runoff, allow more water to percolate into the ground, and filter out contaminants. In addition, weirs, movable dams, and detention areas can provide storage capacity in storm and combined sewer systems, thereby reducing the frequency and volume of combined sewer overflows (NRC, 1993).

Land Disposal of Treated Wastewater

When pit privies were in common use, human excreta were disposed of primarily into the soil. The same approach is used by those who are served today by septic tanks. This procedure changed with the development of municipal sewage collection and treatment systems and the widespread use of chlorine as a disinfectant (beginning early in the twentieth century). In conjunction with these practices, the common approach was to discharge the treated wastewater into rivers, streams, and lakes. The result was a multitude of problems, including localized cases of widespread pollution of the aquatic and marine environment.

In recent years it has again been recognized that land disposal of treated wastewater offers many advantages. Such an approach:

Returns nutrients to the soil where they can be used for agriculture and forests; the nutrients can also support aquaculture.

Reclaims and/or preserves open spaces and existing wetlands and allows the development of new wetlands—parks, golf courses, and recreational areas irrigated with wastewater enhance human activities; wetlands receiving such wastes provide habitats for wildlife.

Creates an ideal environment in which natural biological, physical, and chemical processes can stabilize the wastes; wetlands serve as nutrient sinks and buffering zones to protect streams and other areas (Vassos and Reil, 1994).

Provides a ready means for recharging groundwater sources.

Frequently results in economies in wastewater treatment, thus saving funds for application to other problems.

Experience has shown that a properly developed land disposal system can be operated for 20 or more years. Such a system can also serve as a viable and beneficial alternative to methods commonly employed for the secondary treatment of municipal sewage. The reuse of human waste in aquaculture, in particular, can produce significant benefits and achieve a variety of useful goals. In countries where nutrition requirements exceed food production, aquaculture can assist in closing the gap by using valuable nutrients that would otherwise be squandered. In countries where water quality must be improved, aquaculture can lessen the harmful impact on water courses due to pollution from overpopulation. In arid regions, it can make an important contribution in the conservation of scarce water resources (Edwards, 1992).

Acknowledging that the application of existing drinking-water standards could be unreasonably restrictive, federal and state regulatory agencies have sought to develop separate standards for wastewaters destined for recycling and reuse. Such wastes include those destined to serve either as a raw water source for industrial processes or as irrigation water for agricultural crops and recreational areas (WHO, 1989). Studies of household systems in which effluent from the septic tank was used to irrigate the lawn have shown no increase in coliform bacterial concentrations above those in lawns irrigated with water from the potable supply (Kleene et al., 1993). Studies of salmonella organisms in composted sludge used as a soil conditioner, however, led to the recommendation that limits be set on the permissible bacterial concentrations in such wastes (Skanavis and Yanko, 1994).

Concerns related to the use of wastewater extend beyond disease organisms to include the possible presence in such water of trace metals, organics, and pesticides. In all cases, it is important not only to examine the purity of the treated wastewater under consideration for disposal, but also to develop detailed knowledge on industrial facilities that may discharge toxic materials into the treatment system. It is also vital to analyze periodically the purity of the wastewater leaving the land disposal system, to assure that it is safe for its intended use. Obviously, this entire area needs further review and evaluation.

The Environmental Protection Agency estimates that the liquid waste treatment plants in the United States produce almost 8 million tons of sludge annually. Additional sludge results from the treatment of industrial wastes. Unless this sludge is properly disposed of, many benefits of the treatment process will be lost. In some cases sludge is used as fertilizer; in other cases, as landfill. New York and other cities transport digested sludge out to sea and dump it, an approach that is receiving ever-closer scrutiny.

The manner of disposal should be designed so as not to create additional public health problems. Because of certain characteristics of the process, sludge produced in the biological treatment of municipal wastes tends to concentrate toxic heavy metals. If this sludge is used as a fertilizer, these metals and other chemicals can be taken up by food crops destined for human consumption. A notable example is cadmium, which tends to concentrate in leafy vegetables. If the sludge is incinerated, the release of toxic materials into the air may be a problem. Recognizing these facts, the EPA has developed regulations governing the disposal of sludges produced in the treatment of liquid wastes. Requirements have been specified for sludges that are to be sent to a landfill, applied to land as a fertilizer, or incinerated (EPA, 1992b).

The General Outlook

At the present time, there are almost 16,000 municipal wastewater treatment facilities in the United States. These have a total daily treatment capacity of almost 40 billion gallons and represent a capital investment of about $4 billion. Unfortunately, many are in need of major repair and/or upgrade. One of the immediate challenges is to find the financial resources to meet these needs. Estimates are that the costs of the necessary repairs, plus funds for the construction of facilities to meet future demands, will exceed $100 billion over the next 20 years (Ouellette, 1991).

Another considerable challenge is the range of potential problems resulting from the widespread use of chlorine as a disinfectant for wastewater. As in the case of drinking water, chlorination of wastewater creates a host of potentially carcinogenic compounds. Such compounds have been a special concern, for example, in terms of the wastewater discharged into the Great Lakes. As a result, the International Joint Commission, a U.S.-Canadian advisory group on the control of pollution in these waters, has recommended banning the application of chlorine to wastewaters discharged into the Great Lakes.

Another challenge is the need to protect groundwater supplies from contamination with toxic chemicals. While contamination of surface water may be of relatively short duration, groundwater contamination can persist for decades or longer. Recognizing these problems, the EPA is focusing on a long-term strategy for protecting groundwater. Its underlying principles are that groundwater must be protected to ensure that the nation's current and future drinking-water supplies do not present adverse risks, and that these supplies are preserved for present and upcoming generations. The basic approaches being considered include pollution prevention, source monitoring, siting controls, the designation of wellhead protection areas, and the protection of aquifer recharge areas (EPA, 1991).

On a long-term basis, the most desirable approach for preventing pollution of the world's water resources is to eliminate production of the waste in the first place and, where such generation is unavoidable, to design systems so that the waste can be recycled and reused. Even here, severe problems exist. Water conservation may, in some cases, reduce the volume of the wastes without reducing the mass of pollutants they contain. Furthermore, many of our conventional treatment systems are primarily designed to reduce the organic load in the waste. Their ability to reduce the concentration of bacteria and viruses is extremely restricted. Conventional primary treatment processes, for example, remove only about 50 percent of the pathogenic bacteria; even secondary treatment systems, such as the trickling filter or activated sludge process, remove only about 90 percent. Many viruses survive these treatments, thereby placing restrictions on how the effluents can be used, even when the desire to conserve resources is present (Mara, 1982).

9

SOLID WASTE

UNTIL World War II most solid or municipal waste took the form of garbage, yard waste (leaves, grass clippings, tree limbs), newspapers, cans and bottles, coal and wood ashes, street sweepings, and discarded building materials. Most such waste was not considered hazardous, and it was simply transported to the local land disposal facility or "dump," where it was periodically set on fire to reduce its volume and to discourage the breeding of insects and rodents. Because this practice often led to windblown debris and unsightly disposal facilities, and because people recognized the need for a technically sounder method of disposal, it was gradually replaced by the sanitary landfill, where municipal waste was buried in the ground (Figure 9.1). As long as windblown debris and fires were contained, material was covered over and sealed daily (so that breeding and habitation by insects and rodents were controlled), and contamination of nearby groundwater supplies was avoided, the sanitary landfill was considered an acceptable method of disposal.

With the development of a throwaway society and an unprecedented demand for new products, the volume of solid waste has increased enormously. Within the United States, the quantity of municipal solid waste rose from 2.7 pounds per person per day in 1960 to 4.3 pounds in 1990. Today the average person in this country annually produces 1,000–1,500 pounds of municipal solid waste, including almost 100 pounds of plastics—more municipal solid waste per capita than in any other industrialized nation (Grove, 1994; AIChE, 1993). Vastly larger quantities of similar nonhazardous wastes are produced by industry. At the same time, the composition of

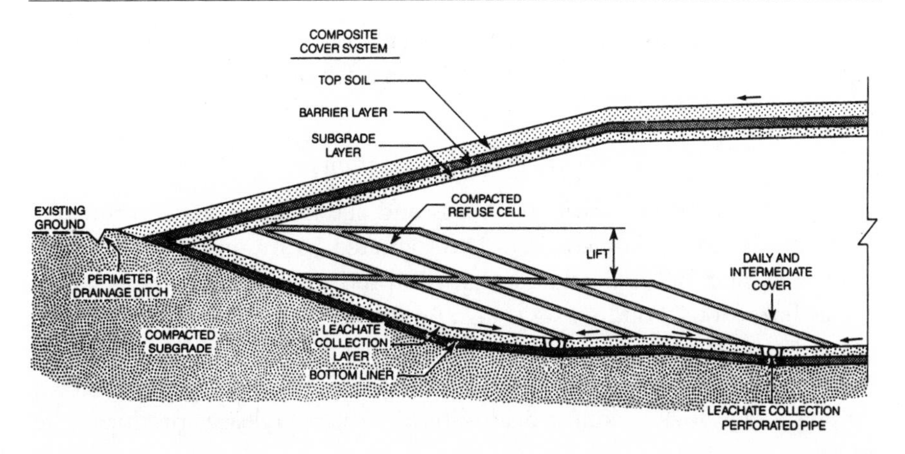

Figure 9.1 Cross section of a typical landfill and leachate collection system

the waste has changed. Solid waste today contains many materials (such as plastics) that are not readily degradable and toxic materials—primarily various types of chemical waste produced by industry—that can contaminate soil and groundwater indefinitely if not properly disposed of.

In a similar manner, the amount of *hazardous waste* generated has been undergoing dramatic change, increasing from about 0.5 million metric tons per year at the time of World War II to some 300 million metric tons in 1993—more than one ton of hazardous waste per person per year. In addition, industry annually discharges some 18 billion pounds (9 million tons) of toxic chemicals directly into the air, water, or land or into deep underground injection wells (NRC, 1995a). The chemical and petroleum industries currently generate more than 70 percent of the hazardous waste in the United States (EPA, 1986a); the rest comes from a wide range of other industries. In total, about 6 billion tons of waste are produced in the United States each year.

About two-thirds of the municipal solid waste produced is sent to sanitary landfills for disposal (AIChE, 1993). However, many existing disposal sites are being filled and phased out. In 1980, there were about 16,000 active sanitary landfills in the United States; by 1995, the number had decreased to about 3,000. In fact, much of the heavily populated East Coast is expected to run out of acceptable landfill space by the year 2000 (Grove, 1994). Although this trend is partially due to the growing difficulty of establishing new facilities, it is also a response to the increasing stringency of federal regulations and a recognition of the benefits of establishing larger, better designed and operated disposal facilities on a regional basis.

Types and Classifications

Solid waste comes in a variety of types and classes. As a general rule, the waste produced by homeowners is designated as municipal nonhazardous waste, and the toxic chemical waste produced by industry is classified as hazardous. Solid waste is also produced by the treatment of liquid and airborne wastes. For example, the purification of drinking water and the treatment of liquid wastes produce sludge, and processes for removing pollutants, such as sulfur oxides, from airborne releases produce solid waste.

Beginning with the Solid Waste Disposal Act of 1965, the U.S. Congress has passed a number of laws that have had a significant impact on the management and disposal of solid waste (Table 9.1). One of the most important was the Resource Conservation and Recovery Act (RCRA) of 1976. The principal programs and goals of this act are summarized in Table 9.2.

One of the most recent acts pertaining to solid waste disposal is the Federal Facility Compliance Act; it requires the Department of Energy to develop technologies for the treatment and disposal of "mixed" wastes,

Table 9.1 Principal federal laws related to management and disposal of solid and hazardous waste

Year	Law	Policy Law number
1965	Solid Waste Disposal Act	89–272
1976	Resource Conservation and Recovery Act (RCRA)	94–580
1980	Comprehensive Environmental Response, Compensation, and Liability Act (Superfund Act)	96–150
1984	Hazardous and Solid Waste Amendments, Resource Conservation and Recovery Act	98–616
1986	Superfund Amendments and Reauthorization Act (SARA)	99–499
1990	Pollution Prevention Act	101–508
1992	Federal Facility Compliance Act	102–386

Table 9.2 Principal programs and goals of the Resource Conservation and Recovery Act (1976)

Solid waste program (directed primarily at management and control of nonhazardous solid wastes)
Primary goals:
To encourage environmentally sound solid-waste management practices
To maximize reuse of recoverable resources
To foster resource conservation
Hazardous waste program ("cradle-to-grave" system for managing hazardous waste)
Primary goals:
To identify hazardous waste
To regulate generators and transporters of hazardous waste
To regulate owners and operators of facilities that treat, store, or dispose of hazardous waste
Underground storage tank program
Primary goals:
To provide performance standards for new tanks
To prohibit installation of unprotected new tanks
To provide regulations concerning leak detection, prevention, and corrective action

those that contain both hazardous chemicals and radioactive materials (U.S. Congress, 1992). Discussion of the full range of laws pertaining to solid waste is presented in Chapter 13.

Included in RCRA is a system for classifying solid waste, with one of the primary objectives being to clarify its definition and to establish a distinction between solid nonhazardous and hazardous waste. According to RCRA, *solid (nonhazardous) waste* is defined as "any garbage, refuse, sludge from a waste treatment plant, water supply treatment plant, or air pollution control facility and other discarded material, including solid, liquid, semisolid, or contained gaseous material resulting from industrial, commercial, mining, and agricultural operations and from community activities." Specifically excluded is "solid or dissolved material in domestic sewage" (EPA, 1986a).

Examples of nonhazardous waste include domestic trash and garbage, such as milk cartons and coffee grounds; refuse such as metal scrap, wallboard, and empty containers; and other materials discarded from industrial operations, such as boiler slag and fly ash. As would be anticipated, munici-

Table 9.3 Composition of typical municipal solid waste

Component	Percentage by weight[a]	Percentage by volume[a]
Paper and Cardboard	37.5	37.0
Glass	6.7	2.3
Ferrous Metals	6.3	8.8
Aluminum	1.4	3.1
Plastics	8.3	18.3
Rubber and Leather	2.4	5.8
Textiles	2.8	5.4
Wood	6.3	5.9
Food Wastes	17.9	9.2
Other	3.7	1.5

a. Due to rounding, numbers do not total 100 percent

pal solid waste is a complex mixture of common household trash, with paper, yard waste, and construction debris constituting the largest fractions. The average distributions of various components of such waste by weight and volume are given in Table 9.3.

RCRA defines *hazardous waste* as any "solid waste, or combination of solid wastes, which because of its quantity, concentration, or physical, chemical, or infectious characteristics may: (1) cause, or significantly contribute to an increase in mortality or an increase in serious irreversible, or incapacitating illness; or (2) pose a substantial present or potential hazard to human health or the environment when improperly treated, stored, transported, or disposed of, or otherwise managed."

In implementing this law, the Environmental Protection Agency specified that a solid waste is hazardous if analysis proves it to be (1) ignitable, (2) corrosive, (3) reactive, or (4) toxic. Title 40, parts 261.31–33 of the Code of Federal Regulations lists almost 300 substances as hazardous wastes. As noted in Table 9.4, these wastes come from a number of segments of business and industry.

Under the existing regulatory structure, commercially generated waste is classified as nonhazardous, hazardous, mixed, or radioactive. Table 9.5 summarizes the characteristics, volumes, and groups responsible for regulating each category.

Table 9.4 Examples of hazardous waste generated by business and industry

Waste generator	Type of waste
Chemical manufacturers	Strong acids and bases Spent solvents Reactive wastes
Vehicle maintenance shops	Paint wastes containing heavy metals Ignitable wastes Used lead acid batteries Spent solvents
Printing industry	Heavy metal solutions Waste inks Spent solvents Spent electroplating wastes Ink sludges containing heavy metals
Leather products manufacturers	Waste toluene and benzene
Paper industry	Paint wastes containing heavy metals Ignitable solvents Strong acids and bases
Construction industry	Ignitable paint wastes Spent solvents Strong acids and bases
Cleaning agents and cosmetics manufacturers	Heavy metal dusts Ignitable wastes Flammable solvents Strong acids and bases
Furniture and wood manufacturers and refinishing	Ignitable wastes Spent solvents
Metal manufacturing	Paint wastes containing heavy metals Strong acids and bases Cyanide wastes Sludges containing heavy metals

Table 9.5 Types, regulation, and characteristics of commercially generated waste

Type of waste	Regulating body	Typical content	Approximate annual volume (1994)
Nonhazardous	State and local governments	Refuse, garbage, sludge, municipal trash	400 billion ft^3
Hazardous	EPA or authorized states	Solvents, acids, heavy metals, pesticide residues, chemical sludges, incinerator ash, plating solutions	4 billion ft^3 (1% of total)
Mixed	EPA and USNRC or states	Radioactive organic liquids, radioactive heavy metals	160,000 ft^3 (0.00003% of total)
Radioactive	USNRC or agreement states	High- and low-level radioactive waste, naturally occurring and accelerator-produced materials	1 million ft^3 (0.0002% of total)[a]

a. Commercial low-level radioactive waste only.

Management and Disposal

For many years, people responsible for protecting the environment accepted the wastes that were generated and tried to develop satisfactory methods for their treatment and/or disposal. This is referred today as the "end of pipe" approach. With the coming of the throwaway society and the rapid development of industry, environmentalists and Congress soon realized that the production of waste was becoming overwhelming, and that new approaches had to be developed, particularly with respect to hazardous waste.

As a result, Congress in 1984 passed the Hazardous and Solid Waste Amendments, Resource Conservation and Recovery Act, which mandated that the reduction or elimination of the generation of hazardous waste at the source (that is, pollution prevention) should take priority over the management of such wastes after they have been generated. Congress further clarified the role of pollution prevention by passing the Pollution Prevention Act of 1990 (U.S. Congress, 1990), which stated that it was the policy of the United States that, whenever feasible, pollutants that cannot be prevented should be recycled, and those which cannot be prevented or recycled should be treated and disposed of in an environmentally safe manner.

WASTE MANAGEMENT

To ensure compliance with these approaches, the EPA announced that it was committed to a policy that places the highest priority on waste minimization (EPA, 1993b). Under this approach, the agency requires generators of hazardous waste to certify on their shipping manifests that they have a waste minimization program in place. The same certification is required for owners and operators of facilities that receive a permit for the treatment, storage, or disposal of hazardous waste on the premises where such waste was generated. Methods for minimizing the production of hazardous waste include:

Separating or segregating waste at its source, to prevent hazardous materials from contaminating nonhazardous waste and thereby making the entire mixture hazardous;

Eliminating or substituting raw materials that generate little or no hazardous waste for those that generate a large amount (for example, the use of non–lead-based paints);

Changing manufacturing processes to eliminate steps that generate hazardous waste, or altering processes so that the waste is no longer produced (for example, using more effective and efficient methods of applying paints).

Table 9.6 summarizes the techniques that can be used to minimize the production of hazardous waste.

Initially, waste minimization was viewed by many industries as simply an additional regulatory burden. Experience has shown, however, that such a program can yield significant benefits: financial savings through

Table 9.6 Techniques for minimizing production of hazardous wastes

Inventory management and improved operations
Inventory and trace all raw materials
Emphasize use of nontoxic production materials
Provide waste minimization or reduction training for employees
Improve receiving, storage, and handling of materials

Modification of equipment
Install equipment that produces minimal or no waste
Modify equipment to enhance recovery or recycling options
Redesign equipment or production lines to produce less waste
Improve operating efficiency of equipment
Maintain strict preventive maintenance program

Production process changes
Substitute nonhazardous for hazardous raw materials
Segregate wastes by type for recovery
Eliminate sources of leaks or spills
Separate hazardous from nonhazardous and radioactive from nonradioactive wastes
Redesign or reformulate end products to be less hazardous
Optimize reactions and raw material use

Recycling and reuse
Install closed-loop systems
Recycle on-site for reuse
Recycle off-site for reuse
Exchange wastes

Treatment to reduce toxicity and volume
Evaporation
Incineration
Compaction
Chemical conversion

increased plant efficiency and reduced waste management and disposal costs, reductions in future liabilities, decreased worker exposure to toxic chemicals, and enhancement of the company's public image (EPA, 1993b).

Where the generating of waste cannot be avoided, the preferred alternative is environmentally sound recycling. That is, every effort should be made to employ the waste as an ingredient in the manufacture of another product, or to process the waste to recover usable materials from it. Appropriate procedures include recovering and reusing the waste through removal of its hazardous constituents, reusing the waste within the process (for example, recycling lead storage batteries), or transferring the waste to another industry that can use it as input to its production process.

Although it is important for industrial organizations to have comprehensive recycling programs, the effort should not end there. Individuals can do a great deal on a personal level to promote this type of activity, especially with respect to the recycling of household solid waste. In fact, many of the public-spirited members of society have seized on this area as one in which they can both exercise control and have a positive impact in helping solve one of our most serious environmental problems. People in communities throughout the country are sorting their trash, filling recycling bins, demanding to be able to purchase products made of recycled materials, and avoiding products with wasteful packaging. Overall, in 1995 more than 80 percent of the cities in the United States had operational curbside recycling programs; 60 percent had programs for collecting recyclables from multifamily buildings (News Briefs, 1995). In addition, over 40 states have established recycling goals—the most ambitious being Rhode Island, which seeks to attain the recycling of 70 percent of its garbage within the next few years (Consumer Reports, 1994). A summary of the degree to which various items are currently being recycled in this country is shown in Table 9.7. The net result has been a 25 percent reduction over the past 15 years in the percentage of municipal waste being sent to landfills for disposal (News Briefs, 1995).

The recycling of household solid waste, however, is not without its limitations, a major problem being the required initial investment in additional collection vehicles and sorting equipment. Whereas normal municipal waste can be loaded onto a truck and compacted for efficient transport, recycled materials cannot. As a result, the amounts of waste that can be hauled by the trucks used for collecting and transporting recyclable materials are far below normal. This leads to increased energy consumption and air pollution. Since the recyclable materials must subsequently be carefully

Table 9.7 Recycling of various materials in the United States, 1994

Product	Annual waste (millions of tons)	Percent—	
		Recovered	Discarded
Paper	73.3	29[a]	71
Plastic	16.2	3	97
Glass	13.2	20	80
Steel	12.3	15	85
Aluminum	2.7	37[b]	63

a. Nearly 50 percent of United States newspapers are recycled—most as newsprint, some as cardboard boxes.

b. About 65 percent of the aluminum cans used as beverage containers are recycled.

sorted, often by hand, the system is labor intensive, further increasing the costs (Consumer Reports, 1994).

Still an increasing number of industries are benefiting from recycling operations. Tissue, cardboard, and boxboard manufacturers have for years used scrap paper as one of their feed materials. The aluminum industry discovered the economies of recycling several decades ago. Recycling is also common practice in the steel industry; depending on the type of furnace used, between 25 and 100 percent of the input consists of scrap metal. Newsprint manufacturers in Canada use up to 50 percent recycled pulp.

At the same time, it is important to recognize that recycling can have negative impacts. In the paper industry, consistency in the quality of a final product is easier to achieve with 100 percent virgin materials. In glassmaking, care must be taken to exclude contaminants that, because of their different melting points, can ruin an entire furnace load. The same is true in the manufacture of plastic and paper (Consumer Reports, 1994). Another area in which contaminants can lead to unfortunate problems is in steelmaking. Radioactive materials, inadvertently present in scrap metal, have led to incidents of major contamination of both the plants and the finished product (Lubenau and Yusko, 1995).

Essential to the success of any recycling program is a demand for products in which the recycled materials are incorporated. Unfortunately, balancing demand with supply has frequently been a problem, especially during the early phases of a recycling program. To overcome these obstacles, certain types of stimuli may be required. Consortia have been formed by leading companies to provide a market for recycled products, such as paper (Consumer Reports, 1994), and the Buy Recycled Business was established in 1992 by 25 blue-chip corporate activists. Today this group includes nearly 700 companies, 80 percent of which are small businesses. In terms of increasing the demand for products made from recycled materials, their success is demonstrated by the fact that in 1993 purchases of second-time-around goods increased to over $10 billion, nearly quadruple the amount the previous year (Grove, 1994). Another step being taken is the establishment of materials exchanges, through which information on available and wanted waste materials is readily made known via catalogs, directories, newsletters, or electronic databases. As a result of these efforts, almost 1,000 waste generators currently "market" several thousand different materials; to date more than 500 transactions, involving the buying and selling of 50,000 tons of reusable waste, have been completed (Melody, 1994).

WASTE TREATMENT

Treatment is defined as any method, technique, or process, including neutralization, that is designed to change the physical, chemical, or biological character or composition of a hazardous waste so as to neutralize it, recover energy or material resources from it, render it nonhazardous or less hazardous; or to make it safer to transport, store, or dispose of, more amenable to recovery or storage, or smaller in volume (EPA, 1993b). Treatment may be either thermal (for example, incineration), chemical, or biological (especially for organic hazardous wastes). Where methods for neutralizing or rendering a waste nonhazardous are not available or are ineffective, immobilization (stabilization) can often be effective (especially for inorganic hazardous wastes).

The general goal is to convert hazardous waste into a solid form for disposal. Treatment may be initiated at any stage prior to or following solidification, for example, in tanks, surface impoundments, incinerators, or land treatment facilities. Special processes such as distillation, centrifugation, reverse osmosis, ion exchange, and filtration can also accomplish the task. Because many of these processes are waste specific, the EPA has not attempted to develop detailed regulations for any particular type of process or equipment; instead, it has established general requirements to assure safe containment. Technologies that can be applied for the treatment of hazardous waste prior to disposal include the following (EPA, 1986a,b):

Biological treatment using microorganisms to degrade organic compounds. Anaerobic bacteria capable of degrading and detoxifying polychlorinated biphenyls exist in nature; unfortunately, most waste disposal facilities are not designed to promote conditions that favor the growth of microbes (Finstein, 1989).

Physical methods such as the application of activated carbon to adsorb organic compounds from liquid waste. The problem of disposing of the contaminated carbon can be avoided if the adsorbed compounds are removed and recovered and the carbon is reused.

Chemical treatment to neutralize acid or alkaline waste, to detoxify chlorinated substances, to oxidize and detoxify waste such as cyanides, phenols, and organic sulfur compounds, or to coagulate, precipitate, and concentrate suspended solids.

Solidification and stabilization of waste to make it less permeable and soluble, and less susceptible to transport by water.

Incineration to destroy certain toxic chemicals or to make waste, particularly that containing organic compounds, less hazardous.

Incineration deserves special mention because it is one of several technologies available both for reducing the volume of solid and hazardous waste and for destroying certain toxic chemicals in it. The increased use of plastics in packaging, however, has created a corresponding increase in the amount of polyvinyl chloride in solid waste. When burned, such plastics produce hydrochloric acid. This extremely corrosive compound can destroy incinerator components such as metal heat exchangers and flue-gas scrubbers, and can threaten human health if released into the atmosphere. Hydrochloric acid can also be produced in incinerators by the combustion of foods and wastes that contain chloride salts. Compounding these problems, incomplete combustion of some organic materials in the presence of chlorides can produce dioxins, a toxic group of compounds. Dioxins may be present in the airborne emissions from the incinerator as well as in the solid residues. However, if the operating temperature is sufficiently high, and if distribution and mixing of the combustion air are adequate, generation of these compounds can be avoided or sharply reduced (Engdahl, Barrett, and Trayser, 1986).

These and other potential threats to human health have led to stringent regulations on emissions from incinerator facilities (Rose, 1994). Although modern technology will provide almost any degree of cleanup required, the economic costs can be high. One response has been to construct and operate centrally located incinerators to serve a group of waste producers. In many communities that (for environmental, political, economic, and other reasons) have a limited capacity for direct disposal of solid waste in landfills, incineration has become the principal method of intermediate treatment.

Finding a place to dispose of the residue from the incineration of hazardous or low-level radioactive waste continues to be a problem, but the ash generally is in a physical and chemical form much more readily disposed of than the original waste; it is biologically and structurally more stable, and many of the compounds it contains are insoluble (Long, 1990). Incineration also produces a waste that minimizes long-term ground subsidence and leaching by rain and groundwater. Another potential benefit of incineration is the energy that can be derived through use of certain types of waste, such as worn-out tires, to supplement the fuel normally consumed in paper mills, cement kilns, and electric utility boilers. Each 20-pound car tire contains an amount of heat energy roughly equal to that in 25 pounds of coal (Lamarre, 1995). But such uses need not stop there. Estimates are that more effective use of the municipal solid waste stream as fuel could generate upward of 100 billion kilowatt-hours of electricity annually—an amount equal to about 4 percent of this country's current electricity demands (EPRI, 1992).

WASTE DISPOSAL

Waste disposal, by definition, means the discharge, deposit, injection, dumping, spilling, leaking, or placing of any solid waste or hazardous waste into or onto the land or water. It has crucial ramifications for environmental health because disposal may permit the waste and/or its constituents to enter the terrestrial environment, be emitted into the air, or be discharged into surface waters. Also of concern is the potential contamination of groundwater.

As previously mentioned, the primary method for managing and disposing of municipal and hazardous waste is burial in the ground. Such disposal includes a range of options (EPA, 1986b):

Landfills—Disposal facilities in which the waste is placed into or onto the land. In most landfills, the wastes are isolated in discrete cells within trenches. As will be noted below, landfills must be lined to prevent leakage and have systems to collect any leachate or surface runoff (Figure 9.2).

Surface impoundments—Natural or man-made depressions or diked areas that can be used to treat, store, or dispose of hazardous waste. Surface impoundments may be of any shape or size (ranging from

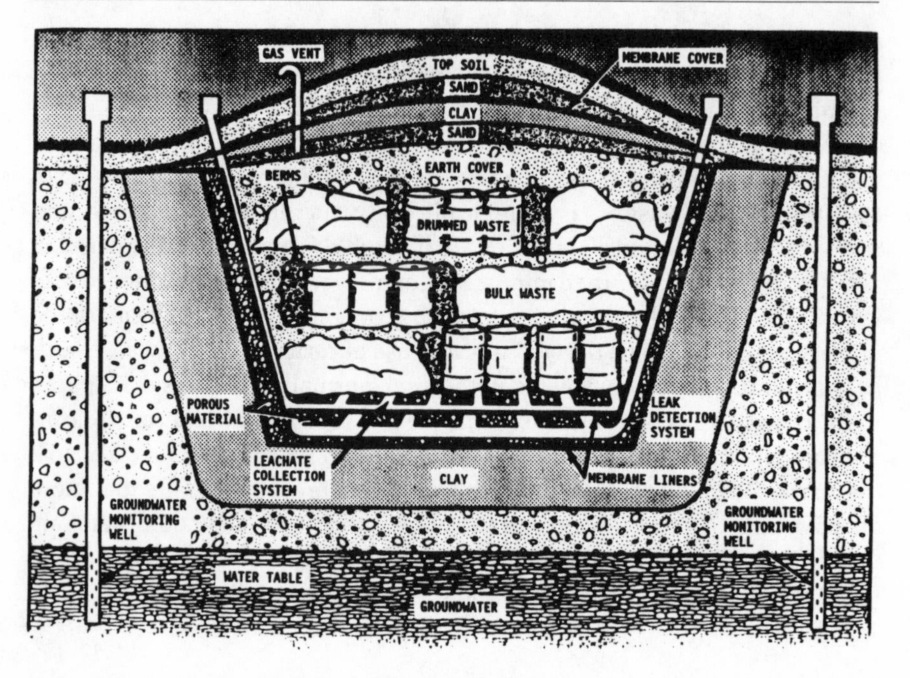

Figure 9.2 Land burial facility for hazardous waste

a few hundred square feet to hundreds of acres). Surface impoundments are often called pits, ponds, lagoons, or basins.

Underground injection wells—Steel- and concrete-encased shafts placed deep in the earth into which wastes are injected under pressure. Hazardous liquid wastes are commonly disposed of in this manner.

Waste piles—Noncontainerized accumulations of insoluble solid, nonflowing hazardous waste. Some waste piles serve as final disposal, many as temporary storage pending transfer of the waste to its final disposal site.

Land treatment—A disposal process in which solid waste, such as sludge from municipal sewage treatment plants, is applied onto or incorporated into the soil surface. Microbes occurring naturally in the soil break down or immobilize the hazardous constituents.

More than 200,000 sites in the United States are now being, or were formerly, used as sites for the disposal of municipal wastes. Although the wastes as buried were classified as nonhazardous, about 35,000 of these sites are known to have received hazardous chemicals and other materials from small-quantity industrial generators. In addition, a certain amount of the waste from most households contains hazardous materials. Through the 1984 amendments to RCRA, Congress mandated that EPA develop new criteria to provide better protection of the public from the potential health risks associated with these facilities (EPA, 1986b). Responding to this mandate, the EPA requires disposal practices for municipal waste that closely parallel those for industrial (hazardous) waste. These requirements, which apply to all aspects of the siting, design, construction, operation, and monitoring of such facilities, can be summarized as follows (EPA, 1993a).

Location. Landfills must not be located on a floodplain; they must not be built on wetlands unless the proposed operator can show that the landfill will not lead to pollution; they cannot be located in areas subject to landslides, mudslides, sinkholes, or major disruptive events such as earthquakes, which could lead to pollution; and they cannot be located near airports, where birds that are frequently attracted to such facilities might constitute a danger to aircraft.

Design. Landfills must be designed to protect the groundwater from being contaminated. Ancillary requirements include lining the bottom of the landfill with clay, covered by an impervious synthetic-material liner, coupled with a system to collect and treat any leachate (liquids) that may collect within the liner and could pose a threat of pollution.

Operation. No hazardous waste should be disposed of in a municipal landfill; the waste must be covered daily with dirt to prevent the spread of disease by rats, flies, mosquitoes, birds and other animals; access to the landfill must be restricted to prevent illegal dumping and other unauthorized activities; the site must be protected by ditches and levees to prevent stormwater flooding; and any runoff that occurs must be collected and controlled.

Monitoring. Generally, landfill owners or operators must install monitoring systems to detect groundwater contamination. If contamination is observed, the concentrations must be reduced so as to assure compliance with federal limits for safe drinking water; methane gas, generated through decomposition of the waste, must also be monitored and controlled, if necessary.

Closure and postclosure care. Upon ceasing operation, the landfill must be closed in a way that will prevent subsequent problems. The final cover must be designed to keep liquid away from the buried waste and, for 30 years after closure, the operator must continue to maintain the cover, monitor the groundwater to be sure the landfill is not leaking, collect and monitor any generation of gases, and perform other maintenance activities.

As would be expected, all approaches to the disposal of solid waste have potential impacts on public health and the environment. Improper discharge or burial of waste on land has led to contamination of many underground aquifers that serve as a source of drinking water. Waste contaminants can also move into streams, rivers, lakes, and other surface waters, killing aquatic life and destroying wildlife and vegetation. Nor are the harmful effects confined to inhalation or ingestion: plastic fishing gear, six-pack beverage yokes, sandwich bags, and styrofoam cups thrown into the ocean entrap and kill more than 1 million seabirds and 100,000 marine mammals every year. In fact, plastics may be as great a source of mortality to marine mammals as oil spills, heavy metals, or other toxic materials (Shea, 1988). To assist in controlling one source of such pollution, the National Research Council has proposed that laws be established to protect beaches and coastal waters from garbage and plastic wastes discharged from ships and that a national commission be established to oversee their implementation (NRC, 1995b).

On the benefit side, landfills, like incinerators, can be operated to serve as a source of energy. An estimated 750 landfills in the United States have the

Table 9.8 Trends in the disposal of municipal solid waste

	Percent	
Disposition	1980	1990
Landfills	81.3	66.6
Combustion		
Waste to energy	1.8	15.2
Without energy	7.3	1.1
Recycled		
Composting	0.0	2.2
Other	9.6	14.9

potential of serving as sources of gas that could be used to replace natural gas in certain applications, such as fuel for internal combustion engines. In fact, a full-scale facility producing and dispensing high-quality compressed natural gas derived exclusively from landfill gas is currently in operation in California (Anderson, 1994).

Over the past decade, the relative use of landfills for disposal of municipal solid waste has been declining, while at the same time the percentage of waste that is being recycled or burned to generate energy has been increasing. A summary of these trends is presented in Table 9.8 (AIChE, 1993).

Cleanup of Disposal Sites

Thousands of waste disposal sites in the United States were improperly designed or operated and have leaked, or have the potential to leak, hazardous waste into the environment. Recognizing the severity of the problem and the urgent need for cleanup of these sites, Congress in 1980 passed the Superfund Act. Under terms of this act, the EPA designates sites for cleanup by including them on a national priorities list. At the present time, about 33,000 hazardous waste disposal sites have been designated as Superfund sites. Of these, almost 1,300 are on the national priorities list (NPL), that is, they have been identified as posing the highest risk and therefore are to be given priority for cleanup. About 200 of the sites on the NPL are former municipal landfills. In addition to the sites contaminated by hazardous waste from industrial activities, a host of sites have been contaminated by operations of the Department of Defense and the Department of Energy.

After placing a site on the NPL, the EPA identifies the potentially responsible parties and gives them an opportunity to implement cleanup. If they fail to do so, the EPA arranges for the cleanup, using Superfund money, then seeks to recover the costs from the responsible parties.

One approach that can be used for the cleanup of hazardous waste sites is to excavate the contaminated material, including the accompanying soil, and transport it to a new burial site. Because in many cases the quantities of material are enormous, various methods for on-site treatment are being developed. Sites containing soils and sediments contaminated with industrial organic chemicals represent a particularly difficult challenge, both in sheer numbers and in complexity. Treatment technologies that are being used involve physical and chemical processes. Either can be applied to the soil in place or after it has been excavated. Which option is selected depends on the types of contaminants and the relevant properties of the soil—for example, its clay and humus content. The most proven separation technology is volatilization of organic chemicals using vapor extraction and/or forcing air through the soil. Many sites are also being remediated by solidification/stabilization, whereby the organic contaminants are not removed or treated, but the contaminated soil is rendered inert to the environment by mixing it with cementitious additives. Thermal desorption has also been found to be an effective technology for separation of the full spectrum of organic compounds from soils (Fox, 1996).

Another method being extensively evaluated is biological treatment, which offers two distinct advantages: it is inexpensive, and it has the unique potential for rendering hazardous constituents nontoxic. Unlike more conventional applications of such processes (for example, in the treatment of domestic sewage and liquid industrial wastes), the biological treatment of contaminated soils is an immature field that, although offering high expectations, is confronted with numerous scientific and engineering challenges (Hughes, 1996).

To encourage the development of innovative treatment techniques, the EPA has established the Superfund Innovative Technology Evaluation (SITE) program, through which such technologies can be both developed and shared throughout the industrial and environmental communities. Even though an array of processes has been proposed, progress has been slow. Obstacles include the unproved nature of many of the proposed methods, the diversity of expertise required, the high costs, and the fact that no two sites are identical. Even within a single site, conditions may vary. Costs of cleanup range up to $1 million per acre, and some sites cover more than 100 acres. Where the groundwater has been contaminated, cleanup may take 20–40 years or even longer. The ultimate cost of cleanup of contaminated sites on the NPL has been estimated to range into the hundreds of billions of dollars. As a result, it is of the utmost importance that the system of identification be based on sound risk assessment procedures (see Chapter 16). It is also crucial to acknowledge that restoration of the sites to a pristine state is not always necessary or technically or financially feasible (Travis, 1993; NRC, 1994).

Management of Radioactive Waste

As is true of hazardous chemical wastes, the management and disposal of radioactive waste is receiving extensive governmental attention. Groups involved at the federal level include Congress, the EPA, the U.S. Nuclear

Regulatory Commission (USNRC), and the Department of Energy (DOE). In general, Congress passes relevant legislation (Table 9.9), the EPA sets applicable environmental standards, and the USNRC develops regulations to implement the standards.

Procedures for developing the necessary disposal facilities, and the associated regulatory requirements, are closely tied to the type of waste under consideration. For low-level radioactive wastes, Congress has mandated that the states, acting either alone or as part of a regional compact, must arrange to design, construct, and operate the necessary disposal facilities. For high-level wastes, Congress has assigned this responsibility to DOE.

Low-level radioactive wastes include those resulting from the operation of nuclear power plants and related industrial facilities, from the decommissioning and decontamination of nuclear facilities, and from the use of radioactive material in medicine, research, and industry—that is, the commercial sector. Low-level wastes also include the tailings generated through the mining and milling of uranium (for military operations and commercial nuclear power plants), and wastes containing transuranic elements (those heavier than uranium) that result from activities related to national defense. On a relative basis, the volumes generated by uranium mining and milling are far larger than those generated by the commercial sector. These wastes are handled "in place," that is, they are stabilized and provided with a cover to protect them from wind and water erosion. A special geologic facility is being constructed in New Mexico for the disposal of transuranic wastes.

Table 9.9 Principal federal laws related to management and disposal of radioactive waste

Year	Law	Public Law number
1954	Atomic Energy Act	85–703
1978	Uranium Mill Tailings Radiation Control Act	95–604
1980	Low-Level Radioactive Waste Policy Act	96–573
1983	Nuclear Waste Policy Act of 1982	97–425
1986	Low-Level Radioactive Waste Policy Amendments Act of 1985	99–240
1987	Nuclear Waste Policy Amendments Act	100–203

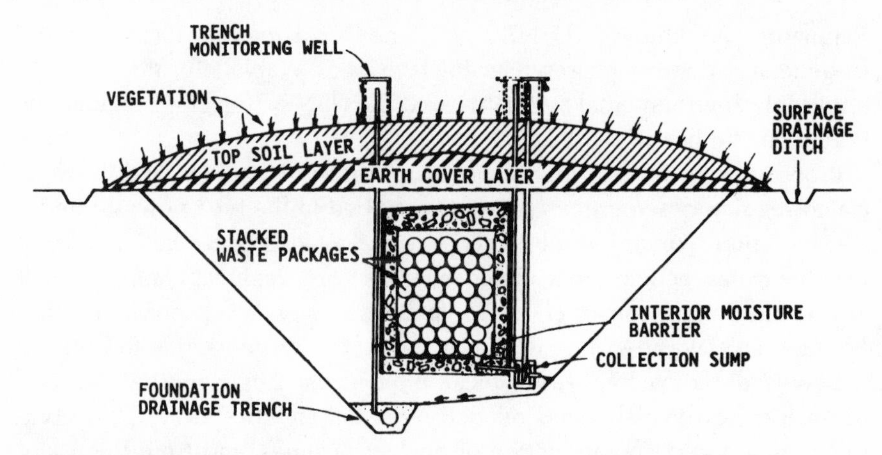

Figure 9.3 Earth-mounded concrete bunker for disposal of low-level radioactive wastes

Over 80 percent of the total volume of radioactive waste generated by the commercial sector is low level. Although it contains much less than 1 percent of the total quantity of radionuclides being disposed of the annual volume is about 1 million cubic feet. Commercial nuclear power plants account for more than half of this volume and for over 80 percent of the total activity in such wastes. In the past, the approach commonly used to dispose of these wastes was shallow land burial, very similar in principle to the disposal of hazardous (nonradioactive) chemical wastes. With the formation of state compacts to develop low-level radioactive waste disposal facilities and the increasing involvement of citizen groups in selecting associated disposal sites, there has been a growing demand that such wastes be disposed of in more robust facilities. These range from above-ground and below-ground vaults to earth-mounded concrete bunkers (Figure 9.3).

High-level radioactive wastes include spent (used) fuel removed from commercial nuclear power plants, and fission products separated from military fuel that has been chemically reprocessed. Removal of the fission products permits the highly enriched, unused uranium remaining in the fuel and the newly formed plutonium to be reclaimed for reuse.

Spent fuel, as removed from a nuclear power plant, is in solid form. In the United States such fuel is not currently being reprocessed; instead, it is being stored on-site until a proposed deep geologic repository for its disposal can be completed. Prior to disposal, the fuel will be enclosed in suitable containers to assure long-term retention of the associated radioactive materials. In countries such as France, Germany, Japan, and the United Kingdom, where spent fuel from commercial nuclear power plants is being chemically reprocessed, plans are to solidify the fission products, initially separated out as liquid waste, prior to disposal.

In evaluating the magnitude of the problem of high-level wastes, it is important to keep in mind the relatively small volumes involved. A single 1,000-megawatt nuclear electricity generating station, for example, will produce only about 30 tons of high-level waste (so-called spent fuel) each year.

In every nation in the world that produces high-level radioactive waste, geologic disposal is the method of choice. Yet no country has adopted the relatively rapid schedule being pursued in the United States for construction and operation of such a facility. Rather, the focus has been on methods for interim storage of such wastes until a repository can be completed. In fact, plans in the United Kingdom call for interim storage of such wastes for periods of 50 years or more (GAO, 1994). Recent indications are that the United States too is considering establishment of an interim storage facility, to relieve the pressure on completing the siting and construction of a repository.

Public Concerns

At the present time, America's "not-in-my-backyard" syndrome extends to the siting not only of waste disposal facilities but also of prisons, airports, mental health centers, and power plants. Unfortunately, the individuals responsible for the design, construction, and operation of such facilities rarely know how to deal with such concerns. It is for this reason that the Superfund Act specifically requires the EPA to conduct community relations efforts, such as public meetings and comment periods, during the cleanup of hazardous waste sites.

According to one group of experts, workable, legitimate sitings of waste treatment and disposal facilities will require: (1) a comprehensive, integrated management strategy giving primary emphasis to waste reduction, detoxification, and recycling; (2) extensive interactive participation by citizens, regulators, and waste managers during planning, siting, and operations; and (3) new institutional arrangements and guarantees that instill public confidence in the need for and safety of proposed facilities. Those attempting to develop such a program should understand the serious concerns of people who oppose waste disposal facilities: the apparent lack of enforcement by regulatory officials; the refusal of public health agencies to negotiate or to take local concerns seriously; the failure of the government to clean up known toxic waste sites that remain hazardous; the "closed" information policies of state organizations and industry; the continual underestimation of the ability of the public to comprehend technical issues; and a lack of good-faith efforts to involve citizens in the process (Peelle and Ellis, 1987).

A variety of tactics can be used to minimize or overcome these perceptions (Connor, 1988). One is to realize that the main sources of public concern are perceived risks and inequities, threats to community integrity, and improper or arbitrary decisionmaking. Information and consultation must begin early in the project and continue through the environmental assessment process. Perceptions of inequity often arise when the benefits of a facility will be highly diffused and the drawbacks concentrated; occasionally some form of compensation for the affected groups may be appropriate. If the proposed facility threatens large-scale changes in the local landscape or an influx of outsiders, involving local residents in the initial development of the project offers a constructive outlet for responding to their anxieties. Perceptions of improper or arbitrary decisionmaking arise primarily when engineers make decisions on issues which they believe to be purely technical and professional, but which the people affected view as issues of political power. When possible, residents should be given a voice, particularly in the development of criteria for site selection. Close, frequent, and suitable interaction with local groups can lead to a successful project; improper interactions will almost assuredly lead to failure (Table 9.10).

Epidemiologic studies have been conducted to determine whether the health of people living near hazardous waste disposal sites is being adversely affected, particularly through an increase in cancer rates. Most such studies have been inconclusive. Even at Love Canal, studies revealed no evidence of higher cancer incidence there than in the rest of New York State (Golaine, 1991). Although certain other studies have shown an apparent association between living near a hazardous waste disposal site and increased risks of certain types of cancer, as well as birth defects, investigators are careful to point out that, owing to limitations in the data, it is too early to reach any definitive conclusions (ATSDR, 1994). As outlined in Chapter 2, such a result is not unexpected, for several reasons. First of all, the earliest recognizable effects of low-level chemical exposures (headache, malaise, minor skin irritation, and respiratory tract complaints) tend to be common to many conditions. In addition, many of the illnesses (such as cancer) that might be expected to result have latency periods of 10–40 years. It is diffi-

Table 9.10 Relation of public participation to success in establishment of waste disposal facilities

Public participation	Project outcomes: Successful	Project outcomes: Unsuccessful
Agency "sells" premade decision		•
Agency redefines public demand or concern		•
No prior public education efforts		•
Public hearings		•
Agency seeks public-attitude data	•	
Agency seeks public-needs data	•	
Workshops or small-group meetings	•	
Prior public-education efforts	•	
Agency exchanges written information with public groups	•	

cult to establish patterns of exposure and equally difficult to gather data on a sufficiently large population group to verify a definitive relationship.

The General Outlook

Most current procedures for disposing of hazardous and radioactive waste involve land burial, ranging from near-surface facilities for hazardous and low-level radioactive waste to deep-underground facilities for high-level radioactive waste. Given that much chemical waste will remain toxic for hundreds of thousands of years or more, and that certain high-level radioactive waste will require thousands of years to decay, it is a matter of urgent necessity to develop methods for reducing or eliminating the generation of this waste or for converting it into less hazardous material. Care must also be taken to assure that society does not lose sight of the temporary nature of the techniques currently being applied. In reality, today's procedures provide at best interim storage, until methods can be found for the ultimate disposal of these hazardous materials.

Recognition of these deficiencies is undoubtedly an underlying cause of the increasing stringency of the regulatory requirements being imposed on disposal facilities. Already cited is the fact that the regulations for landfills designed for the disposal of nonhazardous municipal waste now closely parallel those for hazardous industrial waste. A similar pattern has developed relative to requirements for the disposal of low-level radioactive waste (Table 9.11).

As in the case of air, water, and food contamination, a systems or "life-cycle" approach is essential in evaluation and analysis of the problem of solid waste (Keoleian and Menerey, 1994). That is to say, in order to understand a product's environmental impact, each stage of its life cycle—from design, manufacture, and use to ultimate disposal—must be examined. Only through such an evaluation can industrial organizations avoid product-related and process-related environmental impacts as well as redesign and reformulation of the process or product at a later stage (Abelson, 1994). The necessity of a systems approach can also be demonstrated on a more fundamental basis. For example, airborne emissions from major coal-burning electricity generating plants are being reduced by the installation of scrubbers. However, not only do scrubbers reduce electrical output, but the sulfur dioxide removal process yields large volumes of solid waste. An alternative is to push forward with the development of new, clean, coal-burning processes (such as the integrated gasification combined-cycle technology) that promise to eliminate many of these problems (Abelson, 1990).

Although industrial operations are a principal source of solid waste in the United States, consumer activities continue to make a sizable contribution—in the form of discarded clothing, yard debris, plastic packaging, and an estimated 16 billion disposable diapers, 1.6 billion pens, and 2 billion razors each year (Langone, 1989). The disposal of old automobiles accounts

Table 9.11 Trends in low-level radioactive waste disposal facility requirements

Factor	Early approach	Intermediate approach	Latest approach
Technology	Simple landfill	Advanced landfill with liners	Multiple engineered barriers
Waste containers	Wooden boxes and 55-gallon drums	Metal and high-integrity containers	Metal and high-integrity containers placed within a vault
Container handling	Open facility: random dumping	Open facility: individual placement	Covered facility: individual placement
Record-keeping	Simple records	Detailed records with computer storage	Detailed records using on-line computer

for tons of waste metal (which fortunately can be recycled), lead batteries, brake linings that may be toxic, and air-conditioning refrigerant, whose release to the atmosphere may lead to destruction of the ozone layer (Herman, Siamak, and Ausubel, 1989).

The declining number of municipal landfills highlights the importance of reducing the volume of all types of solid waste. Use of biodegradable plastics may help, but for the long term other approaches are needed. These include changing our industrial processes to eliminate such materials, developing more methods for recycling, and discovering ways to use waste material beneficially instead of burying it. An excellent example of this last approach is the current use of discarded plastics as substitutes for materials such as wood, metal, or glass. Plastics can now be converted into synthetic wood that is resistant to water, corrosion, chemicals, bacteria, and insects. In the final analysis, there is only so much land where hazardous materials can be buried, and groundwater sources, once contaminated, may be lost for thousands of years. The disposal of solid waste is a problem of enormous magnitude. Solving it will require changes both in our technology and in our lifestyle.

10

RODENTS AND INSECTS

SCIENTISTS estimate that there are 10 million insect species in the world (Holden, 1989). Of these, nearly 1 million have been identified, including more than 100,000 species of butterflies and moths, over 100,000 species of ants, bees, and wasps, and almost 300,000 species of beetles. About 4,000 new varieties are discovered each year. Some, such as the honeybee and silkworm, bring financial benefits; in fact, honeybees are responsible for the pollination of $10 billion worth of agricultural products (for example, oranges, apples, alfalfa, and almonds) in the United States each year (Watanabe, 1994). Other insects (such as the butterfly and lightning bug) are aesthetically pleasing. Still others, however (flies, mosquitoes, boll weevils, corn borers, termites, and locusts) are destructive and even dangerous to humans. The mosquito, in particular, is the vector (transmitter) for a wide range of disease agents. The common housefly and the cockroach are also thought to be implicated in a number of human diseases.

Rodents too are known transmitters of disease agents and represent a major challenge to environmental health. It is estimated that in the United States there are 125 million rats, or one for every two people. Table 10.1 lists rodents, as well as various insect and noninsect vectors, and the diseases they can transmit. These vectors and vector hosts have major public health, social, and economic impacts throughout the world.

Rodents

Rodents have been a public health problem for centuries. Most famous is the role of rats in the successive epidemics of bubonic plague, collectively known as the Black Death, that swept Europe in the fourteenth century. One

Table 10.1 Public health impact of various disease vectors and hosts

Vector	Impact
Flies	Diarrhea, dysentery, conjunctivitis, typhoid, cholera, fly larvae infestations, annoyance
Mosquitoes	Encephalitis, malaria, yellow fever, dengue, filariasis, annoyance, bites
Lice	Epidemic typhus, louse-borne relapsing fever, trench fever, bites, annoyance
Fleas	Plague, endemic typhus, bites, annoyance
Mites	Scabies, rickettsial pox, scrub typhus, bites, allergic reactions, annoyance
Ticks	Lyme disease, tick paralysis, tick-borne relapsing fever, Rocky Mountain spotted fever, tularemia, bites, annoyance
Bedbugs, kissing bugs	Bites, annoyance, Chagas' disease
Ants	Bites, annoyance
Rodents	Rat-bite fever, leptospirosis, salmonellosis, rat bites

of the earliest recorded epidemics was launched in 1347 in Genoa, when ships arriving from Black Sea ports brought with them rats carrying infected fleas (Duplaix, 1988). The subsequent spread of bubonic plague depopulated some 200,000 towns and in three years killed 25 million people, or a quarter of the population of Europe. The Black Death remains the greatest calamity in human history. In 1665 another epidemic of plague killed 100,000 people in London; in the 1890s it struck in San Francisco; today it continues to exist in Africa, North and South America, and Asia. A major outbreak in India in 1994 resulted in at least 1,000 cases with almost 100 deaths.

Three species of rodents cause major environmental concern in the United States today: the Norway rat, the roof rat, and the house mouse. Species of importance in other parts of the world include the Polynesian rat *(Rattus exulans)*, which has spread from its native Southeast Asia to New Zealand and Hawaii, and the lesser bandicoot *(Bandicota bengalensis)*, which is predominant in southern Asia, especially India. The main impact of these rodents is their widespread destruction of food, particularly grains (Canby, 1977).

Knowing the characteristics of the rodents that pose an environmental problem is essential to their control. Figure 10.1 highlights the distinguishing characteristics of the most common domestic rodents in the United States.

> The Norway rat *(Rattus norvegicus)* is characterized by its relatively large size and short tail. Norway rats frequent the lower parts of buildings and inhabit woodpiles, rubbish, and debris. They also burrow under floors, concrete slabs, and footings and live around residences, warehouses, chicken yards, and in sewers. They nest in the ground and have a range of 100–150 feet.
>
> The roof rat *(Rattus rattus)* is characterized by its smaller size and longer tail. Roof rats live in grain mills, dense growth in willows, and old residential neighborhoods. They are excellent climbers, frequently occupying shrubbery, trees, and upper parts of buildings. They usually nest in buildings and have a range similar to that of the Norway rat.
>
> The house mouse *(Mus musculus)* is characterized by its small size, including small feet and eyes, and long tail. Mice live in buildings and in fields and their range is limited (10–30 feet).

RATS

Among the fastest-reproducing mammals, rats have a gestation period of 21–25 days and they can reproduce every 60–90 days. A typical litter ranges in size from five to nine offspring. Their life span is 9–12 months.

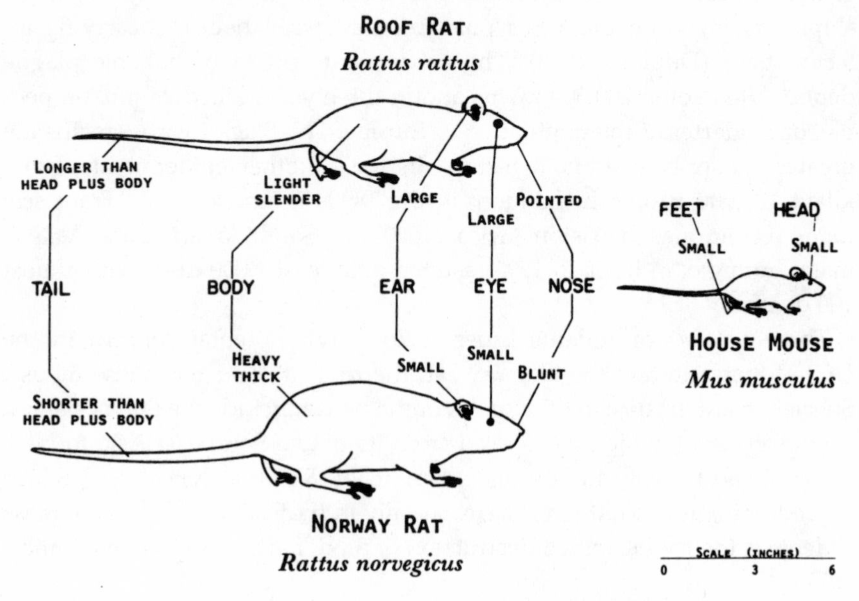

Figure 10.1 Distinguishing characteristics of three species of rodents

Rats are very intelligent and survive in a hostile human environment by means of complex social mechanisms. Although their vision is poor, they have a keen sense of smell; they like the same food as people and prefer it fresh. Yet they can also eat decayed food and consume contaminated water with no apparent ill effects. Rats have a well-developed sense of touch via their nose, whiskers, and hair. They have an excellent sense of balance and can fall three stories without injury. All rats are accomplished swimmers.

Rats have a strict social structure. Although those in neighboring colonies will tolerate one another, those in nonadjacent colonies are openly hostile. Rats are rarely seen in the daytime. Cowardly, they will seldom pick a fight with people. They follow established paths, which are readily identified by the presence of droppings (feces), grease marks (where the rats have rubbed), urine stains (located by using ultraviolet light), and their characteristic odor (Fischer, 1969).

Rats are extremely adaptable and can survive under adverse conditions (Canby, 1977; Lore and Flannelly, 1977). Scientists returning to Pacific islands that were virtually destroyed in nuclear weapons tests in the early 1950s found flourishing colonies of rats. The same species that lives in burrows in the United States and in attics in Europe can live in palm trees in the South Pacific. Other species, finding shortages of food on land, have learned to dive into lakes and ponds to catch fish.

Rats affect the quality of human life in many ways. Although their ability to transmit disease is important, they also have significant economic impacts and social implications.

In addition to bubonic plague, rats can transmit typhus fever through infected fleas, salmonellosis through food contaminated by their urine, and rat-bite fever through a spirochete in their blood. In poor housing conditions infants, paraplegics, and people under the influence of alcohol or drugs are especially vulnerable to rat bites. On babies, the targets for mutilation are often the nose, ears, lips, fingers, and toes.

One rat can eat 10 bushels of grain or 40 pounds of food per year. Rats are estimated to destroy 20 percent of the world's crops annually, including 48 million tons of growing rice and some 30–40 million tons of bread grains and rice in storage. In locations such as India, they compete seriously with humans for food. In the United States, rats are estimated to cause about $1 billion in losses per year. This figure includes fires caused by rodents chewing on electrical wiring. In fact, 25 percent of fires in rural areas are thought to be caused by rodents (Canby, 1977).

To many inner-city residents, the presence of rats is a vivid and gruesome symptom of community environmental degradation, a token of the larger pattern of social and economic breakdown and disorder in the real world of the urban poor. The appalling quality of life in such conditions often becomes clear to others only during urban renewal, when rats from buildings that are being torn down or renovated stream into adjacent neighborhoods.

One of the earliest accounts of attempts at rodent control dates to 1284, when the now-legendary Pied Piper of Hamelin played on his pipe and led all the rats in the town into the river Weser, where they drowned (Guest, 1984). Rodent control is complex because it is so closely tied to human behavior and to large-scale social and economic factors. Effective control programs are possible, however; they consist of three basic elements.

Eliminating food sources. Rats cannot live and reproduce without food. Making food unavailable to them, however, requires control of garbage and refuse, which in turn requires comprehensive public education. Garbage should be stored in metal cans with tight-fitting lids and collected twice a week; otherwise, storage containers will be filled, and residents will switch to plastic bags or cardboard boxes, which rats can easily tear open.

Ratproofing. Ratproofing of buildings is an essential part of a long-range control program. Existing buildings should be modified so that rats cannot enter or leave. All openings should be sealed and a concrete floor or underground shields installed outside the building walls to prevent rats from burrowing underneath. (A young rat can squeeze through a ½-inch opening; a young mouse, through a ¼-inch hole.) Buildings that cannot be ratproofed should be demolished, and strict codes should be enacted to assure proper construction of new buildings.

Traps, fumigants, or poisons. Once all buildings have been ratproofed and food has been made unavailable, the rat population can be reduced by trapping, fumigation, or poisoning programs inside buildings and area-wide kills outdoors.

Traps avoid the use of poisons, but the rats that are caught must be collected and buried or incinerated. In the process, any fleas that survive may transfer to people.

Fumigants consist of gases, such as calcium cyanide and methyl bromide, that are released inside buildings. They provide a quick kill but require care to prevent dangerous exposures of humans during use.

Poisons are generally placed in food (baits) for the rats. Examples include:

Warfarin: a slow-acting anticoagulant rodenticide. Unfortunately, many rats have developed immunity to this chemical.

Red squill: a bitter-tasting red powder that causes heart paralysis. For most animals, red squill is a powerful emetic. Since the Norway rat cannot vomit, this compound has proved to be a selective method of killing.

Zinc phosphide: a fast-acting black powder with a garlic odor that reacts with acid in the rat's stomach to produce phosphine gas. Although most animals are repelled by this poison, rats relish it.

Norbormide: a fast-acting poison that causes shock impairment of blood circulation. It too is lethal only to Norway rats.

Effective use of rat poisons is difficult. Rats have developed efficient feeding strategies that enable the members of a colony to avoid poisoned baits and to adjust to sudden changes in the food supply. Both laboratory and wild rats tend to avoid any contact with novel objects in their environment. Typically they avoid a new food for several days, and they may never sample it if their existing diet is nutritionally adequate. Eventually, small sublethal quantities may be ingested. If feeding animals become sick, the entire colony thereafter avoids the new food (Lore and Flannelly, 1977).

Other approaches being developed or considered for rat control include single-dose chemosterilants that could sterilize both male and female rats, and new rodenticides that are rat specific and thus not hazardous to nontarget animals.

MICE

The house mouse lives in any structure to which it can gain entrance. Mice that live outside during the summer tend to move into buildings with the onset of cold weather and heavy rains. Non-domesticated varieties live outdoors essentially all of the time.

Although the effects of the house mouse are primarily of a nuisance nature, other varieties play a major role in the transmission of serious diseases. One is Lyme disease, in which the white-footed mouse serves as one of the hosts for the ticks that transmit the disease. The deer mouse aids in the transmission of the hantaviruses, which for many years have caused episodes of pulmonary disease and killed thousands of people in East Asia. In 1993 an outbreak occurred in the southwestern United States.

It is virtually impossible to control mice populations living outdoors. As will be noted below, a successful approach in controlling Lyme disease has been to kill the ticks on the deer mice, rather than attempting to control the mice per se.

Insects

As with rodents, the control of insects requires knowledge of their characteristics, including their life cycles and breeding habits, as well as their role as vectors of disease.

Insects are highly specialized. Houseflies, for example, have hundreds of eyes mounted in such a way as to provide them with wide-range vision, coupled with unusual visual powers. Some insects can detect sex attractants more than 15 miles away. One of the unique characteristics of certain insects is their ability to protect themselves from cold weather. Those with dark colors survive by absorbing sunlight; others gain heat by basking on dark surfaces, or have heavy layers of hair or scales that retard heat loss. Still others beat their wings to keep warm while at rest. Some survive subfreezing temperatures by lowering the freezing point of their body fluids, producing compounds that function in a manner similar to the antifreeze used in automobile engine cooling systems. Other insects protect themselves by purging their body fluids of nucleators, tiny specks of dust to which other molecules (such as ice) can attach, before hibernating for the winter (Conniff, 1977).

The life cycles of the group of insects that includes mosquitoes and flies have four stages: egg, larva, pupa, and adult. Some noninsects (for example, the tick discussed below) follow a similar life cycle. Many insects lay hundreds or thousands of eggs, and some pass through their entire life cycle in a matter of days or weeks, producing numerous generations each season. A tiny fraction who survive the winter can quickly multiply to enormous numbers in the spring.

Each year insects infect multitudes of people with diverse agents of disease; some of the more prominent are summarized in Table 10.2. Mosquitoes alone cause millions of new cases of malaria worldwide each year. Many of these cases are in Africa, where the disease causes an estimated 2 million deaths annually; in Central and South America, the number of deaths averages about 25,000 each year. Half of these are children, often under the age of 5 years (Cromie, 1994). In addition, an estimated 90 million people throughout the world have lymphatic filariasis, an infection caused by a parasitic worm transmitted by mosquitoes. In fact, filariasis is one of the most rapidly spreading diseases.

Various kinds of flies can also be major transmitters of disease agents. The World Health Organization estimates that worldwide there are 17 million cases of onchocerciasis, or river blindness, a disease initiated by infec-

Table 10.2 Global impacts of tropical disease infections

Disease	Insect vector	Number of countries affected	Number of people infected (millions)	Total population at risk (millions)
Malaria	Mosquitoes	103	270	2,100
Lymphatic filariasis	Mosquitoes	76	90	900
River blindness	Black flies	34	17	90
Chagas' disease	Triatomine (kissing bugs)	21	16–18	90
Leishmaniasis	Sandflies	80	12	350

tive larvae passed through the bite of the black fly. Most of these cases are in Africa; at least 300,000 people have lost their vision as a result (APHA, 1990).

Although insect-related diseases are not as prevalent in the United States as in most other countries, they are still a serious public health problem. Each year several hundred people die from insect stings or from vector-borne diseases such as encephalitis, Rocky Mountain spotted fever, and Lyme disease; and tens of thousands suffer illness or injury from insect bites. In addition to the health effects, insects have an enormous economic impact on agricultural production. They attack all stages of plant life, eating seeds, seedlings, roots, stems, leaves, flowers, and fruit; after the harvest, they eat the stored product. In the United States, insects destroy an estimated 13 percent of agricultural crops (Holden, 1989); in Kenya, the figure is 75 percent. Insects also eat wooden buildings and woolen clothing.

The impact of insects, however, ranges far beyond the effects listed here. Thousands of species of insects obtain their livelihood from the bodies of animals. There is not a wild or domesticated animal that is not attacked by various kinds of external and internal parasites. The presence of flies and other insects can reduce the yield of milk from dairy cows and eggs from chickens, and can cause cattle to lose weight and work horses to be less efficient. In many parts of the world, the persistent biting of mosquitoes, black flies, and other bloodsucking insects seriously impairs the productive capacity of workers, sometimes even bringing their activities to a standstill.

MOSQUITOES

Essentially every person in the world has heard the buzzing of mosquitoes and suffered their bites. Their characteristics, especially in interactions with humans, may provide the key to their control.

Though seemingly frail, mosquitoes show remarkable abilities in flight: those that fly during the day navigate by polarized light from the sun; those that fly at night navigate by the stars. Their wings move even faster than those of a hummingbird—an estimated 250–600 strokes per second (Conniff, 1977)—and produce the familiar whine that is their mating call. In general, mosquitoes do not fly when the temperature is below 57° F (14° C); they seldom fly at altitudes higher than several thousand feet; and their maximum flying speed is 7 miles per hour. As a result, they cannot maneuver if the wind is stronger than 8 miles per hour.

Mosquitoes generally have a life cycle of one to two weeks. Hardy females lay about 200 eggs every four days. Fortunately, less than 5 percent of the eggs become mature adults, and each fall the initial frost kills most of the adult mosquitoes. Still, some eggs may survive for years to decades, hatching when the right combination of warm weather and moisture occurs.

Only female mosquitoes bite people, and they do so only to obtain the blood they need to reproduce. Mosquitoes can fly with a blood meal two to three times their body weight. Some bite only during daylight; others bite only at dusk or at night (Conniff, 1977). When biting a victim, a mosquito immediately injects an enzyme, apyrase, that acts on the platelets to assure the continuing flow of blood. The victim's immune system responds by releasing histamines and other substances into the tissue around the bite, causing the blood vessels to dilate and produce swelling and itching.

Mosquitoes appear to be especially attracted to certain odors and gases. One of these is carbon dioxide, which humans exhale with every breath. Some mosquitoes also are attracted by lactic acid, a common by-product of muscle contractions and accompanying perspiration. Mosquitoes appear to prefer dark colors.

Mosquitoes are transmitters of diverse agents of disease. *Anopheles quadrimaculatus,* which breeds in swamps, was the principal vector of malaria in the southeastern United States; *Aedes aegypti,* which breeds predominantly in artificial containers (cans, bottles, old tires), is the urban vector of yellow fever and dengue fever, a debilitating viral disease common in parts of Asia, West Africa, and the Americas.

Control of mosquitoes and mosquito-transmitted infections involves two basic steps: (1) reducing the mosquito population by eliminating their breeding habitats—draining land areas in the case of *Anopheles quadrimaculatus*, or applying insecticides or other agents to kill the adult mosquitoes or their larvae; and (2) preventing mosquitoes from biting people and providing medical treatment to individuals who have been, or are subject to being, infected. This second step is only ameliorative; it is not possible to attain a sufficient level of treatment to influence the transmission of a mosquito-borne disease.

Mosquito control is complex because of the large number of species involved and their widely different breeding places, biting habits, flight ranges, and relations to disease. Shoreline towns may be troubled by salt-marsh mosquitoes, inland towns by freshwater ones. Control is also complicated by the mosquito's rapid development of new behavior patterns, such as the shift from indoor to outdoor blood feeding and resting, in response to insecticide control programs. Agencies planning any type of control program would be wise to hire an entomologist to identify exactly which mosquitoes are the problem and to give advice on the most effective control measures.

Whereas earlier programs to kill mosquitoes provided temporary relief at best, the discovery and exploitation of *Bacillus thuringiensis israeliensis* (BTI), a natural enemy of mosquito larvae, have changed this situation dramatically. The bacterium was discovered in the gut of dead mosquito larvae in an oasis in the Negev Desert. It appears to kill only mosquito and black fly larvae; it has shown no toxicity to humans or other nontarget organisms. As a result, BTI is being widely used throughout the world as a larvicide for the control of mosquitoes and is proving spectacularly successful. Its first widespread use in the United States was in the Cape Cod area of Massachusetts in 1991. More recently a second larvicide that incorporates the *Bacillus sphaericus* bacterium, commonly present in the soil, has been developed and approved for use in the United States to control mosquitoes such as the Culex, which breed in municipal wastewater lagoons and stormwater basins.

Another variety of BTI, *Bacillus thuringiensis kurstaki* (BTK), is being used to kill the gypsy moth and spruce budworm larvae. Other varieties of *B. thuringiensis*, in combination with natural predators and fungi, are controlling certain agricultural pests. One example is the potato beetle, where tests in Maine have shown that a combination of the bacillus, a stinkbug, and a fungus were as effective as chemical insecticides in controlling this pest. Unfortunately, some insects are now showing resistance to some of the varieties of *B. thuringiensis* (Spielman, 1995).

Chemical insecticides continue to be widely used to control adult mosquitoes. In the past, the most commonly used insecticide was dichlorodiphenyltrichloroethane (DDT). Because of environmental concerns, its use was banned in the United States in 1972; subsequently, certain other organochlorine compounds (such as aldrin and dieldrin) were also prohibited. The use of DDT has been banned in Europe and many other parts of the world as well. Malathion is now the insecticide of choice. Since *Anopheles* mosquitoes typically bite people indoors at night and then land on the wall to rest, a successful strategy is to apply the insecticide to the inside walls of a house so that the mosquitoes will come in contact with it and be killed. Before the development of BTI, DDT was also applied to water areas to kill mosquito larvae. Another larvicide in use today is Abate.

Other approaches for controlling mosquitoes include elimination of breeding zones by digging drainage canals, preventing construction and other practices that lead to the creation of stagnant water, changing the salinity of existing waters, and raising and lowering the water level in lakes, such as those created by dams, to disrupt the life cycle of the mosquito.

Besides staying indoors except on breezy days or hot afternoons, individuals can prevent or at least reduce mosquito bites by placing screens on doors, windows, and porches; by using protective clothing and mosquito netting; and by applying mosquito repellents to the skin. Trials in Gambia have shown that the use of permethrin-impregnated bed-nets is especially effective in reducing childhood mortality from malaria. A variety of repellents is available commercially (Consumer Reports, 1987); all contain diethyltoluamide (DEET). In deciding which to use, people should try several to determine which is most effective against mosquitoes and most compatible with their skin. Because DEET is harmful to children, only products containing less than 35 percent DEET should be used.

Treatment of people who have been bitten by mosquitoes may range from applying rubbing alcohol on the resulting welt to administering drugs, such as chloroquine in the case of malaria. Unfortunately the malaria parasite in Africa has developed resistance to this drug, so it is no longer an effective treatment there. Chloroquine continues to be effective in parts of South America, however, and new drugs such as artemisinin (a substance extracted from a fernlike plant) are being either developed or refined. In addition, progress is being made in the development of a vaccine for malaria (Chege, 1994). Field tests in humans in Columbia and Venezuela have shown variable results, some of which are promising; similar tests are under way in Brazil, Gambia, and Tanzania (Maurice, 1995).

FLIES

Among the members of the fly family, three have been selected for discussion here. One is the housefly *(Musca domestica),* which is present in many of the temperate parts of the world and may be a carrier of the agents for several diseases. The other two chosen are the screwworm fly *(Cochliomyia hominivorax),* because of its potentially devastating impact on livestock; and the Mediterranean fruitfly *(Ceretitis capitata),* because of the likelihood of its destructive effects on citrus and other fruit.

The housefly. Gray and about a quarter-inch long, the housefly breeds in a variety of decaying animal and vegetable matter, and its larval stage is the maggot. In rural areas, horse, pig, cow, or chicken manure frequently serves as a breeding habitat; human excreta can also be involved where proper disposal methods are not observed.

The housefly's larval stage lasts 4–8 days; the pupal stage, 3–6 days. In warm weather the average time from the laying of eggs to the emergence of the adult is 10–16 days. Flies live 2–8 weeks in midsummer; in cooler weather, up to 10 weeks. Although flies have been reported to travel several miles in one day, most flies present in a given area probably originated nearby.

Although their role in transmitting disease is difficult to document, houseflies pick up and carry a wide range of pathogens (including viruses, bacteria, protozoa, and eggs and cysts of worms), both externally (on their mouth parts, body and leg hairs, and the sticky pads of their feet) and internally (in their intestinal tract). As a rule, pathogens picked up by the larvae are not transmitted to the adult fly, and most pathogens picked up by adult flies do not multiply in them. The germs on the surface of a fly often survive only a few hours, especially if exposed to the sun. In contrast, pathogens can live in the intestinal tract and be transmitted to humans when the fly vomits or defecates (Keiding, 1976). In order to eat, the housefly regurgitates a fluid that dissolves its food. Part of this effluent may remain behind on the food when the fly departs and may contain pathogenic organisms. Specific diseases in which houseflies may play a role include typhoid, dysentery, diarrhea, cholera, yaws, and trachoma. Of these, bacillic dysentery (shigellosis) is probably the most important.

The screwworm fly. The adult has a metallic blue body and three vertical black stripes on its back between its wings. It is about twice as large as the housefly. In contrast to the housefly, it lays its eggs in fresh wounds of warm-blooded animals. Any accidental or surgical wound, a fresh brand mark, or the navel of a newborn animal can serve as the site for initial invasion by screwworm maggots. In warm areas populated by screwworm flies, few newborn calves, lambs, kids, pigs, or the young of larger game species escape attack.

The maggots hatch in 12–24 hours and begin feeding on the flesh head down, soon invading the sound tissue. The feeding larvae cause a straw-colored and often bloody discharge that attracts more flies, resulting in multiple infestations by hundreds to thousands of maggots of all sizes. Death is inevitable in the case of an intense infestation unless the animal is found and treated.

The maggots become full grown in about five days. They then drop out of the wound, burrow into the ground, and change to the pupal or resting stage. The adult flies emerge from the pupal case after about eight days during warm weather, live for two to three weeks, and range for many miles (Knipling, 1960).

Early in the twentieth century, screwworm flies were present in southern Texas and northern Mexico and annually migrated northward into Louisiana and Arkansas. In 1933, screwworm flies appeared in Georgia, presumably introduced through shipment of infested cattle from the Southwest. During that summer screwworm flies spread southward into Florida, where they were a problem in the 1930s and early 1940s. Subsequent outbreaks occurred in the United States in the late 1950s and the 1970s.

Screwworm flies can have devastating effects on livestock growers. Losses from screwworm infestations along the Atlantic seaboard in 1958 were estimated at $20 million (Richardson, Ellison, and Averhoff, 1982). Today losses from a major outbreak would be many times that amount. For this reason ranchers in states such as Texas and Florida gladly pay a tax per head of cattle to finance control programs. The screwworm fly continues to be a problem in Central and South America, and more recently it has appeared in Libya (Palca, 1990).

The Mediterranean fruitfly. Also known as the Medfly, this fly is slightly smaller than the common housefly, has yellowish-orange spots on its wings, and thrives in warm climates. Scientists believe that it originated in West Africa. By 1850 it had spread throughout the Mediterranean region; it was found in Australia in the late 1800s, and in Brazil and Hawaii in the early 1900s. In 1929 it was discovered in Florida.

A Medfly typically lays her eggs in a ripe, preferably acidic fruit by drilling tiny holes in the skin or rind while the fruit is still on the tree. Choice targets are oranges, grapefruit, peaches, nectarines, plums, apples, and quinces. In 2–20 days the eggs hatch into larvae, which eat their way through the fruit, causing it to drop to the ground. The larvae later burrow into the ground, where they pupate. Adult flies emerge after some 10–50 days.

Although quarantines of fruit and other measures have brought the Medfly under control, infestations recurred in Florida and Texas between 1930 and 1979, and again in the 1980s. The Medfly also appeared in California in 1975, 1980, 1987, and 1990. Because the export of fruit is prohibited from any areas where the Medfly has been detected, the economic impact is tremendous and could, if infestations were left unchecked, approach a billion dollars a year in crop damage in California alone.

Several approaches can be used to control flies. The specific technique depends on the habits of the species in question.

Although installing screens in buildings helps reduce contact between houseflies and people, it does nothing to reduce the fly population. That goal calls for other approaches, one of the most important of which is a careful sanitation program. Keeping garbage and excreta covered and disposing of them promptly and properly will eliminate major breeding grounds. Timely disposal of garbage, especially decaying fruit, and prompt removal and disposal of infested fruit that has fallen from trees have proved effective in controlling the Mediterranean fruitfly in Israel and Italy. These measures, however, have essentially no effect on the screwworm fly. One method of controlling that fly is to restrict the breeding of cattle so that births occur only during the winter months, when the screwworm fly population is at a minimum.

Chemical insecticides are widely used for controlling flies. They can be wiped or sprayed on indoor surfaces to kill houseflies, or tapes impregnated with insecticides can be hung from the ceiling. Care must be taken, however, to avoid contamination of foodstuffs. Outdoor control measures include the application of larvicides to breeding areas and the use of bait stations and sprays.

Initially the principal insecticides used for killing adult flies were the organochlorine compounds—DDT, dieldrin, chlordane. All except DDT were found to be extremely toxic to humans. They also proved to be persistent, remaining in the environment for long periods. In addition, because of its persistence and the fact that it was bioaccumulated in the environment, DDT was found to have adverse effects on birds (through decalcification of their eggshells), fish, and bats. This led to the ban on the use of DDT in the

United States and in Europe. Recently there has been a shift to the far less persistent organophosphorus compounds (such as malathion, used to control the screwworm fly and Mediterranean fruitfly), the carbamates (such as Sevin), and the pyrethroids (such as permethrin).

Flies, even more than mosquitoes, have become resistant to virtually all insecticides. Thus insecticides cannot and should not be used as the sole means of control.

Radiation sterilization was developed in the late 1950s and led to the eradication of the screwworm fly in the United States. The technique remains very effective in controlling this insect. It consists of artificially breeding and growing millions of adult males, sterilizing them with radiation, and then releasing them. The result is that eggs of the indigenous female flies with which they mate do not hatch. The technique benefits from the fact that insects generally mate only once; but it is applicable only where the density of the fly population is low, so that sufficient numbers of sterile males can be released to have an impact. Radiation sterilization is now being used in Mexico, Belize, and Guatemala against the screwworm fly and it is being considered for use against the Mediterranean fruitfly in Egypt and the screwworm fly in Libya (Palca, 1990).

TICKS

Ticks are a good example of noninsect pests that can be important vectors of disease. For many years the primary disease of concern relative to this vector was Rocky Mountain spotted fever, but the recent upsurge in Lyme disease has caused renewed interest in these pests.

Ticks are leathery-bodied, eight-legged arthropods with mouthparts that enable them to penetrate and hold fast in the skin and withdraw blood from animals. The female mates while attached to a host and usually feeds for 8–12 days. The tick that plays a major role in the transmission of Lyme disease in the United States differs from one part of the country to another. On the East Coast and in the Northern Great Plains, the disease is spread by the bite of the *Ixodes dammini* tick; in the western states it is transmitted by the bite of the *Ixodes pacificus* tick. Both are common parasites of deer and mice. Because young stages (nymphs) of these ticks are only as large as a poppy seed, they often go undetected on humans.

The first cluster of cases of Lyme disease was reported in Connecticut in 1975; it is now known to be present in more than 40 states (Weinstein, 1990). In fact, Lyme disease is the most common arthropod-borne disease in this country, with over 13,000 cases reported each year (Spielman, 1995). The emergence of this disease as a serious problem is an excellent illustration of the interactive nature of changes that take place within the environment. Factors that have promoted these changes include the shifting of major farming activities to the western portions of the United States, the resulting abandonment of many of the farms in the Northeast, and the subsequent regrowth of trees over the area, providing an ideal habitat for both the deer and its ticks (Spielman, Telford, and Pollack, 1993).

The control of ticks and Lyme disease can be complex. Common methods include the wearing of protective clothing, the application of chemical repellents (for example, DEET) and toxicants, the removal of attached ticks, area control and vegetation management with insecticides, and control of ticks on pets and in buildings (Reed, 1993). Prompt removal of ticks is essential because the Lyme disease spirochete, *Borrelia burgdorferi,* is not likely to infect the patient before the vector ticks begin to engorge (Matuschka and Spielman, 1993).

The effective control of any insect requires a fundamental understanding of its biology. One of the novel methods that has been developed for controlling the spread of Lyme disease is based on just such an understanding. Recognizing that one of the hosts for the ticks that transmit this disease is the white-footed mouse, scientists distribute cotton balls, impregnated with permethrin, in areas foraged by these animals. The mice collect the cotton to use as a liner for their nests, the net result being that the ticks they normally carry are killed (Spielman, 1995).

A unique approach is being employed to develop a vaccine for Lyme disease. The vaccine is designed to engender, through multiple inoculations of humans, a very high complement of antibodies to the parasite. When a tick begins to feed on a vaccinated host, it takes in these antibodies with the blood meal. The antibodies destroy the Lyme disease spirochetes while they are still in the stomach of the tick. Laboratory tests with animals have demonstrated that the vaccine is virtually 100 percent effective. Tests in humans are under way in Rhode Island and Massachusetts (Spielman, 1995; anonymous, 1994).

Trends in Pest Control

Many problems have developed as a consequence of the widespread use of insecticides. For this reason other control methods, such as the application of technologies involving biological control and genetic engineering, are being developed. Also being applied is the integrated approach, involving a combination of methodologies.

IMPLICATIONS OF INSECTICIDE USE

References to the use of pesticides can be found as far back as Homer's mention of the fumigant value of burning sulfur. In the first century A.D., the Roman naturalist Pliny the Elder, in his *Historia naturalis,* advocated the use of arsenic as an insecticide and suggested the use of soda and olive oil to protect legumes. In the sixteenth century, Chinese farmers used arsenicals and nicotine in the form of tobacco extract as insecticides. Subsequent developments included the use of tobacco to combat lace bugs; pyrethrum to kill fleas; and Paris green to control the Colorado potato beetle (Lang, 1993). Today insecticides are widely used for the control of a host of insects. Concerns about the dual character of pesticides—that they control pests but at the same time pose potential toxicity to nontarget species, including humans—did not become widespread until the late 1960s. One of the primary stimulants for this awakening was the publication in 1962 of Rachel Carson's *Silent Spring,* now a classic.

The most widely known and used modern insecticide is DDT. First synthesized in 1874, its insecticidal properties were not discovered until 1939. In 1960 the World Health Organization introduced its use worldwide. The success of DDT prompted the development and introduction of a host of similar chemical derivatives, including chlordane, heptachlor, aldrin, dieldrin, toxaphene, and endrin. Unfortunately, the use of these materials has led to widespread contamination of human food and animal feed, with subsequent absorption of those chemicals into the body. Other adverse results are bioaccumulation within the environment and contamination of groundwater and surface water supplies as a result of seepage and runoff from agricultural lands. In 1984 an accident at an insecticide manufacturing plant in Bhopal, India (see Chapter 18), led to the release of large amounts of a pesticide into the environment and killed between 2,500 and 7,000 people.

Exacerbating these problems has been the dramatic increase in resistance of insects to these chemicals. From 1970 to 1980 the number of arthropod species resistant to pesticides increased from an estimated 200 to almost 450. Today, they number over 500. Thirty-six countries report that 25 species of beetles, caterpillars, mites, and other insects that attack cotton plants are now resistant to pesticides. And 84 countries report that the malaria-carrying *Anopheles* mosquito is resistant to one or more of the principal insecticide groups (Georghiuo, 1985). Although some of these responses are evolutionary in nature, others take place at a relatively rapid pace. Unfortunately, scientists do not yet know how to monitor or control these changes (Gianessi, 1987). If a control program is to be successful, such responses must be anticipated and taken into account (Gould, 1991).

BIOLOGICAL CONTROL

For these and other reasons, many scientists have advocated different approaches to the control of insect pests. These include technologies involving biological control—suppression of pest populations by natural enemies—and the use of microbial pesticides—relatively stable formulations of microorganisms that suppress pests by producing poisons and causing disease (OTA, 1995). Again, potentially harmful ecological effects may ensue.

One such proposal, the introduction of wasps to control grasshoppers in certain rangelands in the western United States, was not approved by regulatory agencies for fear the wasps would kill nontarget grasshoppers that serve beneficial ends, such as controlling the poisonous snakeweed plant. The proposed release of specific organisms has met with similar concern relative to possible impacts on nontarget species (Goodman, 1993a).

Some of the concerns raised by this approach are illustrated by tests conducted in Australia in 1996. The plan was to release a virus that causes rabbit hemorrhagic disease, to eradicate the population of nonnative wild rabbits. The rabbits, imported from Europe some 150 years ago, now number about 300 million and are the source of extensive economic and ecologic damage. Initial studies were conducted on an island off the coast, with the plan being to move the experiments to the mainland if they proved successful. Animal activists, however, worried about the possible effects of the virus on other animals, as well as humans. Their concerns were stimulated by the fact that the virus had escaped quarantine during the earlier tests (Drollette, 1996; Matson, 1996).

Also being investigated are the identification and synthesis of compounds emitted by insects to kill their predators and to attract other insects (Yoder, Pollack, and Spielman, 1992). One outgrowth has been the use of artificially produced pheromones to attract and trap insects such as the gypsy moth. In more novel approaches, pheromones are used to protect crops against the destructive effects of fungi. One of the enemies of the corn plant is the dusky sap beetle. By using pheromones to attract these beetles to devices containing *Bacillus subtilis,* a natural enemy of the fungi that produce aflatoxin in corn, agricultural experts have been able to cause the beetles, in turn, to transport *Bacillus* to the corn plants, where they kill the fungi (Associated Press, 1993).

GENETIC ENGINEERING

The capabilities that scientists have developed to engineer the genetics of plants and animals could have a profound impact on the control of insects and insect-related diseases. One approach would be to introduce into the pest population individuals that have been genetically altered to carry genes that interfere with reproduction (OTA, 1995). Researchers are also evaluating the possibility of replacing natural mosquito populations with populations that are unable to support normal parasite development (Collins and Besansky, 1994). In addition, work is under way to produce mites and insects that are highly effective enemies of crop pests. For instance, a transgenic version of a mite is helping to control spider mites in almond orchards. There are, however, associated concerns. One is that arthropods (in contrast to transgenic crops) have the ability to make sudden changes in their diets and even in their genes, coupled with the fact that they can readily move from one area to another. Furthermore, it may be that the foreign gene from the genetically engineered organism can be transferred to other arthropod species (Goodman, 1993b).

Other researchers are directing their attention to functional analysis of the genes that assure the virulence of various parasites. One of the most vicious malaria parasites, for example, protects itself by manufacturing a protein that expels drugs before they can be fully effective. Having identified and isolated these genes, scientists are attempting to develop a compound that will interfere with the action of the protein. This, in turn, should permit engineering the genes within the parasite to either prevent or reverse the resistance of these organisms to various drugs (Cromie, 1994).

In related studies, genetic engineers have achieved impressive successes in designing agricultural plants with improved resistance to insects and viruses. Initially such efforts were directed to improving plant resistance. Today the aim has been expanded to include protecting the harvested crop. In certain crops (cereal grains, beans, peas) weevils and other insects can cause losses during storage equal to those incurred during the growing season (Schmidt, 1994).

INTEGRATED PEST MANAGEMENT

One of the strategies for the control of pests that is gaining widespread acceptance is an integrated approach. Successful implementation of this approach, commonly termed "integrated pest management," involves acquiring complete information about a given insect (including its physiology, predators, and life cycle), learning the technical measures available for its control, and the related political, industrial, and environmental factors, then applying a combination of strategies and techniques.

Planning of the control program occurs at the local level, with full input from the community. Strategies may include rotating crops to interrupt the cycle of specific pests, interspersing one crop with another to confuse insects, carefully timing control efforts (that is, applying pesticides only when insects appear and using them in carefully controlled amounts), introducing natural predators to combat specific pests, and applying an insecticide developed specifically for a given pest. Mechanical methods of pest control, such as soil aeration, tillage or no-till, tractor-mounted flaming devices, vacuuming machines, and pest barriers, are often an integral part of such a program.

The goal of the integrated approach is not to halt the use of chemical insecticides, but rather to assure that they are used judiciously to avoid the destruction of beneficial insects and the development of resistance among the pests being controlled. Although the integrated approach requires time, especially at the planning stage, once launched it tends to be self-perpetuating.

Disadvantages of the integrated approach include cost and complexity. It is labor intensive and must be structured around the growing cycle. Nonetheless, farmers increasingly are adopting one or more aspects and the method is growing in popularity. In fact, it is estimated that 20 percent of the farmers in the United States are currently utilizing this approach. The goal is 75 percent usage by the year 2000 (Lang, 1993).

The General Outlook

During the past several decades, an array of new infectious diseases has emerged. In addition, there has been a resurgence of many of the older diseases. Infectious diseases currently account for almost one-third of all deaths worldwide (Holden, 1996). One of the most worrisome aspects of the new diseases—acquired immunodeficiency syndrome (AIDS), Ebola disease, and Lyme disease—is their rate of increase; at least 29 such diseases have emerged in the last 20 years (Anonymous, 1995b). Because of related concerns, the Centers for Disease Control and Prevention have developed programs of increased monitoring for emerging infections, including the enhancement of international efforts to control the spread of such pathogens (Pennisi, 1996).

In most instances, the increase in infectious diseases can be directly tied to one or more environmental changes that have facilitated contact between the vectors of disease and their human hosts. The origins of such changes include the unprecedented rate of growth and mobility of populations and the rapid rate of urbanization. Some scientists predict even more dramatic changes if the predicted global warming materializes. An undeniable lesson is that the environment is an integrated system, and changes in one segment can have effects throughout the system.

Negative results of such interactions abound, one being dengue fever, which is now pandemic throughout the world's tropics. Large outbreaks have been reported in a dozen or more Latin American countries and the Caribbean. The *Aedes aegypti* and *Aedes albopictus* mosquitoes, which can carry the virus, are found in Texas, Florida, and Louisiana (Anonymous, 1995a). The cause can be traced to accumulations of discarded automobile tires, as countries have advanced economically. These tires, which retain water following periods of rain, provide an abundance of mosquito breeding sites located close to human residences (Spielman, Telford, and Pollack, 1993). The previously cited epidemic of Lyme disease in the northeastern United States can be directly attributed to widespread increases in deer herds, again largely the result of changes in the activities of people. These and other environmental trends demonstrate the dynamic relationships that link an ever-changing landscape, the vectors that exploit these instabilities, and the pathogens that may thereby affect human health (Telford, Pollack, and Spielman, 1991). Similarly, the continued proliferation of rodents is almost totally a result of urbanization, the deterioration of many of our inner cities, and the lack of proper garbage and refuse disposal.

11

INJURY CONTROL

WORLDWIDE, approximately 3 million people die each year from injuries and about one-third of all hospitalizations result from injuries. In many countries the problems of injury and injury control have assumed an importance equal to that of infectious diseases. In the United States, after heart disease, cancer, and stroke, unintentional accidents are the leading cause of death. In fact, unintentional injuries in this country account for about 90,000 deaths and some 19 million disabling injuries each year (NSC, 1995). These figures include some 43,000 people who are killed in vehicular accidents and another 50,000 who die as a result of injuries suffered in community and recreational activities, fires, and work-related accidents. An additional 30,000 deaths result from suicides, and some 25,000 people are killed in the United States each year through violence, often the result of acts leading to what are now termed intentional injuries. Many environmental and public health officials regard the deaths and injuries associated with these latter events as the most underrecognized public health problem today.

For children and youths aged 1 to 24 years in the United States, unintentional injuries are the leading cause of death, accounting for over one-quarter of the total fatalities. Of these, motor-vehicle accidents account for 65 percent, followed by drownings and fires and burns. For those aged 25–34 years, however, the firearm-related death rate almost equals the motor-vehicle related death rate (NSC, 1995). Firearm deaths are also an increasing problem in the 15–24-year age range, a grim illustration of the impact that violence is having on our society.

Accidents remain a major source of injury and death in all age groups. In fact, the economic cost of fatal and nonfatal unintentional injuries in the United States is estimated at over $400 billion per year. This enormous figure includes associated medical expenses, administrative costs, and employer costs, as well as vehicle damage costs, fire losses, and wage and productivity losses. Because accidents predominantly kill people in the younger age groups, the impacts are particularly destructive in terms of the unfulfilled lives of the victims and the lost quality of life due to the associated injuries. The impact of these considerations is estimated to represent an *additional* loss approaching $800 billion per year (NSC, 1995).

While motor vehicles account for many unintentional injuries and deaths, some 120 million occupational accidents, leading to more than 200,000 fatalities, occur each year worldwide (WHO, 1995). In the United States, more than 5,000 workers were killed and over 3.5 million suffered disabling injuries in 1994. About 28 percent of these injuries resulted from overexertion; 27 percent from being struck by or caught in a piece of equipment; 19 percent from work-related accidents in motor vehicles; and 17 percent from assaults and acts of violence. Almost half of the deaths occurred in the work group aged 25–44 years, and 94 percent of those killed were males (CDC, 1989). Annual fatality rates per 100,000 workers were highest for workers in agriculture, mining and quarrying, construction, and transportation (Table 11.1).

Table 11.1 Number of deaths and injuries of various industrial groups, 1994

Group	Deaths per year		Disabling injuries
	Total	Per 100,000	
Agriculture[a]	890	26	140,000
Mining and quarrying	160	27	20,000
Construction	910	15	300,000
Transportation and public utilities	740	12	260,000
Government	520	3	530,000
Trade	450	2	800,000
Manufacturing	690	4	590,000
Services[b]	640	2	860,000
All industries	5000	4	3,500,000

a. Includes forestry and fishing.
b. Includes finance, insurance, and real estate.

Overall, significant progress is being made in reducing the accidental death rates of workers. The death rate in 1994 was less than one-quarter that in 1960. However, the cost of work-related accidents remains high, exceeding an estimated 120 billion in 1994 (NSC, 1995).

The New Approach

According to Julian Waller (1994), injury control as a public health endeavor began in Germany in 1780, when Johann Peter Frank urged that injury and its prevention be addressed not only by individuals but also by nationwide public health programs. In the mid-1900s, several state and local health departments in the United States initiated modest data-collection efforts and child safety, burn prevention, and other programs. The effects of these programs on behavior, morbidity, or mortality were never fully evaluated, however.

In 1942 Hugh De Haven, an engineer at Cornell University, published a paper (De Haven, 1942) that began a conceptual revolution in injury control. He showed how people successfully survived falls of 50–150 feet, in some cases with only minor injury, through proper dispersion of kinetic energy in amounts as great as 200 times the force of gravity. Through this process he demonstrated that it is possible to disconnect the linkage between accidents and the resultant injuries. De Haven's studies in turn led to the development and introduction of seat belts and other occupant restraints as an effective method of reducing injuries in automobile accidents (Waller, 1987; 1994).

In 1961 J. J. Gibson observed that injury events have only five agents, namely, the five forms of physical energy: kinetic or mechanical energy, chemical energy, thermal energy, electricity, and radiation (Gibson, 1961). Expanding on this new concept of injury causation, in the early 1960s William Haddon, then at the New York State Department of Health, launched a movement to base accident and injury prevention programs on sounder scientific and public health concepts. Instead of relying primarily on attempts to change human behavior, he applied an environmental approach to injury control. With the 1970s came federal programs to deal with threats to health in the environment, but most programs addressing accidental injuries targeted the workplace.

Current efforts combine broadened application of Haddon's concepts and strategies (discussed below) with sophisticated use of behavioral interventions. Since the 1970s there has been an increasing awareness that, for certain types of situations, injuries cannot be reduced without changes in the attitudes and behaviors of the target population. There are exceptions, of course, as in the case of fire-related deaths caused by cigarettes. A fire-safe cigarette, which is currently available, could reduce such deaths without a change in the behavioral patterns of smokers. This is an example of a passive, versus an active, approach. Wherever possible, injury-control people today promote passive prevention technologies.

Haddon developed a generic approach to the analysis, management, and control of accidents, which he treated as fundamentally a result of the rapid and uncontrolled transfer of energy (Haddon, 1970). His approach can be applied to all types of occupational and environmental hazards, ranging from automobile accidents to oil spills to major accidents in nuclear power plants. It can also be applied to controlling acts of violence. Instead of focusing on "educating" people to be more careful, this approach emphasizes taking advantage of the full range of opportunities available for intervening and reducing injuries.

To facilitate an analytic approach, Haddon divided accidents into three phases: the pre-event phase (the factors that determine whether an accident occurs), the event itself, and the postevent phase (everything that determines the consequences of the injuries received). The factors operating in all three phases are the humans involved, the equipment they are using or with which they come in contact, and the environment in which the equipment is operated. Combining the three accident phases and the three factors yields a nine-cell matrix (Figure 11.1) that public health workers can use to determine where best to apply strategies to prevent or control injuries.

Because vehicular accidents account for almost half the deaths resulting from unintentional injuries in the United States, they are used as examples in the following discussion.

Pre-event phase. The goal in this phase is to reduce the likelihood of a vehicular collision. Factors that should be considered include:

Humans involved: driver impairment by alcohol or other drugs; the thoroughness of testing procedures for licensure; the degree of enforcement of traffic rules and regulations, including mandatory use of seat belts; and the availability of mass transportation as an alternative to the use of private vehicles

Equipment: the condition of headlights, tire treads, and brakes (and whether they include antilock features); the size and visibility of brake lights; the speed the vehicle can attain; and vehicular crash tests

Phases	Factors		
	Human	Equipment	Physical and socioeconomic environment
Pre-event	(1)	(2)	(3)
Event	(4)	(5)	(6)
Postevent	(7)	(8)	(9)

Figure 11.1 Matrix for the analysis of accidents

Environment: the presence of barriers and traffic lights to protect pedestrians; the design, placement, and maintenance of road signs for ready comprehension; and the design of bridge abutments to prevent or reduce impact damage

Event phase. The goal in this phase is to reduce the severity of the "second collision," as when the victim hits the windshield or steering column. Factors that can reduce the extent of injuries include:

Humans involved: occupants' use of vehicles equipped with air bags, proper use of seat belts and child-resistant systems, and driver abstention from alcohol (which affects cell membrane permeability, so that even in low-impact collisions people who have consumed alcohol are more likely to sustain severe or even fatal neurological damage)

Equipment: whether the vehicle is equipped with a collapsible steering column, high-penetration-resistant windshield, interior padding (for example, on the dashboard), recessed door handles and control knobs, and structural beams in doors; low bumpers with square fronts to reduce the likelihood of pelvic and leg fractures in pedestrians who are hit; and, on trucks, a bar under the rear end to prevent cars from "submarining" beneath them

Environment: breakaway sign posts, open space along the sides of the road, wide multiple lanes, guard rails to steer vehicles back onto the road, and road surfaces that permit rapid stopping

Postevent phase. The goal in this phase is to reduce the disabilities due to the injuries. Factors that can reduce or limit the effects of injuries include:

Humans involved: rapid and appropriate emergency medical care, followed by adequate rehabilitation; properly trained rescue personnel; and injury severity scores to help medical personnel evaluate multiple traumas and predict outcomes

Equipment: fireproof gasoline tanks to prevent fires after an accident

Environment: public telephones along the roadway for summoning emergency help; "jaws of life" to extract victims from vehicles; helicopters for rapid transport of victims to medical-care facilities; trauma centers equipped to handle injured victims; ramps and other environmental changes to reduce the real "cost" to the victims of being disabled; and rehabilitation of the victims

Vehicular Accidents

Through application of the analytic approach developed by Haddon, as well as other strategies, deaths in the United States caused by most categories of vehicle-related accidents have shown a continuing decrease over the past several decades. Since 1980 the number of deaths of pedestrians, for example, has dropped by over 50 percent; for motorcyclists, the reduction has been comparable. Bicyclist deaths have been reduced by 20 percent since 1975, and occupant deaths in accidents involving large trucks have fallen 55 percent since peaking in 1978. At the same time, deaths of people in passenger vehicles—cars, pickup trucks, utility vehicles, and cargo vans—have not shown comparable declines. In fact, because of the reduction in fatalities among pedestrians and other groups, passenger-vehicle occupant deaths have been a steadily increasing proportion of all motor-vehicle deaths since 1975 (IIHS, 1994e).

Total deaths, however, do not reflect the true story of what is being accomplished, particularly during a period in which the number of vehicles on the road, and the distances they were being driven, were undergoing enormous increases. For example, the death rate per million registered passenger vehicles one to three years old dropped from 258 in 1978 to 157 in

1993. This decrease applied to every type and size of passenger vehicle, with the most dramatic decline occurring in small utility vehicles. The death rate per million such vehicles in 1978 was 1,063; in 1993, it was 257—a 76 percent reduction, accomplished through the development of newer models that are larger and less subject to rollover (IIHS, 1994e).

In spite of these advances, vehicular accidents continue to cause more than 40,000 deaths each year in the United States and account for about 2 million disabling injuries. The net result is that about 14 percent of all hospital emergency room visits are devoted to patients injured in vehicular accidents. The annual economic costs of these accidents are estimated to exceed $175 billion (NSC, 1995). Because so many of the victims of vehicular accidents are relatively young, the loss in years of life is very high.

The development of a program for preventing or reducing injuries suffered in vehicular accidents has political, social, behavioral, and economic aspects. It therefore requires a multifaceted approach involving new technical approaches as well as new policies and strategies.

TECHNICAL ADVANCES

A number of safety-enhancing technical advances have been incorporated into motor vehicles in recent years, a primary stimulus being the increased safety consciousness on the part of the public. In fact, consideration is being given in the United States to a safety rating system for new cars, similar to the fuel economy rating now required. Development of such a system will require the evaluation of a wide range of factors, including the size of the car and the results of crash testing (NRC, 1996). The more common examples of technological advances being utilized are discussed below. In one case—the use of speed-monitoring devices—the advance is being used to circumvent measures being taken to improve safety.

Air bags—Data show that driver deaths in frontal crashes are 29 percent lower in air bag–equipped cars, compared to cars equipped with manual lap-shoulder safety belts only. Moderate to severe injury was 25–29 percent lower among drivers in automobiles with air bags than in cars equipped with automatic safety belts. For all kinds of crashes, driver deaths were 20 percent lower in cars equipped with air bags. Hospital inpatient rates were 24 percent lower among drivers of cars with air bags (IIHS, 1991a, 1992). Although there have been instances in which air bags have caused injuries, the vast majority have been minor except for cases involving infants in the front passenger seat (IIHS, 1992b). Side-impact air bags were first introduced into cars by a Swedish manufacturer in 1995; beginning with the 1997 models, they have been introduced into a variety of cars by U.S. manufacturers (IIHS, 1996b). This feature should further reduce occupant injuries and deaths.

Antilock brakes—One development that has been promoted as a major step in helping drivers avoid accidents is the antilock braking system. Evaluations on the test track indicate that such a system would provide many benefits in emergency braking situations, especially on road surfaces that are wet and slippery (IIHS, 1994b). Yet accident data show that antilock braking systems have reduced neither the frequency nor the cost of crashes. Although there may be several explanations, one is that emergency situations on the road often involve complicated scenarios that differ significantly from test situations. In addition, it may be that individuals driving cars with antilock brakes place too much confidence in the system and take more risks. In contrast to the situation with cars, the benefits of antilock braking systems in large trucks, particularly tractor-trailer units, appear to be far more convincing. Many groups are calling for such systems to be mandatory equipment on all new heavy-duty trucks.

Daytime running lights—Data show that the use of daytime running lights (reduced intensity headlamps) is an effective, low-cost method of reducing car-to-car crashes. In urban areas, where traffic congestion is heavy and demands on driver attention are numerous, such lights are particularly beneficial. The National Highway Traffic Safety Administration in 1993 approved the use of such lights on automobiles in the United States and most manufacturers are equipping new cars with them. They are currently mandatory in Sweden, Norway, and Finland, and are required in Canada for all cars manufactured after 1 December, 1989. Data from Canada indicate that the use of running lights will reduce automobile collision damage claims by about 5 percent (IIHS, 1991b).

Truck visibility—Studies have shown that collisions of other vehicles with large trucks can be reduced by making the trucks more visible. One approach is to add reflective material to the sides and rear of the trailers of large trucks. Such markings are particularly valuable at night and in bad weather. As a result, the National Highway

Traffic Safety Administration now requires that all new truck trailers have improved reflectors or reflective tape markings. The cost of retrofitting old units is estimated to be as low as $100 per trailer, with the estimated reduction in accidents being about 15 percent.

Radar detectors—Many drivers, including operators of commercial trucks, use radar detectors, devices that alert them to the presence of police speed-monitoring units. Studies show that people who use such units are likely to travel at excessive speeds. Because the devices have only one purpose—to alert speeding drivers to slow down—they have been banned in many states. Their use has also been banned by the Federal Highway Administration on all commercial vehicles, primarily trucks, used in interstate commerce (IIHS, 1994a). Ninety-eight percent of those killed in 1991 in two-vehicle crashes involving a passenger vehicle and a tractor-trailer were occupants of the passenger vehicles (IIHS, 1994c).

POLICY ISSUES

A number of policies and strategies have been developed and applied in recent years to improve vehicular safety. Sometimes policies developed primarily with other goals in mind have been found to influence vehicular safety. Examples include the following.

Speed limits—It is well known that vehicles traveling at high speeds allow the driver less time to react to an emergency situation and require greater distances to stop. One of the principal factors in the reduction in U.S. vehicular deaths in the mid-1970s was the speed limit of 55 miles per hour imposed nationwide to conserve fuel. With the increasing availability of fuel in recent years, the U.S. Congress has withdrawn this restriction and many states have increased the speed limit on their interstate highways to 65 miles per hour. This change has resulted in roughly a 25 percent increase in deaths on interstate highways in rural areas (IIHS, 1994f).

Teenage drivers—Sixteen-year-old drivers have eight to nine times the accident rates of drivers 25–65 years old; 17-year-old drivers have rates about six times as high. Data clearly show that teenagers living in states with fewer steps in the licensing procedure have higher crash rates. One approach to reducing these rates is to delay or restrict the licensure of such drivers, or to require longer supervised driving periods, specifically: delay the age at which practice driving is permitted, provide learners' permits that are valid for longer periods of time, and raise the minimum age for licensing (IIHS, 1994g).

Seat belt use—The use of seat belts is a well-proven method for protecting passengers in motor vehicles. For some years, however, many have chosen not to take advantage of this protective device. Surveys show that people who do not use seat belts are least likely to purchase insurance voluntarily and, holding insurance status and other factors constant, are more likely not to reimburse the hospital if they are injured and require medical treatment (Hemenway, 1995). The net result is that the public pays a significant share of the medical-care charges associated with such risk-taking behavior.

For these and other reasons, a number of efforts have been made to encourage more universal use of seat belts. One program that has proved especially effective is that of the state of North Carolina, where police are allowed to ticket motorists solely for failing to buckle their seat belts. Nationwide, over 44 states now have laws of some type concerning the use of safety belts. As a result of these and other actions, seat belt use has gradually but steadily increased in this country. In 1982, the proportion of people in cars who used seat belts was 11 percent; in 1994, it was 58 percent. States achieving the highest rates of use are those, like North Carolina, that have comprehensive programs including strong laws mandating safety belt use, aggressive law enforcement, and vigorous educational programs (GAO, 1996).

Vehicle size and body style—To encourage energy conservation, the U.S. Congress has set miles-per-gallon goals for new motor vehicles. Although many approaches can be employed, one of the primary mechanisms is the design and manufacture of cars that are smaller and less massive. Subsequent experience has shown, however, that such vehicles can be a source of increased injuries and deaths. In fact, among the 11 existing vehicles in the United States with driver death rates at least twice as high as the average, 10 are small and 1 is midsize; none is large. Conversely, 8 of the 12 vehicles with the lowest driver death rates are large, and the other 4 are midsize; none is small (IIHS, 1994i). Also of significance is the poor roll stability of lighter-weight vehicles, particularly small pickup trucks and certain sports utility vehicles. Some are so unstable that the added weight of passengers increases the likelihood of a rollover (Whitfield and Jones, 1995).

Daylight saving time—Because it adds an hour of sunlight to the afternoon commuting time and increases the visibility of both vehicles and pedestrians, the adoption of daylight saving time is a proven method of reducing vehicular accidents. Although this step also eliminates an hour of sunlight in the morning, the increase in accidents at that time is not enough to outweigh the lives saved in the afternoon, when many more pedestrians and vehicles are on the road. Data show that over 900 lives—an estimated 727 pedestrians and 174 vehicle occupants—might have been saved if daylight saving time had been in effect year-round in the United States from 1987 through 1991 (Ferguson et al., 1995).

OTHER CONSIDERATIONS

The previous discussion has been largely devoted to steps that can be taken to reduce the injuries associated with accidents involving automobiles and trucks. Other aspects of vehicular-related injuries must also be addressed. These include unintentional injury–related deaths of motorcyclists and pedestrians, and accidents caused by collisions of motor vehicles with animals.

Motorcycles—Only a small percentage of vehicles on the road are motorcycles. Yet the death rate per registered motorcycle is more than 3 times the rate per car; per mile traveled, the death rate on motorcycles is estimated to be 22 times the rate for cars (IIHS, 1994h). Although the number of deaths due to motorcycle accidents decreased by 46 percent between 1982 and 1992, this resulted in part from a decline in motorcycle registrations. Nonetheless, total motorcyclist deaths in 1993 numbered 2,341 (IIHS, 1993d).

One measure that can help protect motorcyclists from injury is to require that they wear protective helmets. Helmet use is strongly associated with reductions in the probability and severity of injury, as well as in economic impacts and motorcyclist deaths (Rowland et al., 1996). Studies in Texas, Nebraska, and California showed reductions of as much as 40 percent in injuries, particularly of the head and neck, following a mandate that helmets be worn. At least 25 states now have such a requirement (IIHS, 1993b).

Pedestrians—Pedestrian deaths constitute the second-largest category of motor-vehicle deaths, after those of passenger-vehicle occupants. Such deaths are a special problem among the elderly (Figure 11.2). Even though death rates of this group have been decreasing since 1950, people 65 years of age and older still have the highest rates of any group. At age 80 and older the death rate for men is more than twice as high as it is at age 70 and younger. At all ages pedestrian death rates are higher for males than for females. Seven of every 10 pedestrian deaths in 1992 were males. For 1993 the total number of pedestrians killed in the United States was 5,638 (IIHS, 1993d). A key observation is that 37 percent of the pedestrians over age 16 who were killed in 1994 had blood alcohol levels of 0.10 or more—the legal limit for automobile drivers (IIHS, 1995). Steps that can be taken to reduce injuries and deaths to pedestrians include providing walkways well back from the road and requiring that the exterior of motor vehicles have no sharp edges or protrusions.

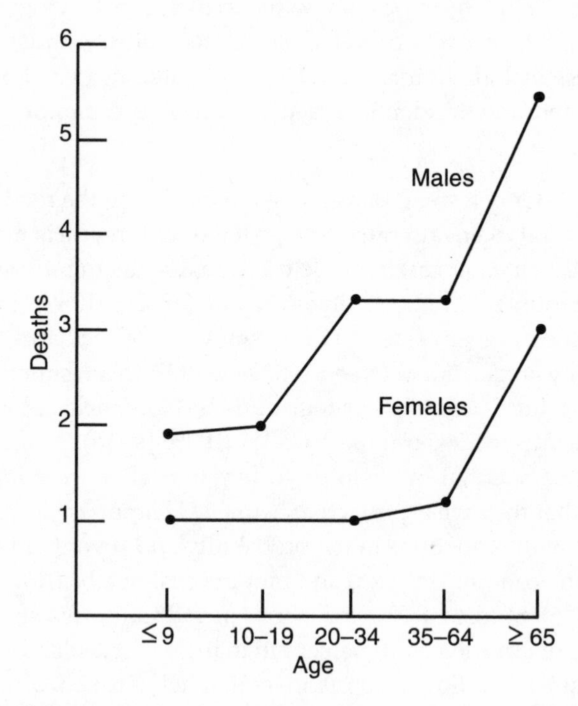

Figure 11.2 Pedestrian deaths per 100,000 people as a function of age, 1992

Collisions with animals—Of surprising importance is the number of accidents caused each year by collisions between motor vehicles and wild animals, particularly deer. During 1991, for example, there were more than 48,000 deer-vehicle collisions in Michigan, over 41,000 in Pennsylvania, and about 40,000 in Wisconsin. The estimated cost of the associated claims in Wisconsin exceeded $70 million, with the average cost per collision about $2,000. Nationwide, such collisions during 1991 resulted in property damage exceeding hundreds of millions of dollars and the deaths of 120 people. The problem appears to be worsening as residential and commercial development spreads farther into formerly rural areas, and the number of deer, vehicles, and miles traveled each year increases.

One countermeasure that has received attention is the deer whistle, a device that is mounted on a vehicle and through which onrushing air creates an ultrasonic sound that is supposed to repel deer. As yet no data confirm the effectiveness of the device. Another approach is to install specially designed roadside reflectors to try to prevent animals from crossing in front of vehicles. Presumably the reflector, illuminated by the headlights of the oncoming vehicle, frightens the deer and causes them to stop (IIHS, 1993a).

A number of people have suggested that an organized effort be undertaken to apply some of the newer developments in computers and technology to reduce the occurrence of vehicular accidents. In response to these suggestions, Congress in 1991 appropriated $659 million to be spent over a six-year period to stimulate development of what would become known as intelligent vehicle highway systems (IVHS). Often referred to as smart cars or smart highways, IVHS encompasses a range of high-technology approaches that proponents claim will revolutionize the U.S. transportation system (IIHS, 1994d).

Although the program has not achieved all that was expected, some of its products are already being applied. These include the previously discussed antilock braking systems, and the installation on highways of electronic message boards that alert drivers to upcoming road conditions. Another tentative step has been to equip cars with navigation systems that provide guidance to drivers, enabling them to select the best route to reach their destination without confusion or error. Further innovative approaches and systems will undoubtedly be developed in the future.

Sports and Recreational Injuries

According to the National Safety Council, sports and recreational activities are another source of unintentional injuries and deaths, particularly to people in the younger age groups. For the United States as a whole, more than 3.5 million people are injured in such activities each year to the extent that they require emergency treatment at a hospital. Of these, over 750,000 involve people playing basketball and almost 500,000 involve people playing baseball or softball. Another 500,000 people are injured playing football, and an additional 200,000 have to be treated for injuries sustained while roller skating or roller blading. Almost 150,000 are injured playing soccer, over 110,000 in accidents involving volley ball, and almost 150,000 in accidents involving swimming.

In terms of deaths, swimming and recreational boating accidents are major contributors, the latest data showing that, for the United States during calendar year 1993, a total of 1,800 people were drowned while taking part in various swimming activities and an additional 800 were killed in boating accidents. More than 80 percent of those in the latter group were victims of drowning, and the deaths were predominantly among people in the age range 20–29 years. Alcohol consumption is believed to be a contributing factor in at least half of all fatal boating accidents (NSC, 1994).

Bicycling is also a significant source of injuries and deaths. During 1994, 600,000 bicyclists were injured and 800 were killed in this country, with almost half of the deaths occurring among children age 16 and younger. The bicyclist death rate per million people begins rising at about age 5 and is highest among 12–14-year-olds, when bike riding is most frequent. At the same time, the number of deaths of adult bicyclists has been increasing over the last decade. Although only 32 percent of the bicyclist deaths in 1983 involved riders 21 years of age and older, by 1992 slightly over half (51 percent) of all bicyclist deaths were of adults, perhaps reflecting the growth of bicycling as a form of adult exercise. The significance of these observations is illustrated by the data presented in Figure 11.3. As will be noted, almost two-thirds of all bicyclist deaths in recent years involve people 16 years of age or older. As is true of many other activities, males are much more involved in fatal bicycle accidents than females, the current ratio being seven to one (IIHS, 1993d, 1995).

The risk of serious head injury can be reduced by as much as 90 percent if the bicyclist wears a protective helmet. The Center for Injury Prevention and Control estimates that if all bicyclists wore helmets, perhaps 500 lives

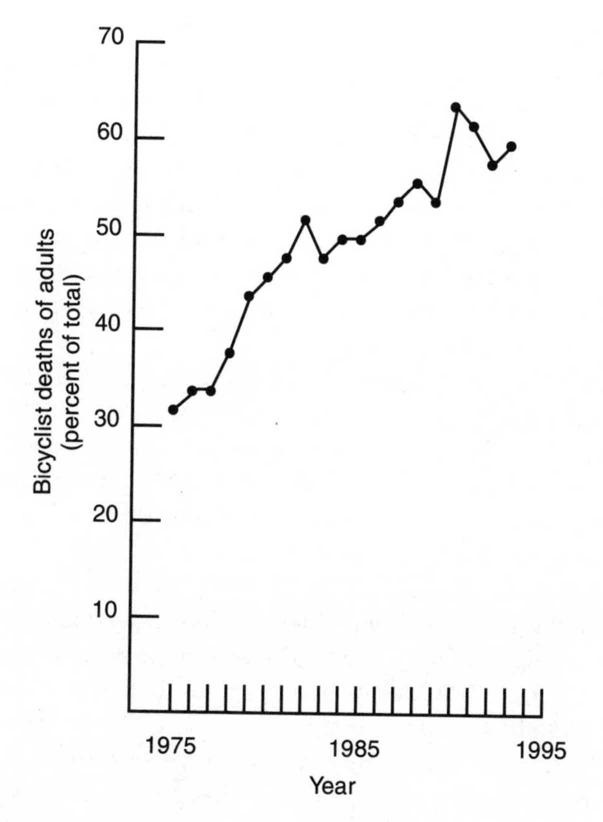

Figure 11.3 Bicycling deaths involving people age 16 and older, 1975–1993

could be saved and 135,000 head injuries prevented in the United States each year.

Intentional Injuries (Violence)

Although the public health community has been slow to acknowledge the importance of unintentional injuries as a public health problem, it has been even slower to recognize acts of violence (suicides, homicides, and assaults) as sources of intentional injuries. For the United States as a whole, events of this sort are the leading cause of injury deaths, exceeding even those of motor-vehicle accidents. On a regional basis, firearm deaths assume even greater importance; in about 15 states, they exceed any other cause of injury-related death.

While motor-vehicle deaths and death rates have been decreasing, for firearms both have been increasing. From 1985 through 1991, firearm deaths increased by 14 percent, with a major share of this increase among adolescents and young adults. In fact, from 1988 through 1991 firearm *deaths* for persons in the 15–24 age group increased 40 percent. This same group showed a 15 percent decrease in motor-vehicle deaths. The firearm-related *death rate* for persons 15–34 increased 8 percent during the same period, while the motor-vehicle death rate for the same age group decreased. In the 25–34 age group, the firearm-related death rate exceeded the motor-vehicle death rate by 4 percent (NSC, 1994).

In the country as a whole, reported firearm-related deaths from accidents, suicides, homicides, and undetermined causes totaled 38,077 in 1991. Expressed differently, on average more than 100 deaths from firearms occurred each day. Suicides accounted for 49 percent of these deaths, nearly 47 percent were homicides, and 4 percent were the result of accidents. Over 86 percent of those who died were males. For firearm-related suicides, the total number of deaths was highest for the 25–44 age group, although the death rate was highest for people 75 and older. For homicides, the total number of deaths was also highest for those aged 25–44, but the death rate was highest for those aged 15–24 (NSC, 1994). The net result is that suicide and homicide today cause about 22 percent of the deaths in children and adolescents, and about 35 percent of all injury deaths. Because firearm deaths occur predominantly among the younger age groups, such events contribute disproportionately to years of potential life lost (Christoffel, 1994).

WORKPLACE HOMICIDE

The problems of violence are not restricted to the community and the home. During 1994, more than 160,000 people in the United States were physically assaulted while at work, and more than 1,000 were killed, making homicide the second leading cause of workplace death; for women, it was the leading cause, accounting for 42 percent of female workplace deaths. In all, women were the victims in 18 percent of the more than 1,000 workplace homicides in 1993. Although homicides accounted for only 11 percent of the occupational injury deaths among men, the homicide rate for males was three times that for females. Workplaces with the highest number of deaths were grocery stores, eating and drinking places, taxicab services, and justice or public order establishments. Occupations with the highest rates were taxicab drivers/chauffeurs, law enforcement officers, gas station or garage workers, and security guards. Factors that increase the risk for homicide among workers include exchange of money with the public, working alone or in small numbers, and working late at night or in the early morning hours (NIOSH, 1993, 1996).

CONTROL OF VIOLENCE

Social scientists have struggled for more than a century to prevent crime by developing activities and other programs for young people. Government and private agencies have spent billions of dollars on law enforcement and on the construction of prisons and correctional institutions. Crime rates, however, have continued to increase. It is evident that the problems of violence and criminal behavior will not be solved until researchers untangle the complex combination of factors that push people down the path toward antisocial behavior. Unfortunately, far too often health authorities are ill equipped, both physically and intellectually, to cope with these problems (Ridley, 1994).

Certain facts are apparent. One is that the personal environment plays a prominent role in determining the extent and nature of violence in a community. Neighborhoods that inspire people to befriend one another, that are protective of local children, and that share resources appear to provide the kind of support that fosters healthy development. These characteristics may explain why some poor urban neighborhoods escape the violence that takes an enormous toll only a few blocks away. In contrast, neighborhoods that are socially and politically disorganized create conditions that contribute to antisocial behavior. When such behavior begins to dominate, there is an exodus of the small businesses that typically provide the glue that holds a neighborhood together. The way is paved for illegal economies, such as drug dealing and gambling (Ridley, 1994).

Another relevant factor is the situation within the home, particularly the relationship between husband and wife or other heads of household. Studies show that more violence is caused by family and former friends than by strangers. Twenty-nine percent of the women murdered in the United States in 1992 were killed by a husband, ex-husband, lover, or suitor. Factors such as the economic dependence of the wife on her husband, her desire to preserve the home, her concern about being separated from her children, and fear for her own safety should she try to leave add to her unwillingness to leave or even to press charges (Koshland, 1994).

Violence is also increasing because of the ready availability of firearms. Furthermore, children and young adults have been led to believe—through music, the media, and the behavior of their associates—that violence works,

Figure 11.4 Poster promoting the "Squash It" campaign aimed at behavioral changes toward confrontation among young people

that it solves problems, and that it is funny (Menninga, 1994). The message in television programs and movies is that problems can be "shot away." These forms of entertainment fail to portray the pain and suffering that accompanies violence (Anonymous, 1991). Lack of education and decreased employment opportunities are other contributing factors. Eighty-five percent of those arrested for violence are high school dropouts; 65 percent of the victims are dropouts.

One obvious flaw in today's reasoning is that 94 percent of current funds for combatting violence is directed to programs for corrections and treatment; less than 6 percent is being spent on development and evaluation of programs for violence prevention. When preventive programs actually are funded, too often they are initiated without first determining if they will work. More research needs to be conducted on causes and on preventive strategies to control violence in our society. Public health authorities need to recognize that, for the most part, violence is not random, it is not uncontrollable, and it can be reduced. Many of the strategies for controlling unintentional injuries are equally applicable here (Menninga, 1994).

One message that has emerged from the current wave of violence in the United States is that many members of the younger age groups do not know how to control their anger. A highly successful program for addressing such groups is being conducted by the Harvard School of Public Health. Entitled "Squash It," a phrase that is already part of the urban street lingo, the program is patterned after the highly successful designated driver campaign that helped draw national attention to methods for reducing the frequency of drunken driving. The new program includes a multimedia effort to assure that the intended audience is reached. It is hoped that a combination of the "Squash It" slogan and a hand gesture that is a more macho version of the T-shaped "timeout" signal used in sports (Figure 11.4) will provide an assertive gesture that young people can use to disengage from confrontations without losing face. In essence, by blocking their fist they will be saying that it is both smart and cool to walk away from escalating confrontations (Winsten, 1995).

Other Types of Injuries

Injuries arise from many sources, many of them covered in the preceding sections. Other types of unintentional injuries, involving specific age groups or having specific sources, are discussed in the sections that follow.

ACCIDENTS INVOLVING CHILDREN

In the developed countries, accidents contribute to more injuries and deaths among children than do communicable diseases. In the United States, approximately one in four children under 17 years of age experiences a medically attended injury each year (Scheidt et al., 1995), and accidents account for 40 percent of the deaths among children aged 1 to 14 years (NSC, 1995). Given these statistics, it seems strange that health agencies spend millions of dollars annually on programs to combat leukemia and muscular dystrophy but very little on the prevention and control of unintentional injuries. One problem is a general lack of data on which to base planning efforts. Fortunately this situation is being corrected. In Massachusetts, for example, studies show that every year 1 of every 5 children is injured and requires treatment in a hospital, 1 of every 50 teenagers is injured in an automobile accident, and 1 of every 12 children under age 5 requires hospital treatment for a fall. Twenty-nine percent of all injuries to teenagers in Massachusetts occur in the workplace. As with adults, the accident rate on farms is especially high. Nationwide almost half of the farm-related accidental deaths among children involve machinery, the most frequent culprit being a tractor (Schenker, Lopez, and Wintemute, 1995). Sports deaths, which gain much publicity when they occur, are minimal in comparison with other causes of death among young people (Gallagher et al., 1984).

Thwarting large-scale programs to reduce deaths and injuries is the fact that the control and prevention of many types of accidents, especially those involving very young children, depend heavily on the awareness of parents and home designers and builders to potential hazards. They need to understand that doors to cabinets for the storage of household cleansers and other toxic agents should be child resistant; that stairs should be equipped with handrails and padding; that play yards should be fenced; and that the ground beneath swings, slides, and other playground equipment should be covered with soft dirt (as opposed to clay or asphalt). Obviously of importance is close parental supervision of small children at all times.

Legislated codes and standards can also do their share to reduce childhood injuries. Examples include the requirement that all toxic materials be sold in containers with childproof caps; that hot-water heaters have temperature limits to prevent scalds and burns; that barriers be installed on upstairs windows of residential buildings; that fences be erected around swimming pools; that playground equipment be designed to prevent entrapment, falls, and contact with protruding parts; that electrical outlets near the floor be covered; that paint used on indoor walls, furniture, and equipment for children be lead free; and that control knobs on stoves be located out of reach of children.

FIRES

Data reported by the Federal Emergency Management Agency (FEMA, 1993) show that the death rate from fires is much higher here than in most other developed countries. Nonetheless, the 4,000 people in the United States who died in fires in 1993 (NSC, 1994) are less than half the number who died in 1963, prior to the availability of low-cost home smoke alarms and other improvements. The high death rate persists because many smoke detectors are incorrectly located or installed, or are nonoperational owing to dead or missing batteries. Data on causes, places of occurrence, and effects again provide useful information for developing possible preventive strategies. For example, residential fires account for about 23 percent of the fires reported to fire departments, but are responsible for 72 percent of the deaths, 66 percent of the injuries, and almost half of all property loss (FEMA, 1993). Furthermore, most fires in residences (23 percent) result from unattended cooking, not equipment failure. The second leading cause of residential fires (19 percent) is a defective local heating system, often installed to supplement or replace a central system (FEMA, 1993).

The leading cause of *death* in residential fires is cigarette smoking (26 percent); 70 percent of the men killed in fires were drunk at the time of death (Berl and Halpin, 1979). For all types of fires, the leading cause of injury (19 percent) and economic loss (18 percent) is arson (NSC, 1990).

For the United States overall, it is estimated that about 1,300 people (including 100 or so children) are killed each year by fires caused by cigarettes. Thousands of additional people are injured (with burns) and the associated damage leads to billions of dollars in losses. Yet each year approximately 500,000 butane-fueled cigarette lighters are sold in this country, two for every member of the population. An estimated 5,000 fires and 150 deaths are caused annually by children playing with such lighters. One obvious need is for lighters that small children cannot operate.

As with most environmental and public health problems, the control of deaths and injuries from fires requires a systems approach. Increased firefighting capabilities, stricter enforcement of building and housing codes, and intensified pursuit of arsonists are all helpful, but these approaches alone will not eliminate fires or the resulting fatalities. These activities must be supplemented by the installation and continued maintenance of smoke detectors and sprinkler systems in buildings; increased attention to the design, installation, operation, and maintenance of heating systems, particularly portable units; the requirement that sleeping garments, especially those worn by children, be fire resistant; the necessity (as previously mentioned) that cigarettes be fire safe, that is, they will not burn hot enough to ignite upholstery; and the requirement that bedding and upholstered furniture be not only fire resistant but also incapable of releasing toxic gases when exposed to heat and flame.

The General Outlook

The annual number of deaths due to injuries in the United States totals approximately 150,000. Of these, about 43,000 (29 percent) are due to motor-vehicle accidents, 50,000 (34 percent) are due to other types of unintentional injuries, and 55,000 (37 percent) are due to suicides and homicides (about 60 percent of which involve the use of firearms).

Nonetheless, progress is being made. From 1984 through 1993, the death rate due to unintentional injuries decreased 14 percent. The death rate for fatalities resulting from motor-vehicular accidents dropped similarly. During this same period, the death rate for workplace accidents declined 31 percent (NSC, 1994).

Not all causes of unintentional injuries are being similarly curbed. The number of deaths due to injuries in the home increased from 21,200 in 1984 to 22,500 in 1993, and the associated death rate declined by only 3 percent. Accidents in the home during 1993 accounted for 6.6 million disabling injuries. In the decade from 1984 through 1993, annual deaths from poisoning by solids and liquids increased by 77 percent (from 3,000 to 5,300). In fact, after falls, poisoning is now the second leading cause of deaths in homes. This increase is entirely due to drug-related deaths, particularly among those between ages 25 and 40. This age group also experienced a 50 percent increase in accidental home deaths in the last decade (NSC, 1990, 1994).

One of the areas of primary concern is the control of deaths from intentional injuries, such as those arising through acts of violence. Deaths due to homicide, for example, increased by 16 percent during the two-year period 1989 to 1991 (from 22,578 to 26,254) (NSC, 1994). Before a comprehensive program can be developed to control such deaths, much work remains to be done. Public health officials are only beginning to recognize the magnitude of this segment of the problem, so the necessary database for planning an effective preventive strategy does not yet exist. Further complicating the situation is the dominant role that personal behavior appears to play.

One basic necessity is that public health officials learn to consider the environment as a totality and constantly be alert to its possible role in all types of public health and social problems. The influence of the home and community environment on the behavioral patterns of people loom as most meaningful. Far too often, this aspect is overlooked. The fact that acts of violence are so frequently caused by family and former friends of the victim confirms this interconnection. Obviously, the personal (home) and community environment play a major role in how, when, and where such acts are committed. Equally obvious is the fact that promotion of a healthful environment must include encouraging strong family ties as well as providing ample opportunities, especially to young people, for gainful employment and education.

12

ELECTROMAGNETIC RADIATION

All human beings are constantly exposed to natural radiation, artificial radiation, or both. Radiation is thus an ideal form of energy with which to demonstrate the principles used by environmental health professionals to monitor, evaluate, and control physical factors in the occupational and ambient environments.

Electromagnetic radiation is propagated through space in the form of packets of energy called photons, which travel at the speed of light (3×10^{10} centimeters per second). Each photon has an associated frequency and wavelength, its energy being directly proportional to its frequency and being expressed in terms of electron volts (eV)—the energy that an electron would acquire in being accelerated across an electrical potential difference of one volt. Higher-energy photons, such as cosmic rays, have frequencies of 10^{21} hertz (Hz, or cycles per second) or more, and energies of 10^7 eV or more; lower-energy photons, such as those associated with electric and magnetic fields, have frequencies of 1–10^3 Hz and energies only a tiny fraction of an eV. Photons in the intermediate energy range (10^{-2}–10 eV), such as those associated with infrared and visible light, have frequencies of 10^{12}–10^{15} Hz. Only intermediate-range electromagnetic radiation can be detected by the human senses. High-energy photons are extremely penetrating and can have effects far from their source; the effects of lower-energy photons are concentrated near the source. Figure 12.1 shows the types and energies of the various parts of the electromagnetic spectrum. The energy ranges for the various types of radiation have not been precisely defined in every case; overlaps are common.

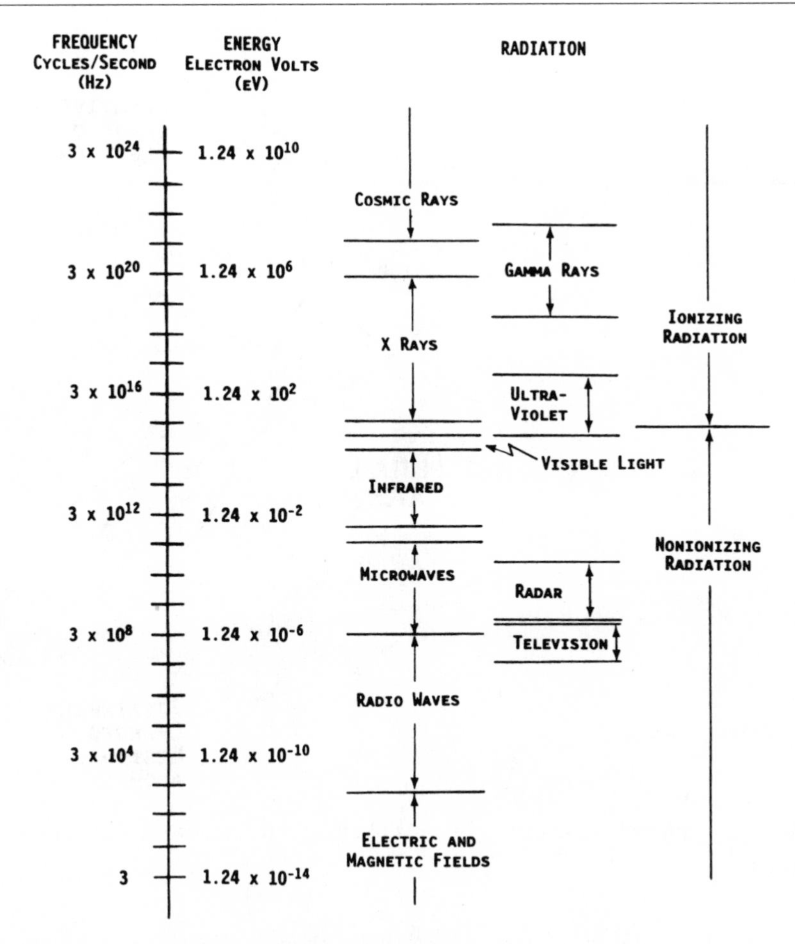

Figure 12.1 The electromagnetic spectrum

As it moves through space, electromagnetic radiation interacts with the atoms of which matter is composed. Only photons in the higher-energy ranges, such as cosmic, X, and gamma rays, have sufficient energy to ionize these atoms by interacting with the orbital electrons and stripping them away. These are referred to as ionizing radiation. Electromagnetic radiation in the lower-energy ranges, such as the lower-frequency range of ultraviolet, as well as infrared, microwaves, and radio waves, do not possess sufficient energy to be ionizing. These are referred to as nonionizing radiation. Once an electron is removed, it exhibits a unit negative charge, and the

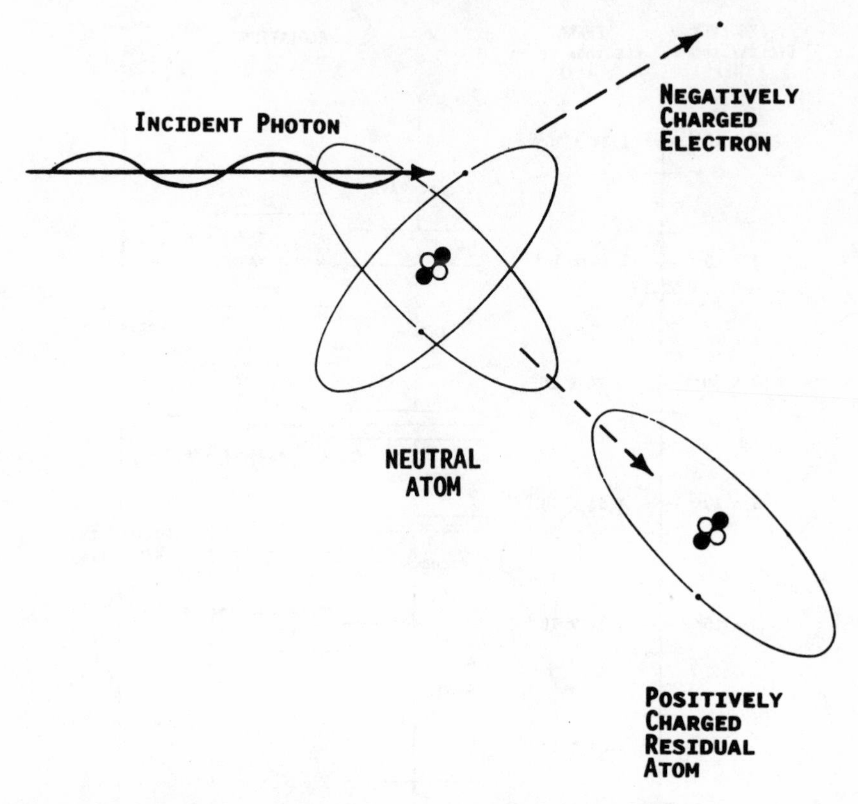

Figure 12.2 Interaction of an X or gamma photon with a neutral atom to produce an ion pair

residual atom shows a net unit positive charge. The two products are known as an ion pair (Figure 12.2). This transfer of energy to atoms can result in chemical and biological changes that are harmful to health.

Nonionizing Radiation

Although the biological effects of ionizing radiation have been recognized and well understood for some time, questions remain concerning the effects of certain types of nonionizing radiation, in particular the photons associated with lower-energy electric and magnetic fields. Techniques for assessing radiation from many such sources are still evolving.

The principal natural source of *ultraviolet radiation* is the sun. Artificial sources include electric arcs used in lights, welding torches, plasma jets, germicidal lamps, and special tanning lamps. The amount of ultraviolet (UV) radiation reaching the earth from the sun depends on factors such as the season, time of day, latitude, cloud cover, and elevation above sea level. At higher elevations there is more UV radiation because the amount of atmosphere between the earth and the sun is less. Interest in the possible effects of ultraviolet radiation has increased as a consequence of the depletion of the stratospheric ozone layer and the accompanying increase in the amount of UV reaching the earth's surface (NRPB, 1993b).

Ultraviolet radiation is unable to penetrate most materials and thus is readily shielded. Ordinary window glass will remove most of the higher frequencies, and a light layer of clothing will eliminate essentially all frequencies (Gies et al., 1994). Because the biological effects of UV radiation do not appear for some time after exposure, people can receive an excessive amount before becoming aware of the extent of the damage. If the exposure is sufficiently high, marked systemic effects, including fever, nausea, and malaise, may ensue. Aging of the skin also appears to be closely related to cumulative exposure to UV radiation. In the areas of highest exposure, premalignant or malignant changes may develop (NRPB, 1996). Ninety percent of skin cancers in humans, for example, develop on the head and neck.

The health effects of *visible light* may be direct or indirect. An example of the former is a retinal burn caused by looking at the sun during an eclipse with inadequate filtration; an example of the latter is injury from an accident caused by insufficient or excessive lighting. Poor lighting can result in a fall; too much lighting, such as the headlights of an oncoming car, can cause a crash.

Other health problems are associated with the use of devices in which beams of light can be focused both temporally and spatially. These devices, called lasers (for *l*ight *a*mplification by *s*timulated *e*mission of *r*adiation), have found a wide range of applications: industrial, consumer, scientific, and medical. These include alignment of tunnels, distance measurement, welding, cutting, drilling, heat treatment, entertainment (laser light shows), and surgery. Lasers are also used in video disc players, supermarket scanners, facsimile and printing equipment, and in certain forms of entertainment. In most such applications, laser units are totally or partially enclosed to prevent exposure to direct or scattered radiation.

Because laser devices have extremely small divergence and their beams are well collimated, they are strongly directional and can be focused on a spot whose diameter is equal to only about one wavelength of the emitted light. As a result, extraordinarily high temperatures can be generated in the small area where the radiation is absorbed. Even minute quantities of laser light can burn a small hole in the retina and impair vision permanently (NRPB, 1991).

All objects emit *infrared radiation* to other objects that have a lower surface temperature. One example is the heat that reaches the earth from the sun; another is the heat produced by a stove or by the radiant heating units used in many dwellings. Fortunately, the sensation of heat is quickly detected and thus provides adequate warning of extreme conditions. Infrared radiation does not penetrate deeply into skin tissues, but if not controlled it can cause burns on the skin, cataracts in the lens of the eye (which has poor heat-dissipating mechanisms), or retinal damage (NRPB, 1991).

Since infrared radiation is readily reflected by polished surfaces, individuals can protect themselves against localized sources by the use of special clothing and shielding. Production of cataracts by intense infrared radiation can readily be prevented through the use of protective glasses. In buildings, air-conditioning and ventilation systems can dissipate excess heat to the outdoor environment.

Microwave radiation sources include radar, radio and television transmitters, satellite telecommunication systems, and microwave ovens. In industry, microwaves are used to dry and cure plywood, paint, inks, and synthetic rubber, and to control insects in stored grain. In medicine, microwaves provide deep-heat therapy for the relief of aching joints and sore muscles, and have even been used to reheat blood rapidly after certain types of surgery (Johnson, 1982).

In microwave ovens, microwaves agitate water molecules and the resulting molecular friction produces heat. The effects of microwaves in living organisms appear to be more complex than simple heating, although most of the changes can be attributed to this phenomenon (Ferris, 1966).

Microwaves have frequencies of 10^8–10^{11} Hz. The human body is largely transparent to the lower frequencies, and microwaves in this energy range produce no biological effects. As the frequency increases, the energy of microwaves is increasingly absorbed, reaching a maximum at about 3×10^8 Hz, in the ultra-high-frequency (UHF) television range. At still higher frequencies ($>10^9$ Hz), less of the energy is absorbed, and above 10^{10} Hz the skin acts as a reflector. The frequency range 10^8–10^9 Hz is potentially the most hazardous because within it there is little or no heating of the skin and the thermal receptors are not stimulated (Ferris, 1966).

Recent years have seen growing concern about two widely used devices that are possible sources of microwave radiation—traffic radar devices and hand-held cellular telephones. Traffic radar devices have been cited as a possible source of increased testicular cancer among police officers, although epidemiological data do not confirm the association. Although traffic radar devices are known sources of microwaves, their power level is less than 0.01 percent of the power of a microwave oven (CDRH, 1992). In the case of cellular telephones, the issue is whether the moderately low levels of radiofrequency energy (in the low microwave range) emitted could contribute to brain cancers or other adverse health effects. Frustrating the efforts to evaluate possible effects are the difficulties in design, execution, and interpretation of related epidemiologic studies, particularly with respect to the identification of study populations with substantial exposure and retrospective assessment of such exposure (ICNIRP, 1996). Although at the present time there is no convincing evidence that cellular telephones are harmful, continuing studies are being pursued to try to resolve the issue (Graham and Putnam, 1995). In the meantime, the manufacturers are examining possible ways to reduce user exposure, including redesign of the phones and provision of instructions on proper use (CDRH, 1993).

Any wire that carries electric current is surrounded by *electric and magnetic fields.* Because they operate at high voltages, electric transmission lines have for some time been considered one of the primary sources of such fields. Household toasters and electric blankets, however, produce electric and magnetic fields that are readily comparable to those near the right-of-way of many such lines (Nair, Morgan, and Florig, 1989).

A number of studies have been conducted to identify and understand the mechanisms through which electric and magnetic fields may influence health. Some research suggests that these fields affect the immune and endocrine systems or growth-regulatory signals, possibly as a result of changes in the flow of calcium through cell membranes. A number of cell types have shown increased cell growth, cell division, and transcription in the presence of such fields. Other studies suggest that magnetic fields in particular may inhibit the secretion of melatonin from the pineal gland (which secretion may play a role in suppressing the growth or preventing the initiation of breast, prostate, and possibly other cancer cells). However, neither the mechanism of action, nor the dose-response relationship associated with magnetic fields, nor the implications of these cellular changes on human health, have been defined (Graham and Putnam, 1994).

As noted in Chapter 3, broader-scale human epidemiologic studies, particularly on possible associations between electric and magnetic fields and childhood leukemia or other childhood or adult cancers, have been inconsistent and inconclusive. Compounding this situation are the difficulties in determining the nature and extent of the exposure. Since the dose is influenced by a variety of factors, questions remain as to exactly which component of the field should be measured: the intensity, the frequency, or the waveform, or possibly its orientation relative to the earth's magnetic field.

Although the risks from exposures to electric and magnetic fields are thought to be small, some public health experts believe that sufficient information exists to mandate a cautious approach. For this reason, the Electric Power Research Institute has initiated a program to develop methods for reducing exposures from electric and magnetic fields associated with both the transmission and the use of electricity (Hidy, 1990).

STANDARDS FOR CONTROL

Recommended limits for exposures of workers and the general public to a wide range of nonionizing electromagnetic radiation have been published by the International Radiation Protection Association (IRPA, 1985a,b, 1988, 1991; Duchene and Lakey, 1990). Specific guidelines for limiting occupational exposures to ultraviolet, laser, infrared, and microwave radiation and for static magnetic fields have been published by the American Conference of Governmental Industrial Hygienists (ACGIH, 1995). Limits for exposures from microwave ovens have been established by the U.S. Department of Health and Human Services (HHS, 1970). Standards for protective eyewear for industrial laser users have been developed by the American National Standards Institute (Tanner, 1990). Guidelines on limits for exposure to magnetic fields have been issued by the International Commission on Non-Ionizing Radiation Protection (ICNIRP, 1994). Because the effects of nonionizing radiation vary widely depending on the specific energy of the source, many of the limits are complex. Frequently the recommendations are based on value judgments, because the only data available were derived from experiments with animals.

Ionizing Radiation

As previously mentioned, ionizing radiation is produced by photons having sufficient energy to ionize atoms. It includes X rays, discovered by Wilhelm Roentgen in 1895, and alpha, beta, and gamma rays, first observed when Antoine Henri Becquerel discovered naturally radioactive materials in 1896. The environment contains a host of naturally occurring radioactive materials, most of which are derived from the decay of uranium. Much lower levels of artificially produced radioactive materials have been discharged into the environment as a result of atmospheric weapons tests and the operation of nuclear facilities.

BIOLOGICAL EFFECTS

Biological effects in living organisms exposed to ionizing radiation develop as the result of a series of events. The first is ionization, which ejects electrons from the atoms in the molecules that constitute tissue. The residual molecule, left with a positive charge, is highly unstable and will rapidly undergo chemical change. One such change is the production of "free radicals," which are extremely reactive chemically. The ensuing reactions may, in turn lead to permanent damage of the affected molecule—or the energy may be transferred to another molecule and the free radicals may recombine. The time required for this chain of physical and chemical events to take place is on the order of a microsecond or less. The subsequent development of biochemical and physiological changes may require hours; the subsequent development of latent cancer may require years (Little, 1993).

All cells are susceptible to damage by ionizing radiation, and only a very small amount of energy needs to be deposited in a cell or tissue to produce significant biological change. For example, if all the deposited energy were converted to heat, a dose of radiation sufficient to be lethal to human beings would raise the temperature of the body by only 0.001° C. Fortunately, ionizing radiation can be accurately assessed, using other methods of measurement, at exposure levels several orders of magnitude below those required to produce measurable biological effects in human cells.

The critical irreversible chemical change responsible for the ultimate biological effects in irradiated cells appears to be damage to deoxyribonucleic acid (DNA), the basic material that controls the structure and function of the cells that make up the human body. Although similar in some respects to other carcinogenic agents, radiation is unique in that it has the ability to penetrate cells and to deposit energy within them in a random manner, unaffected, for example, by the cellular barriers presented to chemical agents. As a result, all cells in the body are susceptible to damage by ionizing radiation (Little, 1993).

The potential effects of radiation on cells can be divided into three categories: (1) at high doses it can cause death, (2) at lesser doses it can inhibit mitosis, and (3) at any dose it can cause alterations in the genetic material of the cell. Because of the effects of radiation on mitosis, the most sensitive tissues are those in which the cells frequently divide—for example, the precursor cells in the bone marrow that give rise to white blood cells and platelets, and the cells that line the stomach and small intestine. Muscular and brain tissues, where cell division is less pronounced, are far less sensitive. At still lower doses, even those insufficient to affect the ability of the cell to divide, radiation may produce mutations or other heritable alterations in DNA metabolism in the genetic material of cells. Presumably, such changes are responsible for the long-term somatic effects of radiation, such as cancer induction. When mutations involve germinal cells in the gonads, they may be passed on to the offspring of the irradiated individual and cause heritable genetic effects (Little, 1993).

UNITS OF DOSE

On the basis of knowledge about the deposition of energy and its associated biological effects, units have been developed for expressing the doses that result from exposures to ionizing radiation (Table 12.1). The one most commonly used today is the unit of equivalent dose, the sievert (Sv), although an earlier unit, the rem, is still in use in some parts of the world. Both units are a measure of the amount, distribution, and resulting biological effects of the ions created by the passage of radiation through tissue. Since both represent larger doses than are usually encountered in the workplace and the ambient environment, subunits have been developed. As would be anticipated, these units express the dose to an individual. To express the societal risk associated with radiation exposures to a large population group, the International Commission on Radiological Protection has developed the concept of collective dose. Expressed in units of person-Sv, it is calculated by multiplying the total number of people exposed (expressed in units of persons) by their average individual dose (expressed in units of the sievert) (ICRP, 1991a).

On the basis of total dose and dose rate, the effects of radiation exposure can be classified as either *deterministic* or *stochastic* (ICRP, 1991). Deterministic effects are those for which the severity of the effect varies with the dose, and for which a threshold may therefore occur. Stochastic effects are those for which the probability that an effect will occur, rather than the severity of the effect, is regarded as a function of the dose, without threshold. Deterministic effects are generally associated with *acute* exposures involving doses in the range of tens of sievert delivered to part or all of the body over a short period. Deterministic effects include cataracts, sterility, tissue damage (for example, erythema), and death (Table 12.2).

Table 12.1 Units of dose for ionizing radiation (historical development)

Unit	Description
Roentgen	The roentgen, now obsolete, was first introduced at the Radiological Congress held in Stockholm in 1928 as the special unit for expressing exposure to ionizing radiation. It was based on the quantity of electrical charge produced in air by X or gamma radiation. One roentgen (r) of exposure will produce about 2 billion ion pairs per cubic centimeter of air.
Rad	The rad was first defined by the International Commission on Radiation Units and Measurements (ICRU) in 1953 as the special unit of absorbed dose. It was developed to reflect the fact that the exposure of soft tissue or similar material to 1 r results in the absorption of about 100 ergs of energy per gram (1 rad). The *r*adiation *a*bsorbed *d*ose served for many years as the standard unit of dose.
Rem	When the biological effects of ionizing radiation were found to depend on the nature of the radiation, as well as on other factors, a unit was needed for expressing the effects of all types of ionizing radiation on a biologically equivalent basis. The *r*oentgen *e*quivalent *m*an, or rem, introduced by the ICRU in 1962, is equal to the absorbed dose in rad multiplied by the appropriate radiation-weighting factor. For X or gamma radiation, the radiation-weighting factor is one. Thus, 1 rad of absorbed dose from X or gamma radiation is equal to 1 rem. Similarly, 1 rad of absorbed dose from beta radiation is equal to 1 rem. For alpha radiation, 1 rad equals 20 rem (the radiation-weighting factor in this case having a value of 20).
Gray	Following the International System of Units, the General Conference on Weights and Measures (CGPM, from the initials of the French name) replaced the rad in 1975 with the gray, the unit of *absorbed dose.* One gray (Gy) is equal to 100 rad.
Sievert	Following the International System of Units, the CGPM replaced the rem in 1977 with the sievert, the unit of *equivalent dose* (often called simply the *dose*). One sievert (Sv) is equal to 100 rem; one millisievert (mSv) is equal to 100 millirem (mrem).
Person-Sv	Although the sievert provided a unit for expressing the dose to an individual, radiation protection specialists recognized the need for a unit to express the societal risk associated with doses to a large number of people. The unit developed for this purpose is the *collective dose,* the product of the number of people exposed and their average dose. Basic to the use of the collective dose is the assumption that the relationship between the dose and the health effects of radiation is linear. Accordingly, a dose of 0.1 Sv to 1 million people (yielding a collective dose of 100,000 person-Sv) has the same societal impact as a dose of 1 Sv to 100,000 people.

Table 12.2 Biological effects on humans of acute whole-body external doses of ionizing radiation

Equivalent dose		
Sievert	Rem	Effects
0–0.25	0–25	No detectable clinical effects; small increase in risk of delayed cancer and genetic effects
0.25–1	25–100	Slight transient reduction in lymphocytes and neutrophils; sickness not common; long-term effects possible, but serious effects on average individual highly improbable
1–2	100–200	Minimal symptoms; nausea and fatigue with possible vomiting; reduction in lymphocytes and neutrophils, with delayed recovery
2–3	200–300	Nausea and vomiting on first day; following latent period of up to 2 weeks, symptoms (loss of appetite and general malaise) appear but are not severe; recovery likely in about 3 months unless complicated by previous poor health
3–6	300–600	Nausea, vomiting, and diarrhea in first few hours, followed by latent period as long as 1 week with no definite symptoms; loss of appetite; general malaise, and fever during second week, followed by hemorrhage, purpura, inflammation of mouth and throat, diarrhea, and emaciation in third week; some deaths in 2–6 weeks; possible eventual death to 50% of those exposed
6–10	600–1,000	Vomiting in 100% of victims within first few hours; diarrhea, hemorrhage, fever, etc., toward end of first week; rapid emaciation; certain death unless heroic medical treatment is available and successful
10–50	1,000–5,000	Vomiting within 5–30 minutes; 100% incidence of death within 2–14 days
>50	>5,000	Vomiting immediately; 100% incidence of death within a few hours to 2 days

Although stochastic effects can occur after acute exposures, they may also result from *chronic* exposures (the type most commonly encountered in the workplace and the ambient environment) involving repeated low doses to all or parts of the body over a period of years. Stochastic effects include cancers (such as leukemia and solid tumors) and genetic damage, manifested in blood abnormalities, metabolic diseases, and physical abnormalities in the descendants of the person who has been exposed.

DOSE-RESPONSE RELATIONSHIPS

Ionizing radiation has sometimes been termed a universal carcinogen, in that it induces cancer in most tissues and most species at all ages (including the fetus). In reality, radiation has proved to be relatively weak in terms of both its carcinogenicity and its mutagenicity. As a consequence, few human data exist on the harmful biologic effects of radiation at low doses. Much of the human epidemiological information derives from observations when those exposed received relatively high doses over a short period of time. A prime example is the ongoing epidemiological study of survivors of the World War II atomic bombings in Japan.

Other studies that have provided extremely valuable data include evaluations of lung cancers in uranium miners exposed underground to airborne radon and its decay products; of bone cancers in young women who ingested radium and thorium while painting luminous markings on the faces of clocks and watches; of breast cancers in women with tuberculosis who had multiple fluoroscopic chest examinations; and of leukemias in patients treated with radiation for ankylosing spondylitis of the spine.

Interpretation of the data is not easy, as exemplified by the survivor studies in Japan. Thus far, the total number of cancer deaths in this population has been close to 6,000, of which only about 100 leukemia deaths and 300 other cancer deaths are estimated to be due to the exposures arising out of the bombings. Even with careful statistical analysis, quantification of the effects is difficult. Furthermore, for most effects of radiation, protracted exposure over days or weeks reduces the biological effects, compared to receiving the same dose in a single exposure. Since the exposures received by the Japanese population involved high doses over a short period of time, epidemiologists must extrapolate these estimates to low doses over long periods of time. Related studies are being pursued through a variety of avenues; subjects include X-ray and nuclear industry workers in a wide range of countries, people living in underground caves and geographical areas with high natural-background radiation, and the large population groups exposed as a result of the accident at the Chernobyl nuclear plant in 1986 (Stone, 1994).

As an outgrowth of these and related studies, a reasonably reliable basis currently exists for expressing the quantitative relationship between the biological effects of ionizing radiation and chronic exposures received at low dose rates. Since the relationships between dose and various biological endpoints differ, several models have been developed (Figure 12.3).

In general, mutations induced by radiation in cultured human cells follow a linear model (graph *A*). In contrast, cell death is related exponentially to dose (graph *B*). Further complicating the situation is that the shape of the curve appears to vary with the nature of the ionizing radiation. After considerable debate, the Committee on the Biological Effects of Ionizing Radiation of the National Research Council (BEIR, 1980) suggested that for X rays and gamma rays, the data appeared to support a relationship combining the linear and quadratic models (graph C). Although newer data appear to support this conclusion with regard to leukemia, it now appears

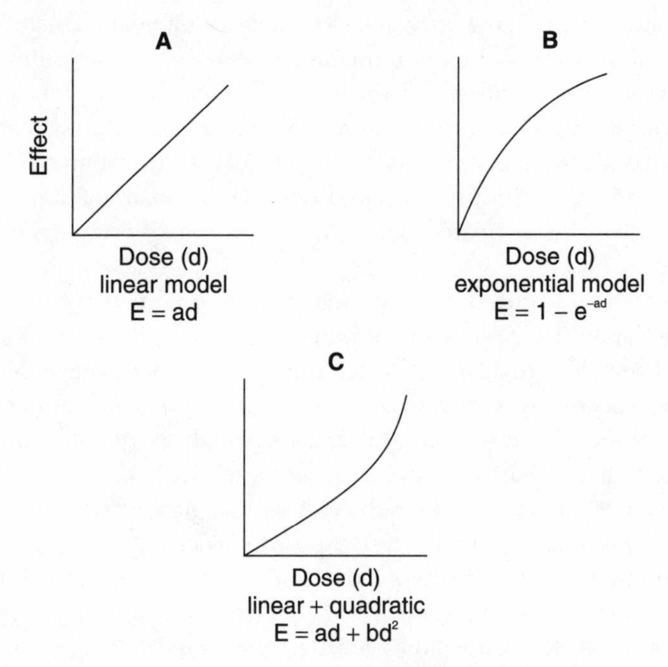

Figure 12.3 Dose-response models for quantifying the effects of ionizing radiation

that the induction of solid tumors (cancers of the thyroid, breast, and bone) follows a linear model (BEIR, 1990).

TYPES OF EXPOSURES

Both external and internal exposures from ionizing radiation are potential threats to human health in the workplace and in the general environment.

Regardless of the assumptions regarding the dose-response model, for sources of ionizing radiation *external* to the body the resulting injury will depend on (1) the total dose, (2) the dose rate, and (3) the percentage and region of the body exposed. In general, the potential for harmful effects increases along with increases in each of these three factors. The principal concern in the protection of both radiation workers and the general public is the probability of stochastic effects, primarily latent cancers.

Internal exposures result from the presence or deposition of radioactive materials in the body through ingestion or inhalation. The potential for harm depends on the types and quantities of material taken in and the length of time they remain in the body. For example, radionuclides that emit alpha particles present a greater hazard than those that emit beta particles. In general, the larger the quantity of radioactive material consumed and the longer it remains in the body, the greater the hazard.

The International Commission on Radiological Protection has developed annual limits on intake for a large number of radioactive materials. These include intake limits for workers (ICRP, 1991b) as well as the general public (ICRP, 1989–95). For radionuclides that distribute rather uniformly throughout the body, the permissible intake is calculated on the basis of the equivalent dose limit for the whole body. For radionuclides that concentrate predominantly in a single organ, the annual limit on intake is based on the concept of effective dose (see Chapter 14). To protect people from airborne radionuclides, the annual limits on intake have been converted to derived air concentrations. In a similar manner, permissible concentrations have been calculated for specific radionuclides in drinking water. Both have been incorporated into the regulations of the U.S. Nuclear Regulatory Commission (USNRC) and the Occupational Safety and Health Administration (OSHA). Techniques for monitoring various avenues of intake to assure compliance with these regulations include analyses of food, water, and air, as well as bioassays and whole-body counting. Techniques also exist for monitoring the movement of radioactive materials within the environment and for estimating doses to members of the public.

Natural-Background Radiation

Human beings have always been exposed to a significant level of radiation from natural sources, in the forms of cosmic radiation from outer space, external radiation from naturally occurring radioactive materials in the earth, and internal radiation from radioactive materials taken into the body in water, food, and air. These sources constitute the overwhelming majority of radiation exposures of people throughout the world (Moeller, 1996a).

The annual dose from *cosmic radiation* at sea level is about 0.3 mSv. Since the atmosphere serves as a shield against this source, the accompanying dose rate increases with altitude. At an altitude of 1 mile, for example, the annual dose from this source is about 0.5 mSv; at 12,000 feet it is about 1 mSv; at 30,000–40,000 feet (where commercial subsonic aircraft operate), the range is 45–70 mSv; at 50,000 feet (where commercial supersonic aircraft operate), the range is 80–90 mSv; and at 65,000 feet and higher (where future supersonic aircraft may operate), the annual dose from cosmic radiation could be in the range of 100–175 mSv or more (NCRP, 1995). Since passengers on flights across the United States in current subsonic aircraft are airborne only a few hours, the actual increase in dose is only a few hundredths of a millisievert. The doses to aircraft crews, however, can be substantial. In fact, many crew members regularly receive doses in excess of those received by radiation workers. Although existing supersonic aircraft operate at higher altitudes, the total dose for a given trip is about the same because the supersonic aircraft make the flight in considerably less time.

When it comes to *terrestrial radiation*, the quantities of naturally occurring radioactive materials in the soil are often far larger than many people realize. Between the surface of the ground and a depth of 5 feet, for example, are an average of 30 tons of uranium in each square mile of earth. Within that same volume are about 3.7×10^{11} becquerels (10 curies) of radium. The resulting worldwide releases of radon into the atmosphere total some 10^{20} becquerels (2.5 billion curies) per year. Because of variations in the quantities of naturally occurring radionuclides in the soil, terrestrial dose rates range from as low as a few tenths of a millisievert to as high as several millisievert per year. Figure 12.4 shows three major regions of variation in the United States. The regions with the highest dose rates are those associated with uranium deposits in the Colorado Plateau, granitic deposits in New England, and phosphate deposits in Florida; those with the lowest rates are the sandy soils of the Atlantic and Gulf Coastal Plain.

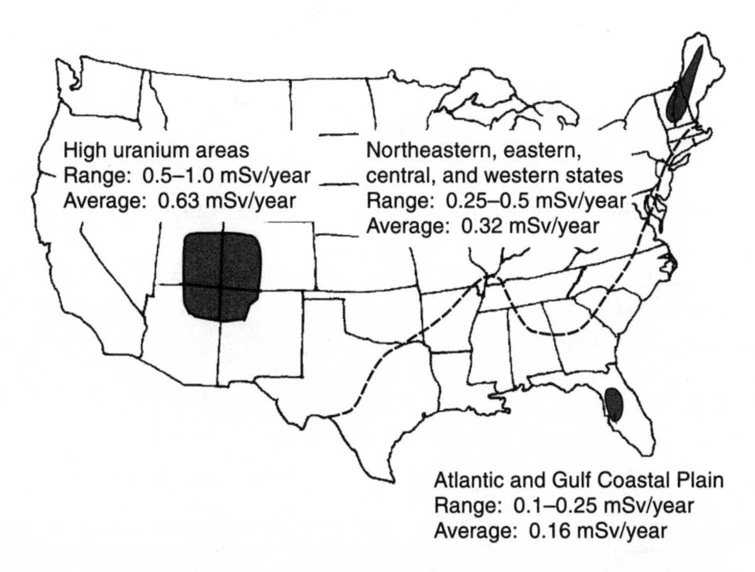

Figure 12.4 Terrestrial dose rates from natural background radiation in the contiguous 48 states

One of the principal naturally occurring radionuclides *within the human body* is ^{40}K. The resulting whole-body dose rate from this source for an average adult ranges from about 0.15 to 0.20 mSv per year. A comparable dose rate is contributed from other naturally occurring radionuclides in the body, such as ^{3}H, ^{14}C, and ^{226}Ra. By far the highest dose rate is that due to the inhalation of radon and its radioactive decay products, which are components of the air inside many homes and other types of buildings. In fact, the annual dose from this source to the lungs of the average member of the U.S. public is about 25 mSv. The associated risk is estimated to be equivalent to a whole-body dose rate of about 2 mSv per year (NCRP, 1987b). Dose rates from this source in certain parts of the country are orders of magnitude higher than these estimates.

For the country as a whole, the total annual dose from natural-background radiation sources is estimated to range from 1 mSv for people living in radon-free houses at sea level on the Atlantic and Gulf Coastal Plain, to 10 to 15 times this value or more for those living at high elevations on the

Table 12.3 Average annual effective dose of ionizing radiation to the U.S. public (exclusive of contribution from cigarettes)

Source	Annual Dose	
	mSv	mrem
Natural sources		
Radon	2.0	200
Cosmic, terrestrial, internal	1.0	100
Medical		
X-ray diagnosis	0.39	39
Nuclear medicine	0.14	14
Consumer products	~0.1	~10
Occupational	~0.01	~1
Miscellaneous environmental sources	<0.001	<0.1
Nuclear fuel cycle	<0.001	<0.1
Total	3.6	360

uranium-bearing lands of the Colorado Plateau and whose houses contain high concentrations of radon. Overall, the average annual dose rate from natural radiation sources to a member of the U.S. public is estimated to be 3.0 mSv. The total annual dose to an average member of the U.S. public from all sources of ionizing radiation is about 3.6 mSv (Table 12.3). Of this total, indoor radon contributes about 55 percent; natural background, including radon, about 80 percent; and artificial sources about 18 percent (Figure 12.5) (NCRP, 1987a,b).

Worldwide, the average dose rate from the natural-background radiation in "normal" areas is estimated to be about 2.4 mSv per year. One-third of this is caused by external exposure to cosmic radiation and terrestrial sources; two-thirds results from exposure to radionuclides internally deposited within the body, with the largest component (about half) coming from radon and its decay products.

Artificial Sources

The principal artificial sources of ionizing radiation are radiation machines, and radioactive materials produced in particle accelerators and nuclear reactors.

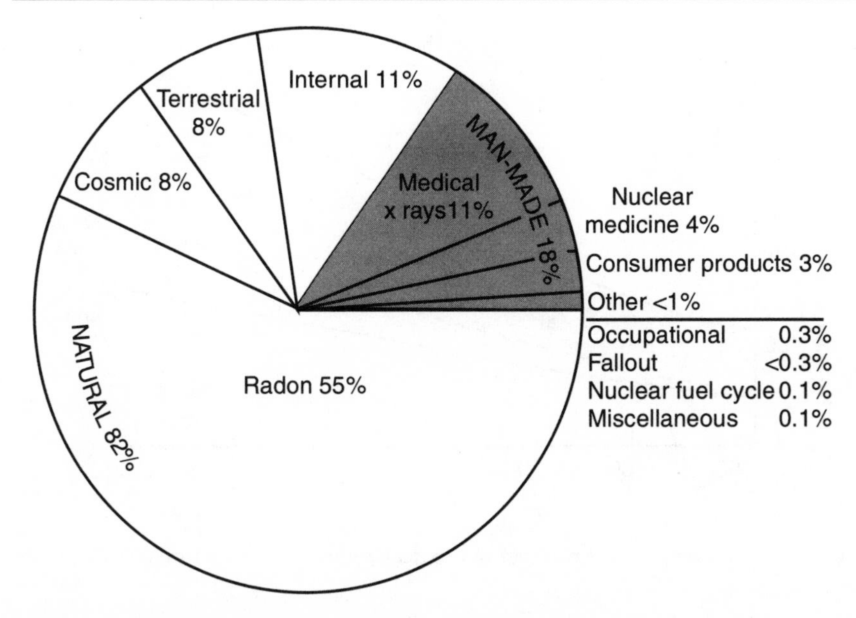

Figure 12.5 Relative annual dose contributions to the U.S. population of the principal sources of ionizing radiation

RADIATION MACHINES

X-ray machines are in widespread use in industry, medicine, commerce, and research. All are potential sources of exposures.

Medical and dental uses. Estimates are that more than 1 billion diagnostic medical X-ray examinations, more than 300 million dental X-ray examinations, and about 4 million radiation therapy procedures or courses of treatment are performed worldwide each year (Mettler et al., 1987). In the United States about 160,000 X-ray units are used by physicians, chiropractors, and veterinarians for medical diagnosis; about 210,000 additional units are used by dentists, for a total of 370,000 units. The rapid growth in applications of such units is illustrated in Figure 12.6, which shows the increase in the number of medical and dental X-ray units over the past 50 years. During the period 1980–1990, the accompanying use of X-ray film increased from 865 million to 1,135 million square feet, a 31 percent increase (Mettler, 1993).

Applications of medical and dental X-ray units result in exposures of both operators and patients. Several million medical and dental personnel, for example, are occupationally exposed in the operation of these units in

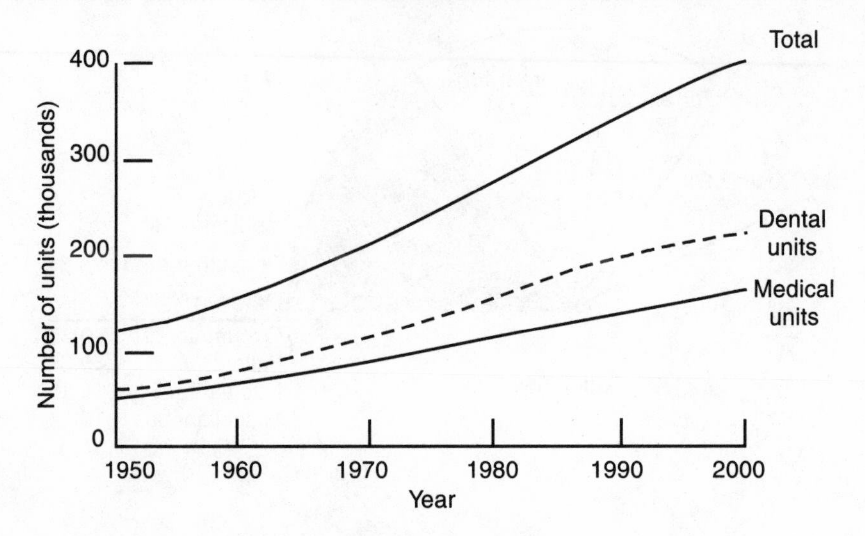

Figure 12.6 Increase in the number of medical and dental X-ray units in the United States, 1950–2000

the United States. Procedures for protecting them include limiting the time of exposure, maintaining an adequate distance between the X-ray beam and the operator, and providing adequate shielding. Because of the effective use of various combinations of these and other control measures, the average occupational exposure of medical and dental X-ray personnel in this country is well below 5 mSv per year, and essentially all of them receive an annual dose less than 10 mSv (Mettler et al., 1990).

In terms of the dose to the patients, about 65 percent of U.S. residents visit their doctor or dentist annually and undergo some type of X-ray procedure. Vigorous programs launched in the early 1960s to control patient exposures (similar to those used for X-ray unit operators) have kept the increase in doses far below the increase in the use of X rays: the federal government promulgated manufacturing standards for medical and dental X-ray machines; and federal, state, and local regulatory agencies developed ongoing inspection programs to assure that X-ray units are operated at the proper voltage, that X-ray beams are properly filtered and collimated, and that physicians and dentists use faster (lower-dose) films and proper processing techniques.

Progress in reducing the doses from medical and dental X-ray machines was particularly significant during the first two decades of the federal and state control programs. In the years 1964–1983, the average X-ray beam dose at the skin entrance for chest X-rays was reduced 25 percent, from 0.28 to 0.21 mSv; and for dental bitewings, the skin entrance dose was reduced 76 percent, from 11.5 to 3 mSv. In the same period the beam size of most units used for taking chest X rays was reduced to the size of the film being used, and the diameter of the beams in over 90 percent of dental X-ray units was reduced to 2.75 inches or less, the size considered acceptable for making a dental X-ray film (Johnson and Goetz, 1985).

Reflecting the success of these efforts, the estimated mean dose to the bone marrow of the average U.S. adult increased by only 13 percent from 1970 to 1980, a period in which the number of medical and dental X-ray machines and the sales of X-ray films were increasing dramatically (NCRP, 1989). Overall, the annual dose to average members of the U.S. population, expressed in terms of an equivalent whole-body dose, was estimated to be 0.39 mSv (NCRP, 1987b).

Industrial uses. Industrial X-ray devices consist primarily of radiographic and fluoroscopic units used to detect defects in castings, fabricated structures, and welds; and fluoroscopic units used to detect foreign material in items such as food products. Today there are about 16,000 active industrial radiographic installations in the United States; some 40,000–50,000 people are occupationally exposed in their operation. The primary concern is the control of exposures to the X-ray machine operators. The same techniques of filtration, coning, shielding, and limiting the time of exposure apply here as in the use of medical X rays.

Commercial uses. Since the 1970s, X-ray machines have been used increasingly to inspect luggage at airports as a security measure against aircraft hijackings and bombings; more than 1,000 such units are in operation in U.S. airports today. Although travelers often pass close to these units when entering the boarding area, their advanced design keeps the doses extremely low, in the range of a few thousandths of a millisievert. The metal detectors used for checking passengers are not a source of radiation exposure.

Research uses. High-voltage X-ray machines and particle accelerators are common equipment in the laboratories of universities and research organizations. More than 1,200 cyclotrons, synchrotrons, van de Graaff generators, and betatrons are in operation in the United States, along with about 3,000 electron microscopes and 10,000 diffraction units. Modern electron microscopes are shielded to protect the operators, but diffraction units still account for a significant number of radiation injuries (primarily burns on the hands).

RADIOACTIVE MATERIALS

More than 22,000 hospitals, academic, industrial, and research organizations in the United States have been licensed to use radioactive materials. Such materials are used in more than 10 million medical diagnostic procedures, 200,000 medical treatment procedures, and 100 million medical laboratory procedures each year. The laboratory procedures include radioimmunoassay tests on blood and bodily fluids from patients. For the industrialized world as a totality, it is estimated that 25 percent of hospital patients undergo a nuclear medicine procedure during diagnosis or treatment. Artificially produced radioactive materials are also utilized in universities and other institutions for teaching and research; they are used by industry in both portable and fixed devices, such as thickness, level, and moisture-density gauges, static eliminators, and gas chromatographs. Artificially produced radioactive materials are also employed in the manufacture of smoke detectors, and sealed gamma and neutron sources log wells during explorations for oil and gas (USNRC, 1991).

The use of artificially produced radioactive materials in the United States is increasing at the rate of 10–15 percent per year. Although from 1980 to 1990 the number of nuclear medicine examinations appears to have remained reasonably stable, increases are still occurring in research and industry. Seventy to 80 percent of all research at the National Institutes of Health, for example, involves such materials. The same is true for the research being conducted at many universities. Of the 15 Nobel prizes awarded in physiology and medicine from 1975 to 1989, 10 were based on research using radioactive materials.

The population undergoing diagnostic nuclear medicine examinations using artificially produced radioactive materials is, in general, much older than the average patient having diagnostic X rays. At the present time, 75 percent of all nuclear medicine procedures are performed on people aged 45 or older; more than 30 percent are performed on patients 65 or older. The annual per capita whole-body equivalent dose from these procedures in the United States in 1982 was estimated to be 0.14 mSv (NCRP, 1987b). In addition to doses to patients, exposures can occur during the preparation, handling, use, and transport of these materials, and through the release of used materials to the environment. Since it is permitted under current regulations, more than half of patients to whom these materials have been administered dispose of their bodily wastes through direct discharge into sewers. This procedure is considered justifiable for three reasons: (1) most of the radionuclides are very short lived; (2) the resulting exposures to members of the public and sewage treatment plant operators are extremely low; and (3) collection and handling of the wastes by hospital personnel or family members would result in far more exposure than otherwise occurs.

Because of these and other uses, approximately 3 million packages containing radioactive materials are shipped in this country each year. The number, although large, is less than 3 percent of the more than 100 million shipments of hazardous materials made annually. Depending on the nature, chemical form, and amount of radioactive material, certain shipments (an estimated 200,000) must use USNRC-certified packaging. Since such packaging is designed to withstand severe accidents, the likelihood of a release is minimal (USNRC, 1993).

Nuclear Power Operations

As of 1996, more than 100 electricity-generating stations powered by nuclear reactors were operating in the United States (Figure 12.7). These units had a generating capacity in excess of 100,000 megawatts and were producing over 20 percent of the electricity consumed (USNRC, 1995). In addition, about 50 reactors were being used for training and research; another 50 were operating at various facilities of the Department of Energy; and almost 200 were in operation or being built for military use as propulsion units in submarines, cruisers, and aircraft carriers.

About 450 nuclear power plants, with a total generating capacity of over 350,000 megawatts, are operating worldwide, producing about 17 percent of the world's electricity. Thirty-one nations produce electricity using nuclear power, in several cases as a major source: France—over 70 percent; Sweden—over 40 percent; and Germany and Japan—about 25 percent. The worldwide installed capacity for nuclear electricity generation is expected to reach 500,000 megawatts by the year 2000 (Mettler et al., 1990).

Sources of radiation exposure from nuclear power operations include the mining, milling, and fabrication of new fuel; the removal and storage or processing of spent fuel; and the reactor itself.

Mining, milling, and fabrication. Several thousand people are employed in uranium mining and milling operations in the United States, mainly in the Colorado Plateau region. At least half of the uranium is now ex-

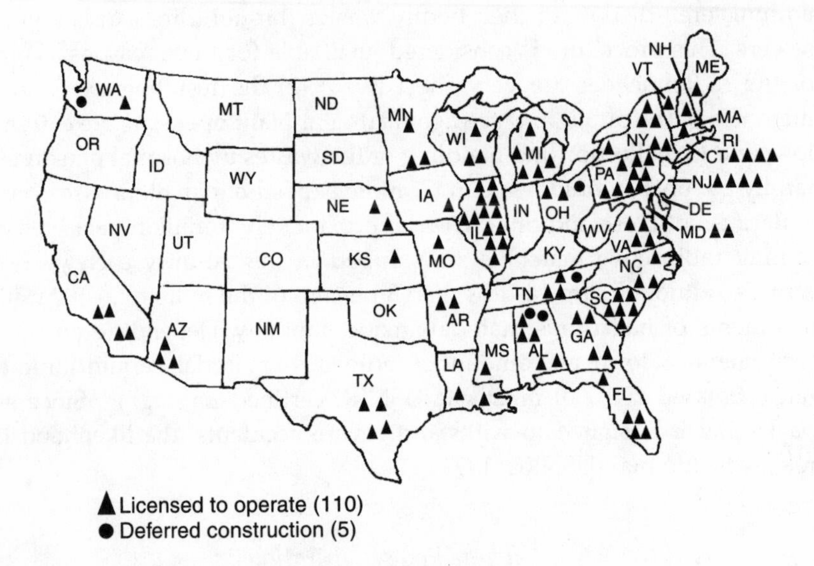

Figure 12.7 Nuclear power plants in the United States, 1995 (none in Alaska or Hawaii)

tracted through strip mining, rather than underground mining, resulting in much lower radiation exposures to workers. Such operations produce vast amounts of uranium mill tailings; about 86,000 tons are produced each year in acquiring the fuel for operating each 1,000-megawatt nuclear power plant (BRWM, 1990). The net result is that today more than 100 million tons of uranium mill tailings are in storage at sites in the United States; about 80 percent of these are located in New Mexico and Wyoming and cover a land area of about 2,000 acres. To retard radon release, uranium tailings stored on the ground surface have been capped with earth covers and stabilized to prevent wind and water erosion and to control exposures of the nearby population groups.

Treatment of spent fuel. Each 1,000-megawatt nuclear plant produces about 30 tons of spent fuel annually. In the United States spent fuel from commercial nuclear plants is being held for ultimate disposal in a high-level waste repository. Other countries, such as France, Germany, and Japan, reprocess their spent fuel so that the unused uranium and the plutonium produced in the fuel by the bombardment of uranium with neutrons can be recycled in nuclear power plants. Reprocessing operations discharge low-level radioactive wastes into the atmosphere and the aquatic environment and produce high-level liquid wastes that must subsequently be solidified for disposal. Although spent fuel from commercial U.S. nuclear power plants is not currently being reprocessed, millions of gallons of high-level liquid wastes from reprocessing of spent fuel from military reactors are now being stored and await solidification prior to emplacement in an underground repository. Through congressional and presidential action, Yucca Mountain in Nevada has been selected as the potential location for the first such repository. If the proposed site is found to be acceptable, the first emplacement of high-level wastes is expected to occur in about 20 years.

Commercial nuclear power plants. Workers at nuclear power plants receive exposures through radiation from the reactor and from the radioactive materials released from it. Data for calendar year 1995 show that the collective occupational dose per nuclear power plant in the United States ranged from about 1.5 to 2.6 person-Sv, depending on the type of plant. No worker at any of the plants in 1995 exceeded the annual dose limit of 50 mSv; in fact, the latest data show that the average dose to each of the approximately 100,000 people who worked at U.S. nuclear power plants and received measurable doses was far less than 10 percent of the limit. In addition, the industrial accident rate at these plants for calendar 1995 was 0.55 per 200,000 work-hours, which makes it one of the safer industrial work environments (INPO, 1996). These results reflect the regulatory efforts of the USNRC; the supporting activities of the Institute of Nuclear Power Operations, an organization supported by the nuclear industry; and the guidance provided by organizations such as the National Council on Radiation Protection and Measurements (NCRP, 1994).

Although the spent fuel contains the bulk of the radioactive material produced in the operation of a nuclear reactor, some radioactive materials leak from the fuel into the cooling system and result in low-level airborne and liquid waste releases. In addition, low-level solid radioactive waste results from various plant operations, as well as from the medical, industrial, and research applications of artificially produced radioactive materials. In 1995 each nuclear power plant in the United States produced an average of 30–110 cubic meters of low-level solid waste, the amount depending on the type of plant. These volumes represent a fivefold to tenfold decrease from the amounts produced in 1980 (INPO, 1996).

Environmental releases from nuclear power plants operating in the United States are limited by USNRC regulations. For example, Title 10, part 50, of the Code of Federal Regulations limits the dose rate to the maximally exposed person offsite to about 0.1 mSv per year. In 1987 the total collective dose to the 150 million people living within 50 miles of nuclear power plants in the United States was less than 0.8 person-Sv (USNRC, 1990). The average 50-year dose commitment to each person in this group for releases occurring in 1987 ranged from 2×10^{-8} to 9×10^{-5} mSv. The maximum dose during that year for any member of the U.S. population was estimated to be less than 0.01 mSv.

Nuclear Weapons Testing

From the late 1940s through the early 1960s, the United States and the former Soviet Union conducted a large number of tests of nuclear weapons in the atmosphere, most of them in 1952–1958 and 1960–1962. Lesser numbers of weapons were tested by the United Kingdom, France, and the People's Republic of China. In accordance with the test ban treaty of 1961, subsequent tests were, in the main, conducted underground and radionuclide releases to the atmosphere were minimized. The earlier tests, however, released into the atmosphere enormous quantities of radioactive material that spread throughout the world. The last atmospheric test was conducted by China in 1980; France, however, continued to conduct underground nuclear tests as recently as early 1996.

The legacy of these tests will remain for many years. In Nevada, for example, there are large amounts of radionuclide contamination near and on the ground surface at the test site as a result of the atmospheric tests, and even larger quantities of radionuclides lie in deep caverns in the soil at the site as a result of the underground tests. The collective dose to the population of the world as a result of all the tests conducted by all the countries involved is estimated to be about 30 million person-Sv (UNSCEAR, 1993). This is by far the largest contributor of dose to the public from any artificial source.

Consumer Products

A number of natural and artificially produced radioactive materials used in consumer products can lead to exposures of members of the public. Such products are vivid demonstrations of the fact that radiation is part of everyday life in the modern world. Consumer products are used in the home, at work, and on a personal basis (Moeller, 1996b).

Sources within the home include the materials used in constructing the home, the cooking and heating systems, the water supply, and other items such as television sets and smoke detectors. Brick, concrete, and/or stone contain relatively large quantities of naturally occurring radionuclides. As a result, homes constructed of these materials yield external dose rates to the residents of about 0.07 mSv per year. Residents of the 40 million homes served by unvented natural-gas cooking stoves receive an estimated annual dose to their lungs from radon of about 0.05 mSv per year. An additional 16 million people may be exposed to radon and its decay products through the use of unvented natural-gas heaters. Radium in domestic water supplies, such as those from groundwater sources, can produce average dose rates approaching 1 mSv per year. Television sets and smoke detectors can also be sources of ionizing radiation; the dose rates, however, are extremely small.

Sources of radiation in the workplace include static eliminators and thickness gauges. The former are widely used in industry to reduce the electrical charge buildup on materials. The 30,000–50,000 people occupationally exposed in the use of such devices receive an annual whole-body dose of 0.003–0.004 mSv. Over a thousand commercial and industrial organizations in the United States use thickness gauges that incorporate radioactive materials, primarily for quality control in the manufacture of sheet steel, aluminum, and other metal products. Average annual doses to the 15,000–20,000 people exposed in the use of these units are estimated to be less than 1 mSv.

A host of consumer products, used or consumed on a personal basis, can be sources of radiation exposure to individual members of the public. Luminous watches and clocks, particularly older units that contain radium, yield external dose rates of 0.005–0.03 mSv per year. Localized doses can range up to 3 mSv per year. Certain eyeglasses and eyepieces used in optical instruments and tinted with uranium or thorium can yield dose rates to the germinal cells of the cornea of up to 40 mSv per year. False teeth, which in the past were glazed with uranium, can yield average dose rates to the basal mucosa of the gums of about 7 mSv per year; it is estimated that 45 million people in this country wear this type of dental prothesis. Cloisonné jewelry glazed with uranium and fashioned as bracelets, earrings, and necklaces can yield dose rates to the skin of up to 0.04 mSv per *hour.* Stimulated by concern over this problem, the Nuclear Regulatory Commission in 1983 banned the use of uranium in enamel products sold in the United States. Gas lanterns include a mantle containing thorium; during initial curing of such mantles, a large portion of the thorium is released into the air. Dose rates to the lungs of users range up to 0.06 mSv per year.

The major consumer-product source of radiation exposure to the U.S.

public, however, arises through the smoking of cigarettes. Tobacco contains relatively high quantities of ^{210}Po, a naturally occurring radionuclide. When a cigarette is lighted, this material is volatilized and taken into the lungs. Estimates of the annual dose to the lungs of the 50 million smokers in the United States, assuming an average of 1.5 packs of cigarettes per day, is 160 mSv. The whole-body equivalent dose is more than ten times the recommended limit for an individual member of the public.

Other products used in our daily lives can release radioactive materials into the general environment. Coal, for example, contains relatively large quantities of uranium and its various radioactive decay products. During the combustion process, a portion of these radioactive materials become airborne and can be inhaled and deposited within the bodies of those living nearby. Although many people are subject to such exposures, the average dose is estimated to be well below 0.01 mSv per year. Millions of tons of fertilizer are used in the United States each year, and many of the phosphate fertilizer products contain naturally occurring radioactive materials such as uranium and thorium, as well as their radioactive decay products. The average dose rate to people consuming agricultural products grown on lands to which such fertilizers have been applied is estimated to range from 0.005 to 0.05 mSv per year (NCRP, 1987c).

Control and Benefits of Radiation

To be effective, control measures must be practical. Although the natural background is the greatest source of radiation exposure, some natural sources are not amenable to control. Take the dose received from radioactive potassium in the body: potassium is essential to health, and there is no practical way to avoid the accompanying dose. Much the same can be said for cosmic radiation; it would be impractical for people living in cities such as Denver to move to lower altitudes simply to avoid the dose from this source.

Although airborne radon and its decay products in dwellings represent the largest natural source of radiation exposure (exclusive of the dose to the lungs of cigarette smokers), people have been slow to adopt control measures. Many of the control measures recommended by the Environmental Protection Agency require relatively expensive systems that can be installed only by an outside contractor. Until homeowners become aware of simpler approaches, such as circulating the air within a home using a fan alone or in combination with an ion generator (Maher et al., 1987), little progress can be expected.

The success of control efforts in the medical field results from several factors. One is that the Department of Health and Human Services (HHS) has established standards for the manufacture of diagnostic X-ray machines so that they will cause minimal exposure to patients. A second is the vigorous manner in which state health and regulatory agencies, in cooperation with HHS, have pursued programs for the inspection of medical X-ray machines on a regular and continuing basis. Joint federal and state approaches have also been effective in controlling exposures from radioactive materials in medicine and research. All users of significant quantities of such materials must be licensed, and part of the licensing procedure involves a demonstration that applicants have the training, equipment, and facilities to handle such materials safely.

The control of radioactive materials in industrial, medical, and research applications is largely a function of their form. Control measures for materials sealed in a capsule (for example, for use in radiography) are very similar to those for X-ray machines. Materials in "loose" form, used in laboratory research and nuclear medicine, require careful handling and monitoring. Usually they must be processed only inside a hood (particularly if there is a chance of airborne release), the handler must wear gloves and a laboratory coat or other protective clothing, and smoking and consumption of food or drink in the work area is not permitted. In some instances, as in decontamination operations at nuclear facilities, workers may require extensive protective clothing (boots, head coverings, and respirators).

Although nonionizing and ionizing radiation sources have potentially harmful effects, they also improve our standard of living in both minor and major ways. Our households contain nonionizing sources in toasters, irons, electric blankets, home computers, and microwave ovens (which not only heat and cook foods quickly but also use far less electricity than conventional ovens). Radar promotes safety in airline transport. Although questions remain about the health effects of the electric and magnetic fields associated with high-voltage lines, transmitting electricity at higher voltages has reduced losses significantly and made electricity and its benefits widely available.

Applications of sources of ionizing radiation are bringing benefits to people throughout the world. Nuclear power, despite its problems, holds promise of generating electricity without airborne release of sulfur, nitrogen, and carbon dioxide. The use of X rays in medicine has not only improved the diagnosis of disease but has also, in the case of radiation therapy, provided a means for treating cancer. Radiation is also used to sterilize medical supplies, such as surgical gloves and syringes, thus protecting patients and medical personnel against the transmission of infection in hospitals and clinics.

Similarly, the use of artificially produced radioactive materials has led to dramatic improvements in medical diagnosis and therapy and related research in physiology and medicine. Approximately 24,000 nuclear scanners or cameras are in use worldwide in hospitals, and nearly 24 million imaging studies are conducted annually. An estimated 18,000 radiotherapy machines are in use, treating about 5 million patients each year (Blix, 1989).

Controls incorporating radioactive materials for monitoring the operation of industrial machines are used throughout the world. These include the previously cited thickness gauges, which have led to improved quality control of manufactured goods. Nucleonic systems now control entire processes in the steel and coal industries. In the paper industry, such systems control both thickness and moisture content, saving materials and energy.

The use of radionuclides in agricultural research has brought many benefits. For example, an isotope of nitrogen has identified the best way of applying nitrogen fertilizers in rice production, reducing the quantity required by up to 50 percent. In addition to economic savings, fewer surplus nitrogen compounds run off into groundwater, rivers, and lakes. Nuclear techniques such as gamma or neutron irradiation have induced mutations in seeds and thus created new varieties of food crops. In 1960 only 15 radiation-induced mutant varieties were available to growers; today more than 1,300 have been bred for higher yield, better quality, and improved resistance to disease and pests. Sixty percent of the durum wheat grown in Italy to produce pasta is a gamma-induced mutant variety. In China about 8 million hectares (almost 10 percent of China's cultivated land) are planted with strains of rice produced through radiation-induced mutations, yielding crops valued at an estimated $1 billion annually (Blix, 1989). And, as previously described, radiation is being used to control insects such as the screwworm fly, the Mediterranean fruit fly, and the tse-tse fly.

The General Outlook

The public urgently needs a more realistic understanding of the major sources of radiation, particularly those of an ionizing nature, in daily life. All too often, the potential benefits have been denied or not fully attained owing to misconceptions. A notable example is the use of radiation for the preservation of food; many of the same misconceptions exist with respect to the use of nuclear power.

One of the first steps in rectifying this situation is to emphasize the overwhelming dominance of natural background radiation as the principal source of the public's exposure. The worldwide annual collective dose from this source, currently estimated at about 12.5 million person-Sv, dwarfs all other sources. Equally important is for individuals to realize the large variations from country to country in the dose rates from natural background. Variations among European countries serve as an excellent example. The combined annual doses from natural sources (cosmic radiation, terrestrial sources, and internally deposited radionuclides) range from about 2.5 mSv in the United Kingdom, 4.0 mSv in Germany, 4.5 mSv in Italy, 6.0 mSv in France, and 6.5 mSv in Sweden, to over 8 mSv in Finland, more than a threefold difference overall (NRPB, 1993a).

The public also needs a fuller understanding of the relative magnitude of various artificial sources of ionizing radiation. At the present time, the largest such contributor of dose, particularly in the more industrialized countries, is the use of X rays in medical diagnosis. For the world as a whole, the annual collective dose from this source is estimated to be about 1.8 million person-Sv. For individuals, the average annual whole-body equivalent dose is about 0.3 mSv, ranging from about 1.1 mSv in countries with the highest level of medical care to 0.05 mSv at the lowest level. The resulting annual collective dose is less than 15 percent of that from natural background (UNSCEAR, 1993).

From the standpoint of the release of artificially produced radionuclides into the environment, the largest contributor over the past half-century has been the atmospheric testing of nuclear weapons, with a total worldwide collective dose estimated at 30 million person-Sv. While this source of radiation exposure is clearly unfortunate and unnecessary, it is important to recognize that this dose is equivalent to the amount received by the same population from natural background over a period of about 2.5 years (UNSCEAR, 1993).

Another significant outcome of the discussion in this chapter is that, in spite of the many environmental and occupational radiation sources that can potentially cause exposures, the primary concerns arise through people's personal environment. For both ionizing and nonionizing radiation sources, those who desire to keep their dose rates to a minimum should avoid overexposure to the sun, should minimize their exposure to magnetic fields by not sleeping under electric blankets, should monitor their homes for radon and take corrective measures if necessary, should ensure that the doctors examining them use X-ray equipment that is up-to-date and well maintained, and should avoid smoking cigarettes.

13

ENVIRONMENTAL LAW

ALTHOUGH laws providing for protection of the environment date back to the late 1800s, modern U.S. environmental law had its birth with the enactment of the National Environmental Policy Act (NEPA) in 1969 (P.L. 91-190). This law declared that it was national policy to: (1) encourage productive and enjoyable harmony between people and their environment, (2) promote efforts that will prevent or eliminate damage to the environment and the biosphere and stimulate the health and welfare of man, and (3) enrich our understanding of ecological systems and natural resources important to the nation. In enacting this legislation, Congress recognized the profound impact of human activities—particularly of population growth, high-density urbanization, industrial expansion, resource exploitation, and new and expanding technological advances—on the interrelation of all components of the natural environment.

The basic goal of the NEPA was to administer federal programs in the most environmentally sound manner, consistent with other national priorities. Through this act Congress made it mandatory that each federal agency prepare an environmental impact statement (EIS) if a proposed action is (1) federal, (2) qualifies as "major," and (3) will have a significant effect on the environment. The primary purpose of the EIS is to disclose to the agency and the public the environmental consequences of a proposed action, as well as the repercussions of the alternative courses of action. Projects or activities to which this act applies cover the construction of new facilities (bridges, highways, airports, hydroelectric and/or nuclear power plants), the dredging or channeling of a river or harbor, large-scale aerial spraying of pesticides, and disposal of munitions and other hazardous materials.

Included are not only projects undertaken directly by federal agencies but also those that are supported in whole or in part by federal contracts, grants, subsidies, loans, or other forms of funding assistance.

A significant initiative of the NEPA was the establishment of a Council on Environmental Quality (CEQ) within the Executive Office of the President. Responsibilities assigned to the CEQ included developing new environmental programs and policies, coordinating the wide array of federal environmental efforts, ensuring that officials responsible for federal activities take environmental considerations into account, and assisting the President in assessing and solving environmental problems. Among the duties of the CEQ was the preparation of an annual report on the quality of the environment.

Although the National Environmental Policy Act was far-reaching and all-encompassing, it focused primarily on process rather than substance. Nonetheless, in many cases proposed projects were halted after initiation of the NEPA litigation. In other cases, federal agencies have refrained from taking controversial actions, to avoid the expense and delay of litigation. Undoubtedly some environmentally unjustifiable projects have either been abandoned or were never formally proposed because of the NEPA requirements. In addition, a host of projects have been modified to reduce their environmental impact. The unanswered question is whether the benefits thus attained are sufficient to justify the expense and delay created in instances when the project has ultimately proceeded (Findley and Farber, 1992).

Interestingly, the NEPA contains no requirements relative to how clean the air and water must be or to how much pollution can be discharged into the environment. To meet these needs, Congress has passed a wide range of laws that address air and water pollution, solid waste disposal, the purity of food and drinking water, and problems of the occupational environment (Table 13.1). The increasing pace of such legislation in recent years is illustrated in Figure 13.1. Many of the laws were passed prior to the NEPA; some of the more important are discussed in the sections that follow.

The Clean Air Act and Amendments

Originally passed in 1955, the Clean Air Act (P.L. 84-159) was designed to protect and enhance the quality of the air resources of the nation. This act was replaced by the Air Quality Act of 1967, which in turn was amended in 1970, 1977, and 1990 (P.L. 91-604; 101-549). Under the 1955 act, it was recog-

Table 13.1 Purpose and/or scope of various environmental and occupational health laws

Law	Date	Purpose and/or scope
Environment		
Solid Waste Disposal Act (Resource Conservation and Recovery Act)	1965 1976 1984	"Cradle-to-grave" coverage of wastes; prohibition of land disposal of nontreated hazardous waste
National Environmental Policy Act	1969	General protection of the environment; requirement of environmental impact statements
Clean Air Act and Amendments	1970 1990	Consideration of public exposures to airborne contaminants; specification of required treatment and ambient standards
Toxic Substances Control Act	1976	Requirements for EPA notification on use, testing, and restriction of certain chemical substances
Clean Water Act and Amendments	1977	Restrictions on pollution discharges into rivers and streams
Comprehensive Environmental Response, Compensation and Liability Act (Superfund)	1980 1986	Requires cleanup of existing disposal sites; establishes financial responsibility for sites
Emergency Planning and Community Right-to-Know Act	1986	Establishes requirements that sources must assess and annually report to EPA their emissions to air, land, and water; establishes emergency release notification system
Pollution Prevention Act	1990	Requirements for prevention, reduction, or treatment of pollution at its source, with disposal to the environment being a last resort
Radioactive Waste Disposal		
Uranium Mill Tailings Radiation Control Act	1978	Requirements for remediation of former uranium mill processing and disposal sites
Nuclear Waste Policy Act	1982	Requirements for disposal of high-level radioactive wastes
Low-Level Radioactive Waste Policy Amendments Act	1985	Requirements for disposal of low-level radioactive wastes

Table 13.1 (continued)

Law	Date	Purpose and/or scope
Food and Water		
Federal Food, Drug, and Cosmetic Act	1938	Protection of foods against pesticides and harmful additives
Federal Insecticide, Fungicide, and Rodenticide Act	1972	Registration of chemicals used for pest control
Safe Drinking Water Act and Amendments	1974 1986	Provision of maximum limits for contaminants in public drinking water and techniques for their removal
	1996	Requirements for specific analyses for emerging contaminants, such as *Cryptosporidium;* provides a revolving fund to help state and local authorities improve drinking-water systems
Food Quality Protection Act	1996	Repeal of Delaney clause; eases regulation of processed foods and tightens regulation of raw foods
Worker Safety		
Occupational Safety and Health Act	1970	Assurance of safe and healthful conditions for U.S. workers
Energy		
Atomic Energy Act	1954	Assurance of safe handling of radioactive materials and safe management of nuclear facilities
Energy Policy Act of 1992	1992	Requirements for improvements in the energy efficiency of a wide range of items, including transportation vehicles and industrial and home appliances, and promotion of the use of renewable resources

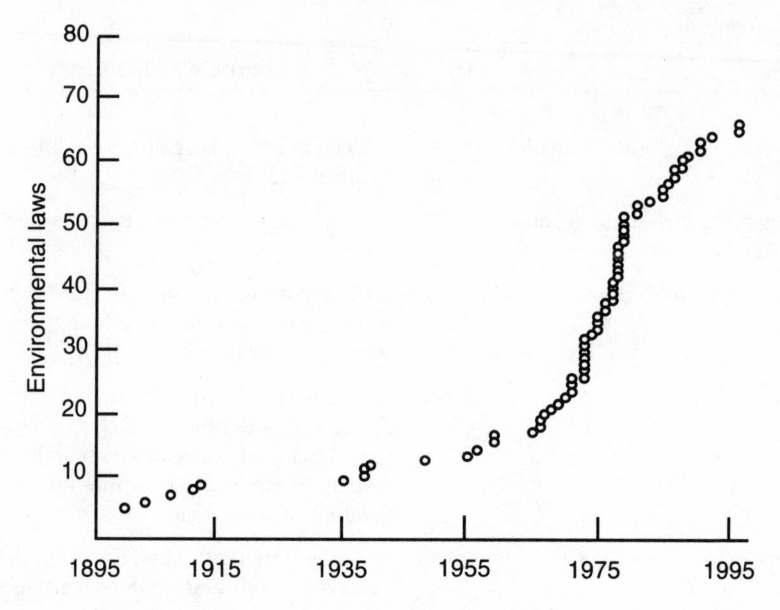

Figure 13.1 The growth of environmental laws in the United States, 1899–1996

nized that the individual states have primary responsibility for ensuring the quality of the air within their borders. The 1970 amendments required the Environmental Protection Agency to set national ambient air quality standards (NAAQS) that will provide "an ample margin of safety to protect the public health." To assure that the NAAQS were achieved, the EPA was also required to set limits for the control of emissions from both stationary and mobile sources. The 1977 amendments required the EPA to apply different levels of stringency to airborne emissions in areas that met the existing air quality standards, the so-called attainment areas, versus those that did not, the so-called nonattainment areas.

To ensure that levels of pollution in attainment areas would not increase, the 1977 amendments incorporated a "prevention of significant deterioration" requirement. New stationary sources in such areas were required to use the best available control technology (BACT). At the same time, the amendments required the EPA to take into account the costs of compliance. Existing sources in nonattainment areas were required to use "reasonably available control technology" (RACT) which, as the wording implies, represents a lesser level of control that can be achieved at lower cost. In essence, the objective of RACT was to provide a minimum level of control in nonattainment areas. To "drive" industry to achieve better controls, the EPA in some cases specified the percentage reductions to be achieved.

One group of emissions receiving special attention were the "toxic air pollutants," or "air toxics." These are primarily substances that tests have shown to be carcinogenic. The establishment of standards for these pollutants has been much more difficult than anticipated. One of the primary problems is whether they have threshold concentrations below which they have no biological effects and, if not, how much risk to the public should be considered acceptable.

One of the important contributions of the 1990 amendments to the Clean Air Act (U.S. Congress, 1990b) was the separation of air pollution standards into several classes. These included harm-based standards designed to protect public health, technology-based standards requiring application of various levels of control technology, and technology-forcing standards designed to ensure that industry develop and apply the very best control technology (NAE, 1993). In many respects, these changes amplified the requirements mandated under the 1970 amendments. At the same time, Congress changed the ozone and carbon monoxide classifications and pollutant control strategies for urban areas; tightened vehicular emission standards, mandated the regulation of almost 200 toxic air pollutants, required reductions in power plant sulfur dioxide emissions and the development of emission standards for nitrogen oxides, called for the establishment of a new permit system consolidating all applicable emission control requirements, and mandated a production phaseout by the year 2000 of the five most destructive ozone-depleting chemicals. The 1990 amendments also required that the statutory reach include airborne particulates 10 micrometers in size or less; that is, it required that primary attention be directed to particles in the respirable range.

Traditionally, the approach relative to mobile sources, for example, automobiles, has been to require that pollution control mechanisms be incorporated into the product by the manufacturer. The EPA could thereby be assured that every vehicle sold in the United States (whether manufactured domestically or imported) would meet the standards. In contrast, the primary approach for limiting airborne emissions from stationary sources (for example, major manufacturing facilities) has been to apply a combination of controls, such as RACT or BACT, and to enforce the requirements through the granting of operating permits. To assure successful control, Congress mandated that each state develop an implementation plan describing how the federally specified standards would be met. Unfortu-

Table 13a Major environmental legislation

1899	River and Harbors Act
1902	Reclamation Act
1906	Pure Food and Drug Act
1910	Insecticide Act
1911	Weeks Law
1934	Taylor Grazing Act
1937	Flood Control Act
1937	Wildlife Restoration Act
1938	Food, Drug, and Cosmetic Act
1947	Federal Insecticide, Fungicide, and Rodenticide Act
1954	Atomic Energy Act
1955	Clean Air Act
1958	Fish and Wildlife Coordination Act
1958	Food Additives Amendment
1964	Wilderness Act
1965	Solid Waste Disposal Act
1965	Water Resources Planning Act
1966	National Historical Preservation Act
1967	Air Quality Act
1968	Wild and Scenic Rivers Act
1969	National Environmental Policy Act
1970	Clean Air Act Amendments
1970	Occupational Safety and Health Act
1970	Resource Recovery Act
1972	Water Pollution Control Act
1972	Marine Protection, Research and Sanctuaries Act
1972	Coastal Zone Management Act
1972	Home Control Act
1972	Federal Insecticide, Fungicide and Rodenticide Act Amendments
1972	Parks and Waterways Safety Act
1972	Marine Mammal Protection Act
1973	Endangered Species Act
1974	Deepwater Port Act
1974	Safe Drinking Water Act

Table 13a (continued)

1974	Energy Supply and Environmental Coordination Act
1975	Energy Policy and Conservation Act
1975	Federal Insecticide, Fungicide and Rodenticide Act Amendments
1976	Energy Conservation Act
1976	Toxic Substances Control Act
1976	Federal Land Policy and Management Act
1976	Resource Conservation and Recovery Act
1977	Clean Air Act Amendments
1977	Clean Water Act Amendments
1977	Safe Drinking Water Act Amendments
1977	Saccharin Study and Labeling Act
1977	Surface Mining Control and Reclamation Act
1977	Soil and Water Resources Conservation Act
1978	Energy Tax Act
1978	Uranium Mill Tailings Radiation Control Act
1978	Federal Insecticide Fungicide and Rodenticide Amendments
1978	Endangered Species Act Amendments
1980	Environmental Education Act
1980	Comprehensive Environmental Response, Compensation, and Liability Act
1980	Federal Insecticide, Fungicide and Rodenticide Act Amendments
1982	Nuclear Waste Policy Act
1984	Resource Conservation and Recovery Act Amendments
1984	Environmental Programs and Assistance Act
1985	Low Level Radioactive Waste Policy Amendments Act
1986	Safe Drinking Water Act Amendments
1986	Superfund Amendments and Reauthorization Act
1987	Clean Water Act Amendments
1987	Nuclear Waste Policy Amendments Act
1988	Federal Insecticide, Fungicide and Rodenticide Act Amendments
1990	Clean Air Act Amendments
1990	Pollution Prevention Act
1992	Energy Policy Act
1996	Food Quality Protection Act
1996	Safe Drinking Water Act Amendments

nately, these requirements continued to emphasize remediation of existing emissions; little attention was directed to pollution prevention.

One of the interesting aspects of the Clean Air Act amendments of 1990 is that they permit the buying and selling of air pollution emission allowances. The goal of such an arrangement is to encourage those industries that can remove pollutants at minimal cost to sell their polluting allowances to industries whose costs are higher.

Two approaches can be used to implement such a system. One is to set a limit on the total amount of pollution that can be released, enabling firms to trade their emission allowances at market prices. The other is to set a fixed cost per unit of pollutant that is discharged, with the total quantity released then adjusting accordingly. A strong advantage of the first approach, preferred by a majority of policymakers, is that it defines how much pollution can be released into the environment (for example, chemicals that lead to acidic deposition). In the case of pollutants that have worldwide implications, such as those that can cause depletion of the ozone layer or lead to global warming, implementation on an international basis is being considered. In fact, the United States has made such a proposal for controlling emissions of the greenhouse gases (Sun, 1990). Although the trading of discharge allowances is also permitted under the Clean Water Act (see below), the practice has seen fewer applications in this field.

The Clean Water Act

The Clean Water Act was passed in 1977 and amended in 1987 (P.L. 95-217; 100-4). This act was designed to restore and maintain the quality of waters throughout the nation, the ultimate goal being to provide "fishable and swimmable" rivers and streams. Like the Clean Air Act, the Clean Water Act required the development of both harm-based and technology-based standards. At the present time, technology-based standards predominate.

One of the basic requirements of the Clean Water Act was establishment of standards for the quality of water within the ambient environment. The act also set up a permitting mechanism, the National Pollution Discharge Elimination System (NPDES), through which the discharge of pollutants could be regulated to ensure compliance with the standards. The authority to establish the standards and to issue NPDES permits could be delegated by the EPA to individual states, provided they had developed a program substantially the same and at least as stringent as the federal program.

Through this process, authority for enforcement of the Clean Water Act has now been delegated to some 40 states.

For these states, one of the first steps in implementing the act is to designate the intended use of each body of water, that is, whether for drinking, swimming, fishing, or boating. The state then sets standards that ensure compliance with the intended use based on "criteria" established by the EPA. To assure compliance, the state also issues NPDES permits for individual "point" sources discharging into each water body. These permits include numerical limits for each specific discharge, and are designed to ensure that the aggregate of all permitted discharges into the given body will not yield pollution levels in excess of the water quality standards.

The Clean Water Act also requires that organizations having the potential for major accidental releases into the environment develop and maintain a plan describing the actions to be taken in the event such a release occurs. Operators of these plants must identify the people who will handle the incident and outline their specific duties and responsibilities as well as the associated reporting and record-keeping requirements. The act requires further that effective application of the plan be guaranteed by regular training.

The 1987 amendments to the Clean Water Act significantly changed the thrust of enforcement. Increased attention is now paid to the monitoring and control of toxic constituents in wastewater; and to discharges of polluted runoff from city streets, farmland, mining sites, and other "nonpoint" sources. The 1987 amendments also changed the NPDES program by requiring much stricter discharge limits and expanding the number of chemical constituents that must be monitored in pollutants that reach waterways.

Unlike the Clean Air Act, the Clean Water Act emphasizes a permitting system for the release of pollutants into lakes and streams. Polluters can apply for discharge permits that specify the required control technology, as contrasted to specifying limits on the concentrations of specific pollutants in the receiving bodies of water. Technology-based, as opposed to harm-based, standards are emphasized.

Safe Drinking Water Act

Closely tied to the control of the discharge of liquid wastes into rivers and streams is the provision of a safe drinking-water supply. The Clean Water Act, however, has been generally interpreted as applying only to surface waters. From this perspective it did not provide protection for one of the most important drinking-water sources—groundwater. This situation was partially remedied through passage in 1974 of the Safe Drinking Water Act (P.L. 93-523), the general objective being to ensure that public drinking-water supplies are free of potentially harmful materials and that they meet minimum national standards for protection of public health. The act includes regulations pertaining to the underground injection of liquid wastes.

Under the Safe Drinking Water Act, the EPA was authorized to establish national drinking-water quality standards, including the specification of maximum contaminant levels (MCLs) for specific substances in water. The MCLs must be set at a level that allows no known or anticipated adverse health effects. In complying with this law, the EPA has accepted the premise that there is no safe level of exposure to a carcinogen. At the same time, the agency has recognized that the costs of controlling human exposure must be reasonable (Merrill, 1986). As a result, some limits are harm based and others are technology based. The agency may either establish a maximum contaminant level for a specific substance or prescribe a technique for its control. In the former case, the limiting technology may be the ability to detect the contaminant in the water. When the technology for monitoring the level of a contaminant at the required sensitivity is readily available, establishment of an MCL is generally the preferred approach. Since the feasibility of achieving a specified MCL or of implementing a given treatment will change with advancing technology, the act requires the EPA to revise and update the regulations on a continuing basis.

Through passage of the 1977 amendments to the Safe Drinking Water Act (P.L. 95-190), Congress recognized the finite nature of the nation's water supplies and the need to assess present and future supplies and demands. It included requirements for an analysis of the projected demand for drinking-water, the extent to which other uses would compete with drinking water needs, the availability and use of methods to conserve water or reduce demand, the adequacy of present measures to assure adequate and dependable supplies, and the problems (financial, legal, or other) requiring resolution in order to assure the availability of adequate quantities of safe drinking water for the future. Of particular interest is the emphasis on conserving water and reducing demand; this approach is part of the growing recognition of the need to manage the nation's limited natural resources in a sustainable manner.

Also reflected in the Safe Drinking Water Act are concerns about the potential health effects of by-products that occur in drinking water as a result of the use of disinfectants such as chlorine. To address these concerns, the EPA was directed to study the reactions of chlorine with humic acid (a common natural ingredient in surface waters), and to evaluate the potential health effects, including any possible carcinogenic nature, of the new chemical products that result. Attention was directed to polychlorinated biphenyl contamination of actual or potential sources of drinking water; contamination of such sources by other substances known or suspected to be harmful to public health; the effects of such contamination; and means of removing, treating, or otherwise controlling the contamination.

The Safe Drinking Water Act was amended in 1986 and 1996 to specify additional contaminants to be regulated and acceptable treatment techniques for each such contaminant. The amendments further required disinfection of all drinking-water supplies, prohibited the use of lead products in drinking-water conveyances, and emphasized once again the need for protection of groundwater sources. A major stimulus for the 1996 amendments (P.L. 104-182) was the recognition that many water suppliers were not analyzing for some of the emerging contaminants, such as *Cryptosporidium.* The new amendments require that such analyses be performed, and that consumers be provided with data on the concentrations of contaminants in their supplies. The amendments also established a revolving fund to help state and local authorities pay for water-system improvements.

Pollution Prevention Act

As noted above, both the Clean Air Act and the Clean Water Act emphasize treatment or remediation, rather than prevention, of pollution. Recognizing this deficiency, Congress passed the Pollution Prevention Act of 1990 (P.L. 101-508). This act established a national policy to assure that pollution is prevented or reduced at the source, recycled or treated in an environmentally safe manner, and disposed of or released into the environment only as a last resort (BNA, 1993). It also created a clearinghouse to encourage the sharing and transfer of source reduction technology and provided financial assistance to states to promote pollution prevention. By advocating the substitution of less hazardous substances for those used in a wide variety of industrial processes, the law has led to significant reductions in worker exposure.

Solid Waste Disposal Act

The original law passed by Congress to address the problem of solid waste was the Solid Waste Disposal Act of 1965 (P.L. 89-272). It was supplemented shortly thereafter by the Resource Recovery Act of 1970 (P.L. 91-512), designed to promote the demonstration, construction, and application of waste management and resource recovery systems. This law was soon found to be inadequate, and Congress followed up by passing the Resource Conservation and Recovery Act (RCRA) of 1976 (P.L. 94-580; EPA, 1986). In essence, the RCRA was designed to regulate the storage, transfer, transport, and disposal of hazardous substances. The goals were quite sweeping: (1) to protect human health and the environment; (2) to reduce waste and conserve energy and natural resources; and (3) to reduce or eliminate the generation of hazardous waste as expeditiously as possible. As with the air and water pollution laws, implementation of the requirements imposed by the solid waste laws was to be handled at the state level.

To achieve these goals, Congress incorporated into the RCRA requirements for the implementation of three distinct yet interrelated programs. The first, outlined under subtitle D, encouraged states to develop comprehensive plans for the management of solid waste, with emphasis on those of a "nonhazardous" nature, for instance, household wastes. The second program, outlined under subtitle C, established a system for controlling hazardous waste from the time it is generated until its ultimate disposal (the cradle-to-grave approach). The third program, subtitle I, was designed to regulate certain underground storage tanks through the establishment of performance standards for new tanks and the development of methods for detecting, preventing, and mitigating leaks in existing facilities (EPA, 1986).

Recognizing the need for increased control over the types and forms of solid waste being sent for disposal, Congress passed the Hazardous and Solid Waste Amendments Act of 1984 (P.L. 98-616). One of its stipulations was that hazardous wastes must be treated in a prescribed manner prior to disposal, such treatment being designed to limit the releases of hazardous chemicals should water inadvertently come into contact with the wastes (Becker, 1991). The act also closed loopholes in previous rules that allowed toxic wastes to be burned in industrial and apartment furnaces (BNA, 1993). In addition, it made more stringent the requirements for the siting of new hazardous waste disposal facilities, and for the continued operation of existing facilities.

While these laws created a framework for the proper management of hazardous and nonhazardous waste, they did not address hazardous wastes encountered at inactive or abandoned sites or spills requiring emergency response. These problems were addressed by the next law to be considered.

COMPREHENSIVE ENVIRONMENTAL RESPONSE, COMPENSATION, AND LIABILITY ACT

The Comprehensive Environmental Response, Compensation and Liability Act (CERCLA) of 1980 (P.L. 96-510), better known as the "Superfund" law, was primarily directed to the cleanup (or restoration) of existing disposal sites that were producing unacceptable environmental releases. A major stimulus for its passage was the Love Canal episode, in which homes in New York State were constructed on lands that had previously served as a disposal site for toxic chemicals. At first it was thought that only a few hundred hazardous waste sites in the United States represented a problem; subsequent investigations, however, revealed that the total was in the thousands.

Under the CERCLA, the Environmental Protection Agency is required to collect data on sites subject to this statute through generation of a preliminary assessment report, followed by a site investigation. Based on the resulting information, the sites are ranked, using risk assessment procedures (see Chapter 16), according to their potential for causing human health impacts or environmental damage. The sites with the highest ranking are placed on what is called the national priority list (NPL) and are subject to mandatory cleanup actions, funded either by potentially responsible parties or by the allocation of Superfund monies derived through taxes on chemical feedstocks and crude oil supplies.

Certain aspects of the CERCLA, particularly its liability provisions, have proved to be extremely controversial. Liability is referred to as "retroactive, strict, and joint-and-several." This means, among other things, that both past and current property owners are liable for the cleanup of contamination created years before. A single producer who shipped a small quantity of waste to a site may be held liable for the entire cost of cleanup, including remediation of the larger volumes of wastes disposed of there by other producers, if the responsible parties cannot be found. Responsible parties may include current facility owners and waste generators and transporters, without any requirement that they be at fault (Boyd and Macauley, 1994). Section 107 of the Act however, establishes an "innocent purchaser" defense that can allow purchasers to escape liability. Under this provision a prospective buyer who conducts a reasonably careful inspection of the property prior to purchase and does not uncover any evidence of contamination, is not liable. And those who owned the property prior to its contamination are also not liable.

EMERGENCY PLANNING AND COMMUNITY RIGHT-TO-KNOW ACT

Signed into law as Title III of the Superfund Amendments and Reauthorization Act of 1986 (P.L. 99-499), the Emergency Planning and Community Right-To-Know Act is a freestanding law designed to address concerns about the effects of chemical releases on communities. The law includes three distinct programs. One, the emergency planning function, provides funds for communities to establish local emergency planning councils to work with industry officials in the development of disaster preparedness programs to cope with accidental releases of toxic chemicals. A second, the toxic substances registry, contains community right-to-know provisions that grant local emergency response personnel and the general public access to information on the chemicals present in local facilities. A third program, the toxic release inventory, requires companies to provide EPA and state officials with an annual accounting of toxic chemicals that are routinely released into the environment (BNA, 1993).

Radioactive Wastes

For purposes of classification, radioactive wastes in the United States have been divided into two categories: high level and low level. Each of these has been addressed through separate laws and regulations.

NUCLEAR WASTE POLICY ACT (HIGH-LEVEL WASTE)

Through the Nuclear Waste Policy Act of 1982 and the Nuclear Waste Policy Amendments Act of 1987 (P.L. 97-425, 100-203), Congress took what it considered to be positive steps to solve the high-level waste disposal problem. The program initiated by these laws has three essential components: (1) the design and construction of a geologic repository for permanent disposal of spent fuel from nuclear power plants and other high-level waste; (2) the establishment of a monitored retrievable storage (MRS) facility for temporary storage and packaging of spent fuel prior to placement in a repository; and (3) the development of a transportation system for moving the waste from its source to the MRS facility and ultimately to the repository. Congress subsequently specified that studies for a high-level waste repository be limited to the Yucca Mountain site in Nevada.

The Department of Energy (DOE) is conducting studies to determine whether the proposed site meets the basic technical and regulatory requirements. If the data indicate that Yucca Mountain can safely isolate high-level waste, the Nuclear Waste Policy Act requires that the DOE, with presidential approval, apply to the Nuclear Regulatory Commission (USNRC) for a license to construct the repository. The USNRC will then determine whether the proposed site meets federal regulations; if it does, construction will begin. As would be anticipated, many Nevadans oppose locating the proposed repository in their state. Similar controversies have surrounded the proposed construction of an MRS, including the fact that several Indian tribes have expressed interest in having the facility located on their lands.

LOW-LEVEL RADIOACTIVE WASTE POLICY AMENDMENTS ACT

For several decades only three facilities were available in the United States for the disposal of low-level radioactive wastes. These were located in Nevada, South Carolina, and Washington. None of the other 47 states had an active disposal site. As a result, the governors and the general public in these three states began to question whether they should continue to be used as a "dumping ground" for wastes from the other states.

In an effort to alleviate this situation, Congress passed the Low-Level Radioactive Waste Policy Amendments Act (LLRWPAA) of 1985 (P.L. 99-240). Each of the states, either individually or as part of a regional compact, was made responsible for the disposal of low-level radioactive wastes (LLRW) produced as a result of commercial or federally sponsored activities within its borders. To achieve this objective, the act established a series of milestones, penalties, and incentives to encourage additional states to move forward with the siting, construction, and operation of LLRW disposal facilities. Incorporated into the act were restrictions on the availability to the states of the existing disposal facilities, including limits on how long LLRW from outside states could be shipped to these facilities, penalties for unmet milestones, and volume restrictions and surcharges on wastes shipped to these facilities after certain intermediate dates.

In spite of these developments, progress in the establishment of new LLRW disposal facilities has been slow. The Nevada facility was closed on 31 December 1992, and only one new facility, in Utah, had been approved and placed in operation by 1995. Although several other states appear to be making progress, obstacles continue to arise. One of these is the public's opposition to the establishment of any new disposal facility. As a result, many waste generators are faced with the necessity of retaining (storing) their LLRW on-site until new disposal facilities can be established. This situation, coupled with the increasing fees being charged for disposal, has led to some perhaps unexpected benefits. One is an accelerated effort to minimize the volume of waste through prevention strategies and technologies as well as through incineration and compaction. As a result, the quantities of LLRW generated in recent years are far below what they were a decade ago.

Toxic Substances Control Act

Prior to 1976, toxic substances were regulated under a haphazard array of statutes that focused on other issues. Through passage of the Toxic Substances Control Act (P.L. 94-469), Congress sought to provide a comprehensive system of control. Though cumbersome and rarely used, the TSCA is a powerful tool in the regulatory arsenal. One of its provisions was to provide federal agencies with adequate authority to prevent unreasonable risks of injury to health or the environment. Having said this, however, the act went on to caution that this authority should be exercised in such a manner as "not to impede unduly or create unnecessary economic barriers to technological innovation while fulfilling the primary purpose of this Act." Obviously, the success of its implementation depends heavily on the relative weights given to these conflicting goals (Findley and Farber, 1992).

Under terms of the act, any company planning to manufacture or import a new chemical must submit to the EPA a premanufacturing notice containing information on the identity, use, anticipated production or import volume, and disposal characteristics of the substance. Also to be reported are any hazards to which workers may be subjected in its manufacture and handling. In addition, the act gives the EPA broad powers to collect information about chemicals, including their production and use. Chemicals used exclusively in pesticides, foods, food additives, drugs, and cosmetics are exempted, since these are covered by other federal laws.

The EPA is permitted to restrict the manufacture, processing, distribution, use, or disposal of any chemical substance if there is a "reasonable basis" to conclude that such activity poses an unreasonable risk of injury to health or the environment. If the EPA has reason to suspect that a chemical may pose such a risk, but lacks sufficient data to take action, it can require the manufacturer to develop the necessary data. The act also stipulates that manufacturers provide the EPA, 90 days prior to initial production of a new chemical, with data relevant to its potential impact on health and the environment (Merrill, 1986).

The requirements of the Toxic Substances Control Act with regard to the export of toxic chemicals are particularly interesting. A chemical intended for export can be regulated by the EPA only if it presents an unreasonable risk to the health of people or to the environment of the United States. The EPA is responsible, however, for notifying the governments of countries importing the chemicals of any associated U.S. regulatory restrictions. In contrast, no chemical substance, mixture, or article containing such materials may be imported into the United States if it fails to comply with any U.S. rule or is otherwise in violation of the Toxic Substances Control Act. These types of issues have had important implications in recent years in the deliberations on international agreements such as the North American Free Trade Agreement.

With the passage of time, those responsible for enforcement of the TSCA have had to balance two competing concerns: (1) that exposure to toxic substances can reduce the quality of public health and the environment, and (2) that federal legislation, if improperly enforced, could seriously discourage industry from introducing new chemicals and implementing new concepts and approaches. From a risk perspective, the overall objective is society's optimal use of toxic chemicals (Graham and Gray, 1993). The primary goal should be to emphasize early identification of hazards, before the chemicals are introduced for widespread use (Greaves, 1992). Such an approach implements the basic principles of public health practice, that is, to prevent problems before they develop rather than to exercise control only after they have reached a crucial stage.

Federal Food, Drug, and Cosmetic Act

Enacted in 1938, the Federal Food, Drug, and Cosmetic Act is one of the oldest of the major health regulation laws. It replaced the Pure Food and Drug Act of 1906 and remains a cornerstone of U.S. food safety policies. The act covers foods for humans and animals, human and veterinary drugs, medical devices, and cosmetics (Merrill, 1986). The original act included two prohibitions that addressed foods containing hazardous substances, restrictions that continue to apply today. The first forbade the marketing of food containing "any added poisonous or deleterious substance which may render it injurious to health," a provision that the Food and Drug Administration (FDA) has interpreted as barring any foods that present a serious risk. The second forbade the marketing of foods containing naturally occurring toxic substances; however, permission was granted to continue the marketing of foods that, even though known to contain such substances, have traditionally been a part of the American diet. Premarketing approval was not required by either provision (Hutt and Merrill, 1991).

The Food, Drug, and Cosmetic Act was revised with the passage of the 1958 Food Additives Amendment, which required that the safety of substances classified as "food additives" be demonstrated prior to marketing. The critical standard for approval was confirmation that the substance is "reasonably certain to be safe." No inquiry into the benefits of an additive was undertaken or authorized. The amendment, however, specifically exempted substances that are "generally recognized as safe" (GRAS) by qualified scientific experts. In the main, such substances are those that have been in use for many years without observed adverse health effects. If new evidence raises doubts about the safety of a GRAS ingredient, it becomes a "food additive," whose use requires approval (Merrill, 1986). The 1958 amendment also exempted food ingredients that had been "sanctioned" by the FDA or by the Department of Agriculture prior to 1958. These included some controversial substances such as sodium nitrite, traditionally used in curing certain meat products such as bacon.

Any review of laws regulating food would be incomplete without mention of the Delaney clause, which Congress added to the Federal Food, Drug, and Cosmetic Act in 1958. Recognizing that cancer was a major and widely feared health problem, this clause stated that "no additive shall be deemed safe if it is found to induce cancer when ingested by man or animal." The result was a multitude of subsequent debates on how much of a chemical shown to be carcinogenic when administered in large quantities to laboratory animals should be considered acceptable as an additive to food. As exemplified by the experience with saccharin, these debates extended well beyond those of a technical or scientific nature. In accordance with the Delaney clause, the FDA proposed to ban the use of this chemical, only to have Congress pass the Saccharin Study and Labeling Act of 1977, which permitted this potential carcinogen to be added to food provided it was accompanied by a label warning the consumer of the potential health risk.

These and similar controversies prompted the passage in 1996 of the Food Quality Protection Act (P.L. 104-170), which repealed the Delaney clause and now permits residues of carcinogenic pesticides in food as long as there is "reasonable assurance of no harm" to consumers. This law also requires pesticide manufacturers to demonstrate that their products are safe for infants and children. Although this law eased regulations on processed foods in some instances, it tightened regulations on raw foods.

Federal Insecticide, Fungicide, and Rodenticide Act

Closely related to the Food, Drug, and Cosmetic Act is the Federal Insecticide, Fungicide, and Rodenticide Act (FIFRA), first enacted in 1947 and substantially amended in 1972 (P.L. 92-516). The FIFRA was again amended in 1975, 1978, 1980, and 1988.

Under this act, no pesticide may be marketed unless it has been registered by the EPA, that is to say, that the agency has reviewed and evaluated data on the pesticide and has determined that it does not present an unreasonable risk to health or the environment. As part of this process, the EPA must establish tolerance levels for any residue in or on both the raw agricultural commodity and the processed foods. Such decisions are to take into account both economic factors and consumer safety—unless a carcinogenic pesticide residue concentrates in processed foods, in which case the EPA must follow a strictly health-based standard (Hutt and Merrill, 1991).

The EPA engendered considerable controversy in the early 1970s by canceling registration for a number of pesticides that had previously been approved for use on agricultural crops. These included DDT, aldrin, and dieldrin; the cancellations were based primarily on studies suggesting that these chemicals were carcinogenic in animals. The criticisms that accompanied these decisions led to important changes in the FIFRA and in the way in which it was implemented (Merrill, 1986). Amendments adopted in 1975 required EPA to submit proposed pesticide cancellations to a scientific panel for review (BNA, 1993). As part of this process, the EPA established a procedure for public comment on the risks and benefits of specific pesticides prior to initiation of the formal cancellation process (Merrill, 1986). Additional amendments passed in 1978 assigned the states primary responsibility for enforcing the law in cases involving violations in the use of pesticides.

Occupational Safety and Health Act

The basic law governing conditions in the workplace is the Occupational Safety and Health Act of 1970. Its goal is "to assure so far as possible every working man and woman in the Nation safe and healthful working conditions and to preserve our human resources." Two of the basic advances provided were the establishment within the Department of Labor of the Occupational Safety and Health Administration (OSHA), and the establishment within the Department of Health and Human Services of the National Institute for Occupational Safety and Health (NIOSH). Among the responsibilities assigned to OSHA was the authority to conduct workplace inspections and to issue citations and impose fines where serious violations are observed. In addition, OSHA was assigned the task of working with the states to develop and implement improved occupational health programs. Accordingly, it established a program to review and approve the occupational safety and health programs within the states; the result has been a shifting of responsibility for primary enforcement from the federal to the state level. As in the control of air and water pollution, OSHA requires that the regulations imposed by the states keep pace with, and be as effective as, those adopted at the federal level.

The tasks assigned to NIOSH were to conduct research on safety and health problems, to provide technical assistance to OSHA, and to recommend standards for adoption by OSHA. As part of this assignment, NIOSH was instructed to make workplace investigations, gather testimony from employers and employees, and require that employers measure and report employee exposure to potentially hazardous substances (Bingham, 1992). It was also empowered to require employers to provide medical examinations and tests to determine the incidence of occupational illnesses among their employees (OSHA, 1980).

Once OSHA has evaluated the recommendations of NIOSH in the setting of standards and has developed plans to propose, amend, or delete a standard, these intentions are published in the Federal Register as a "Notice of Proposed Rulemaking." This format, mandated by the Administrative Procedures Act, was designed to provide all interested parties, including employers, employees, and members of the public, an opportunity for input into development of the standards. Unfortunately, emphasis during the early years was narrowly focused on standards for worker safety; only in recent years has OSHA begun to pay attention to standards for the control of occupational illnesses.

Although considerable progress has ensued since the Occupational Safety and Health Act became law, there is a limit to what can be accomplished through the minimal enforcement efforts being conducted under this act (Bingham, 1992). A mere 2,000 inspectors at the federal and state level are responsible for overseeing almost 6 million work sites. Although inspections, followed by citations when violations are observed, will remain an integral part of the program, continuing improvements in worker safety and health will require cooperative efforts by employers and employees. In the long run, improved worker health and safety will be to their mutual benefit.

Atomic Energy Act

The Atomic Energy Act (AEA) of 1954 established the Atomic Energy Commission, which was later replaced by the Nuclear Regulatory Commission and the U.S. Department of Energy (DOE). Under this act, the USNRC is responsible for licensing the transfer, manufacture, acquisition, possession, or use of any nuclear facility, and for regulating radioactive materials and their by-products (USNRC, 1991). Through the AEA, the USNRC is authorized to enter into agreements with any state whereby the state will regulate nuclear materials within its borders, with certain exceptions such as the construction and operation of nuclear power plants. At the present time, more than half of the states have entered into such agreements.

The USNRC is required by law to ensure that radioactive materials and related facilities are managed so as to protect the public health and the environment. Regulations developed by the USNRC to meet these responsibilities must be in accord with standards developed by the Environmental Protection Agency. Another law closely related to the control of radioactive materials is the Uranium Mill Tailings Radiation Control Act, which requires the DOE to designate and assign priorities for the remediation of wastes at former uranium mill tailings processing sites. Under this act, the DOE selects and performs such remedial actions with the concurrence of the USNRC and in accordance with standards set by the EPA. Additional relevant laws include the previously cited Nuclear Waste Policy Act and the Low-Level Radioactive Waste Policy Act (USNRC, 1991).

Other Environmental Laws

A number of other laws pertain to environmental protection, for instance, those concerning energy policy, flood plains and wetlands, and endangered species.

In the 1970s, Congress passed several laws designed to encourage energy conservation. The Energy Policy and Conservation Act of 1975, for example, promoted disclosure of efficiency ratings for appliances. The Energy Conservation Act of 1976 encouraged energy conservation measures in new buildings, and the Energy Tax Act of 1978 provided tax incentives for energy-efficient automobiles and residences. Many states have adopted similar measures (Findley and Farber, 1992).

A recent piece of legislation that holds promise for significant improvement in the quality of the environment is the Energy Policy Act of 1992 (P.L. 102-486). One of its major provisions is the establishment of energy-efficiency standards for government buildings and for offices and private residences. Such standards apply to lighting, appliances, heating and air-conditioning systems, and plumbing products (such as showerheads and toilets). Other components of the act require new approaches to improved efficiency in transportation, including the development of alternative-fueled and electric vehicles, and increased use of such vehicles by federal agencies. Also covered are promotion of renewable energy sources, including those of solar and geothermal origin and advanced applications of nuclear fission; development of waste-to-energy technologies; improved control of pollutants from coal-burning industries; and reduction of emissions that could potentially deplete the ozone layer and/or affect long-range climate change.

Executive Orders 11988 and 11990, issued by the President in 1977, outline federal policies related to managing floodplains and the protection of wetlands. The former requires federal agencies insofar as possible to avoid adverse impacts associated with the occupancy and modification of floodplains. Any impacts must be fully assessed and documented as required under NEPA. Before a project proposed for construction in a floodplain can be implemented, the agency must demonstrate that there is no reasonable alternative to the given location.

Executive Order 11990 requires that federal agencies identify and reduce potential impacts on wetlands resulting from proposed activities. Where impacts cannot be avoided, action must be taken to minimize the damage by repairing the harm or replacing the wetlands with an equal or greater acreage of man-made wetlands that are as similar to the original as possible.

The Endangered Species Act of 1973 is designed to protect plant and animal resources from the adverse effects of development. Under its provisions, no one may kill, capture, or harm an endangered species. Federal and state agencies, industrial organizations, and related groups are required to assess proposed projects to determine if any threatened or endangered species, or critical habitat of these species, exists and will be affected. If no such species or habitat is present, that fact must be documented in a letter to the Fish and Wildlife Service. If such species or habitat is found to exist, the Fish and Wildlife Service is to be notified; a series of consultations and studies will then be carried out to determine the extent of the impact and any special actions that must be taken to minimize this impact.

In 1978, Congress added an exemption procedure establishing a Cabinet-level review board to rule on the impacts of federal projects. The board can permit a project to cause a species to pass into extinction if: (1) that project is of regional or national significance; (2) there is no "reasonable and prudent alternative"; and (3) the project as proposed "clearly outweighs the alternatives." One recent outgrowth of the Endangered Species Act has been a move to enact legislation to create a National Biological Survey, which would assign to the Department of the Interior the responsibility to inventory every animal and plant species in the United States.

Closely related to the Endangered Species Act are the Marine Protection Act, which makes it unlawful, except under certain special circumstances, to take, possess, or trade a marine mammal or marine mammal product; the Bald and Golden Eagle Protection Act, which makes it a criminal offense to pursue, wound, capture, kill, or disturb these birds; and the Wild Free-Roaming Horses and Burrow Act, which places these animals under the jurisdiction and protection of the Department of the Interior.

Summarized in Table 13.1 earlier in the chapter are some of the more important laws that relate to environmental and occupational health.

The General Outlook

In spite of the efforts of Congress to provide laws covering all facets of the environment, voids remain. One is the lack of requirements for preliminary assessment of the exposures of workers to new chemicals. While the FDA and EPA can authorize premarket approval of food additives, drugs, and pesticides, no employer is obligated to obtain prior approval of new processes or materials, or to conduct tests to assure that changes in operations will not jeopardize worker health. Only if it is discovered that a material already in use threatens the health of workers may the Occupational Safety and Health Administration take action (Merrill, 1986).

Another void is the lack of comprehensive legislation on indoor pollution, particularly inside the home. Although Congress has passed laws relating to radon and lead paint (P.L. 102-550), only in recent years has this body begun to address such problems on a broader basis. Only comparatively recently were the problems of indoor pollution recognized by the environmental health profession itself. The long history of the sanctity of the home has conspired too against such regulation.

In recent years Congress has increasingly called on federal agencies to develop risk-based systems for the assessment and control of environmental and occupational health hazards. Regulations developed by various agencies to implement federal laws, however, frequently convey different levels of concern about risks to human health from various environmental stresses and the weight that should be given to the costs of control. The problem is being addressed by interagency task forces seeking to unify the acceptable levels of risk, but progress has been slow.

Responsibility for implementing and enforcing many of the environmental laws passed by Congress rests with the individual states. In most cases, regulations adopted by the states will be approved by federal agencies as long as the state requirements are at least as stringent as those endorsed by the federal government. In this regard it is interesting to consider the regulation of airborne emissions from automobiles in California. Because of the unique nature of air pollution problems in the Los Angeles basin, environmental regulators in that state imposed requirements far more stringent than those proposed by EPA. Although automobile manufacturers complained vociferously at the time, the net result was that the proposed standards were met, and ultimately the whole country benefited.

It is encouraging that Congress has addressed as many environmental concerns as it has. The problems cited above relate primarily to the inner workings of the laws, not to the basic goals they are designed to achieve. Another heartening sign is Congress' increasing recognition of the need to support programs designed to minimize the production of waste and to conserve our natural resources. This is exemplified by key features in RCRA, the Clean Air Act, and the Clean Water Act—and by the emphasis being placed on pollution prevention. In many cases pollution prevention/waste minimization is a win-win-win situation for government, industry, and the public. Beyond mere protection of the environment, society as a whole benefits through reduced capital and operating costs, decreased liabilities, cleaner and safer working conditions, conservation of energy and material resources, and the opportunity for government and industry to work together cooperatively.

14

STANDARDS

MANY organizations have developed guidelines, recommendations, and standards for limiting exposures to a variety of occupational and environmental contaminants. The Environmental Protection Agency, through its national ambient air quality standards, has established limits for airborne contaminants in the outdoor environment; it has also set standards for various contaminants in drinking water and in rivers and streams. Standards for limiting exposures to contaminants in food have been developed by the Food and Drug Administration. Guidelines for limiting exposures to chemical and physical stresses in the workplace have been developed by the American Conference of Governmental Industrial Hygienists. Standards for environmental and occupational exposures to ionizing radiation have been set by the EPA, and the Nuclear Regulatory Commission and the Occupational Safety and Health Administration have promulgated regulations to assure that the standards will be met. Guidelines for protection against nonionizing radiation have been set out by the International Radiation Protection Association and the International Commission on Non-Ionizing Radiation Protection.

Unfortunately, because of the multitude of organizations involved, there is no uniform methodology for dealing with the host of considerations that must be taken into account in the development of occupational and environmental standards. Such considerations include the approaches used to develop the standards, the underlying scientific bases, the associated goals for limiting the risks to the public and the environment, and the application of the standards, once developed.

Factors to Be Considered

CURRENT APPROACHES

The basic goal of many environmental standards is to protect human health. The associated guidelines are generally referred to as primary standards. Sometimes other affected entities must be considered as well; those associated guidelines are called secondary standards. In the case of air pollution, the secondary standards are aimed at protecting agricultural crops and property, such as buildings and statues; in the case of drinking water, the secondary standards assure the aesthetic qualities of the product, such as temperature, color, taste, and odor. Interestingly, secondary standards for many contaminants in air and water are more stringent than the primary standards.

One of the first requirements in the development of primary standards is to identify or define the exposed member who is to be protected. Is it an adult, a child, an infant, or a fetus? Aware of this need, the International Commission on Radiological Protection in 1994 published age-dependent guidance for use in protecting members of the public (ICRP, 1994a). A specific consideration in the development of occupational exposure standards is the need to protect pregnant women.

Once these types of decisions have been made, the next step is to decide whether the standard should protect the maximally exposed or the most susceptible member of a given group. In the case of the general population it may be difficult to identify such an individual and determine his or her exposure. For this reason some organizations have recommended that standards be based on protecting an average member of the "critical group," that is, the group that because of its location or living habits will be most heavily exposed. This group may be real, in which case the living habits of the members are known or predictable, or the group may be hypothetical, in which case the habits may be assumed, based on observations of similar groups elsewhere. The dose to an individual within the critical group is then assumed to be that received by a typical member of the group. This approach has the advantage of ensuring not only that members of the public do not receive unacceptable doses but also that decisions on the acceptability of a given practice are not prejudiced by a small number of individuals with unusual habits.

Once the basic standards have been developed—for example, by placing limits on the permissible doses to members of the public—derived or tertiary guides must be established for determining through monitoring programs whether the basic standards are being met. These guides may include limits on the intakes of individual contaminants by members of the exposed population, and/or limits on the concentrations of individual contaminants within various environmental media (air, water, food). Derived guides can also be developed on the basis of allowable releases of specific contaminants into the environment via the airborne or liquid pathway.

Refinements and specificity can be added by setting limits on the inhalation of contaminants based on the size of the airborne particles: lower limits may be specified for certain sizes within the respirable range. This is the approach used today for setting limits on airborne particles in the ambient environment. Where the health effects of a contaminant tend to depend on the short-term concentration (as is the case for ozone), it may be necessary to set limits on hourly concentrations rather than on cumulative intake (Lefohn and Foley, 1993).

In certain cases it is difficult, expensive, and/or time-consuming to quantify the presence of a specific contaminant in a given medium (say, viruses and other disease organisms in drinking water). To circumvent this problem, current standards for drinking water are based primarily on limits on the concentrations of coliform organisms, which serve as surrogates. Although not normally a source of disease, these organisms are frequently present in the human intestines, and their identification in drinking water is an indicator of the possible presence of fecal matter. In a similar manner, it is sometimes easier to assure containment, as in a waste disposal facility, through requirements directed to features other than those in question. For example, one of the concerns of the Nuclear Regulatory Commission is that interacting water will leach radionuclides from wastes in a disposal facility. Rather than specifying limits on leachability, the USNRC sets specific requirements for the structural stability of the wastes, the assumption being that radionuclides in such wastes will not be readily leached.

LIMITATIONS

The underlying assumptions and scientific bases for occupational and environmental standards are subject to a host of limitations. The following are some of the more prominent.

1. In only a very few instances have dose-response data been developed with a view to setting standards. Most data are by-products of

descriptive and analytic studies designed to test specific scientific hypotheses (AAEE, 1983).

2. Even where such data are available, they contain a range of uncertainties—not only those commonly encountered in the study of any biological system, but also the uncertainties involved in extrapolating data from animals to humans.
3. Few standards take into account differences in the weight, size, diet, and lifestyle of various population groups (for example, Japanese versus Americans). One solution might be to express the total intake limit for a given contaminant in terms of body weight.
4. Despite a consensus that standards for the general public should be much more stringent than those for workers, neither the standards nor the agencies responsible for setting them coordinate with one another. Moreover, standards limiting concentrations of airborne contaminants inside dwellings and office buildings are essentially nonexistent.
5. Some standards do not apply to all sources of a given contaminant or physical factor. For example, guidelines for acceptable radiation doses to workers and the general public do not include exposures from natural background radiation and from medical applications.
6. Standards for the ambient environment and for protection of the public are commonly set for individual contaminants in specific environmental media—air, water, food, and soil—even though it is the total intake of the contaminant that is critical.
7. In some instances, such as exposures to electric and magnetic fields, definitive data or sound epidemiologic evidence on which to base standards have been lacking. Yet public pressure and the fact that exposures are occurring make standards necessary even when their basis is suspect. Too often the public neither appreciates nor accepts the tentative nature of such standards.
8. Many standards, such as those for controlling airborne contaminants in the workplace or nonionizing radiation in the occupational and ambient environments, assume the existence of a threshold for associated health effects, even though many public health experts believe that most environmental stresses have some impact, however low the dose.

9. Usually the risks associated with exposures as expressed by the limits in the various standards have not been quantified. Unless they are, it is not possible to compare the stringency of, or the protection afforded by, standards developed for different environmental contaminants or stresses.
10. Except in the case of air and water, few limits have been derived for protection of the natural environment, property, or aesthetic features. Similar considerations need to be applied to other environmental stresses.
11. Few standards consider the effects of exposures to a *combination* of occupational and environmental contaminants and stresses. Simultaneous exposure to two contaminants often has synergistic effects; that is, the combination has a greater effect on health than the sum of the two exposures independently.
12. Certain of the standards developed for radiation protection include limits on the exposures to individuals as well as on the total integrated dose that the population as a whole can receive. Limits are placed on both the individual and the societal impacts of the given radiation source. A similar approach might usefully be applied to other occupational and environmental contaminants.

OTHER PROBLEMS AND DIFFERENCES

As an outgrowth of these limitations and the lack of a uniform system for establishing environmental standards, various federal agencies use different technical assumptions, different calculational methods for estimating dose and risk, different approaches on the degree of conservatism to be incorporated into the calculations, different depictions of how human exposure may occur, and different assumptions on the period over which protection is to be provided. Furthermore, federal agencies incorporate different concepts of how the standards will be interpreted and applied.

Two examples are the "top-down" regulatory strategy adopted by some agencies and the "bottom-up" strategy adopted by others. The former involves setting an upper bound or limit, then reducing the limit (on the basis of site-specific compliance) to a reasonably achievable lower level. The Nuclear Regulatory Commission and the Department of Energy have consistently favored this approach in the development of radiation protection standards (GAO, 1994). Conversely, the bottom-up strategy has been used to control a wide variety of other environmental exposures. It involves initially setting a lower, relatively stringent dose or risk goal, the understanding being that the goal is to be considered a desirable target, not a limit. If the goal is not achievable on the basis of technical feasibility, cost, and other factors, the regulatory agency may decide to accept a less stringent level (Robinson and Pease, 1991). This strategy is reflected in certain EPA regulations, such as those pertaining to toxic air pollutants and to the cleanup of contaminated soil and groundwater (Buonicore, 1991). With two such opposite strategies, standards that in the end result in comparable risk control may, on the surface, appear to be quite different.

In deciding on acceptable human exposure levels, federal agencies have increasingly supplemented numerical dose and risk limits with considerations of economic, social, technical, and other factors. The incorporation of economic factors illustrates the complexity and potentially controversial nature of such evaluations. Cost-benefit analyses may require consideration not only of the risks and costs of serious health effects but also of less quantifiable social factors: ethical concerns, equitable sharing of costs and benefits, perceived public aversion to the given contaminant at any exposure level, and the costs and benefits that could accrue to those defined as outside the at-risk population. Agencies making such evaluations are apt to be criticized if their cost-benefit analyses are perceived to be inequitable or to have unfairly placed a monetary value on human lives.

Taken together, the existing environmental and occupational standards reflect a lack of interagency consensus on the amount of risk that is acceptable to the public. Because the standards have different regulatory applications and are based on different technical methodologies, the estimated risks vary substantially. These problems, coupled with the differences in acceptable limits for human exposure set by various federal agencies, raise questions about the precision, credibility, and overall effectiveness of federal standards and guidelines in protecting public health. Clearly, interagency guidance is needed to derive a structured approach to the process of incorporating cost and benefit considerations into various protective strategies (GAO, 1994).

Radiation Protection Standards

To provide insights that may be useful in developing occupational and environmental standards, the current approaches relative to protection against ionizing radiation are reviewed in the sections that follow. A framework has been established for coordinating radiation standards throughout the world, and a risk-based approach employed to establish dose limits. The resulting system permits partial and whole-body exposures to be compared on the basis of relative health risk (morbidity and mortality).

THE BASIS FOR DOSE LIMITS

Shortly after the discovery of X rays in 1895, radiation injuries began to be reported. Recognizing the need for protection, physicists recommended limits on the allowable doses from X-ray generators. Their primary concern was to avoid direct physical symptoms. As early as 1902, however, scientists suggested that radiation exposures might also have latent effects, such as the development of cancer. This hypothesis was confirmed for external sources during the next two decades, and for internally deposited radionuclides by 1930, when bone cancers were reported among workers who applied luminous paints containing radium to the dials of clocks and watches (Eisenbud, 1978).

As methods to control radiation were formulated, and as people learned more about the potential health effects, dose limits inevitably were reduced. Initially, recommendations were developed on an informal basis. This system changed in 1928 with the establishment of the International X-Ray and Radium Protection Committee (known today as the International Commission on Radiological Protection, or ICRP) and the U.S. Advisory Committee on X-Ray and Radium Protection (known today as the National Council on Radiation Protection and Measurements, NCRP). The international committee and its successor have provided a forum in which radiation protection experts from throughout the world can meet regularly, discuss the latest information on the biological effects of ionizing radiation, and formally propose appropriate radiation protection standards. The ICRP and its scientific committees continue to formulate guidelines and recommendations on a wide range of occupational and environmental radiation protection matters. Since there are firm ties between the ICRP and the NCRP, their recommendations tend to be in close conformity (Moeller, 1990; NCRP, 1993).

At about the same time as these two organizations were formed, Hermann J. Muller's report on his experiments with *Drosophila* flies (Muller, 1927) aroused concern about possible hereditary effects of radiation exposure in humans. This consideration shaped radiation protection guidelines and standards from the end of World War II until about 1960. The initial focus was on dose limits for occupational exposures. As public concern increased about worldwide exposures through fallout from atmospheric weapons tests, attention shifted to dose limits for the public. In 1959 the Federal Radiation Council was created to develop U.S. policy on human radiation exposure and to establish dose limits for the public (FRC, 1960).

Epidemiologic studies were showing an increase in leukemia among the survivors of the World War II nuclear bombings of Japan, but failed to demonstrate the anticipated hereditary effects. As a result, somatic effects, primarily leukemia, became the basis for radiation protection standards. In 1970 the Committee on the Biological Effects of Ionizing Radiation (BEIR) reported that solid tumors (cancers of the lung, breast, bone, and thyroid), not leukemia, were the dominant effects of human exposures to ionizing radiation (BEIR, 1972). Ever since, solid tumors have remained the primary basis for the development of radiation standards (BEIR, 1980, 1990).

In 1977 the ICRP broke new ground by proposing a mathematical system that permitted radiation protection standards to be based on what was considered to be an acceptable level of risk (ICRP, 1977). The approach was refined in 1991 and the ICRP recommendations were later endorsed by the NCRP (ICRP, 1991; NCRP, 1987, 1993). The evolution of radiation protection standards is summarized in Table 14.1.

The ICRP recommendations for occupational whole-body external dose limits for the years 1934–1990 are summarized in Table 14.2. Many of these

Table 14.1 Evolution of the basis for dose limits for ionizing radiation, 1900–1990

Approximate period	Protection criteria
1900–1930	Avoidance of immediate physical symptoms
1930–1950	Avoidance of longer-term biological symptoms, plus concern for genetic effects
1950–1960	Concern for genetic effects (and leukemia)
1960–1970	Concern for somatic effects (primarily leukemia)
1970–	Concern for somatic effects (primarily solid tumors)
1980–	Application of a risk-based approach to radiation protection standards, taking into account latent cancer mortality
1990–	Expansion of risk approach to include additional effects, such as loss in life expectancy and effects of morbidity

Table 14.2 ICRP recommendations for occupational whole-body equivalent dose limits, 1934–1990

	Dose limit		
Year	Per day	Per week	Per year
1934	0.2 roentgen		72 roentgens
1950		0.3 roentgen	
1958		0.1 rem	5 (N − 18) rem[a]
1965			5 rem (maximum)
1977			50 mSv (5 rem); based on acceptable risk[b]
1990			50 mSv (5 rem); maximum of 100 mSv (10 rem) in any 5 years

a. N is age of worker receiving exposure.
b. Average dose to all workers recommended not to exceed 10% of limit for individual workers.

recommendations, and those of the NCRP, have been incorporated into USNRC regulations (USNRC, 1991). As the table shows, the ICRP currently recommends a dose-rate limit of 50 mSv per year; however, no individual worker should receive more than 100 mSv over any five-year period. This allowance is comparable to the NCRP's recommendation that the occupational dose never exceed 10 mSv times a worker's age (NCRP, 1993): under either system, a worker who was first exposed to ionizing radiation at age 20 would be permitted to have accumulated 0.4 Sv by age 40.

Over the years the levels of occupational radiation exposure considered acceptable have been reduced significantly (Figure 14.1). Except in the very early years, these reductions have occurred as a result not of observed health effects but of improved control technology and better understanding of the risks associated with radiation exposure.

CURRENT STANDARDS

The present system of radiation protection is based on the following general principles (ICRP, 1991): the application must be justified, that is, it must have a positive net benefit; the application must be optimized, that is, all exposures must be kept as low as reasonably achievable, with economic and social factors taken into account; and doses to individuals must not exceed established limits.

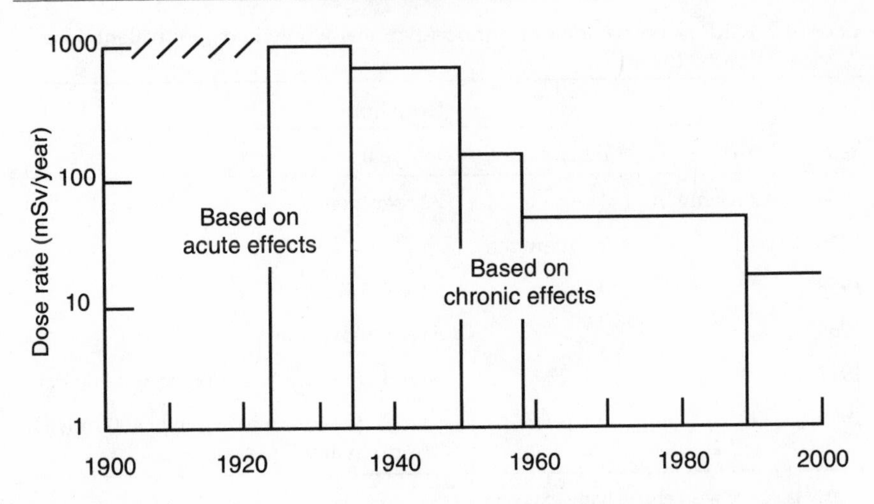

Figure 14.1 Change in limits for occupational exposures to ionizing radiation, 1900–2000. The limits since 1925 are based on recommendations of the ICRP and its predecessors

DOSE LIMITS

Column 2 of Table 14.3 summarizes the current recommended dose limits for radiation workers. Dose limits are provided for short-term (one-year) as well as for longer-term (five-year) exposures, and they apply to the sums of the doses received from both external and internal exposures. The standards are specified in units of the effective dose, a quantity that expresses partial-body doses in terms of equivalent doses to the whole body. This approach permits the addition, on an equivalent-risk basis, of partial and whole-body doses.

For a variety of reasons, dose limits for the general public are set lower than for radiation workers (NCRP, 1971). The population includes pregnant women, infants and children, the sick, and the elderly, each of whom may represent a group at increased risk; for example, their metabolism and breathing rates may be different from those of an adult radiation worker. Furthermore, members of the public may be exposed 24 hours a day, 7 days a week, for their entire lifetime, whereas exposures of workers are generally limited to their working (adult) lifetime and presumably only to the time they are on the job. Members of the public also have no choice about their exposure to most environmental sources, and they may receive no direct benefit from that exposure.

Table 14.3 Occupational and population dose limits

		General population	
	Radiation workers	Individuals	Total population
Annual limit	50 mSv	5 mSv	
Cumulative limit	20 mSv per year averaged over any 5 years	1 mSv per year, averaged over any 5 consecutive years	<< 1 mSv per year
Relative magnitude	100%	5%	<<5%

Many individuals are exposed to risks in their own occupations; those who are radiation workers may already be receiving significant exposures on the job. Finally, members of the public are not subject to the selection, supervision, and monitoring afforded radiation workers. Even when doses are sufficiently low that the risk to the individual is extremely small, the collective risk in the population may justify additional limitations.

Current radiation dose limits recommended for the general population are summarized in Table 14.3, columns 3 and 4. Although the NCRP would permit individuals to receive up to 5 mSv in a single year, it concurs with the ICRP that over the longer term the whole-body dose rate should not exceed 1 mSv per year. The average annual dose to the population as a whole is expected to be well below these values.

THE CONCEPT OF THE EFFECTIVE DOSE

The ICRP has used the concept of the effective dose to develop a system that provides a unit for expressing radiation protection standards for both whole-body and partial-body exposures on an equal-risk basis (ICRP, 1977, 1991). One of the benefits is that internal exposures can be converted to equivalent whole-body doses and summed with external exposures to calculate total dose.

In developing this system, the ICRP based the limits on the total risk (including both mortality and morbidity) to all tissues and organs and on any associated hereditary effects; it considered, in the case of internally deposited radionuclides, not only the dose occurring during the year of intake but also the dose resulting from the continuing presence of this material in the body. This is referred to as the committed dose. For workers, the time span over which the committed dose is assessed is 50 years, based on the assumption that the exposed person begins work at age 20 and dies at age 70. For members of the public, the committed dose is assessed over a time span of 70 years (NCRP, 1993).

Parameters taken into account in developing this system include the probability that radiation will induce a given type of cancer or hereditary effect, the probability that that cancer or effect will be lethal (the cancer lethality fraction), the years of life that will be lost as a result of the death, and the associated health-care and psychological impacts. These parameters were subsequently incorporated into what are called tissue-weighting factors, which provide a mechanism for converting doses to single organs or tissues into whole-body (effective) doses that carry with them an equivalent risk.

CALCULATION OF RISK OF DEATH

On the basis of epidemiological and biological studies, the risk of a cancer fatality (the fatal cancer probability coefficient) can be estimated for known doses to specific body organs. Basic to this calculation is the concept of collective dose, a unit developed to express the integrated dose to a given population group. For example, if 10,000 people each received an average whole-body dose of 1 Sv, the collective dose would be 10,000 person-Sv (1 million person-rem). Similarly, if 20,000 people each received an average dose of 0.5 Sv, or if 5,000 each received an average dose of 2 Sv, the integrated or total population dose would again be 10,000 person-Sv.

Although the concept of collective dose is useful in evaluating the public health impact of radiation exposures, it should be applied only if the following assumptions are valid (NCRP, 1995):

The relationship between the dose and its resulting biological effects is linear;

The potential effects of the rate at which the doses are received (and whether they are fractionated or protracted) are unimportant.

The individual doses and dose rates are sufficiently low that only stochastic (latent) effects need be considered;

The doses are sufficiently high to be statistically significant.

Thus, the concept of collective dose is not applicable if the doses to individuals in the target population group are either very high or very low. For example, if 1,000 people each received 10 Sv, the collective dose would be

10,000 person-Sv. Yet such a dose received over a short period would be fatal to all members of the exposed group. At the other extreme, if 1 billion people each received 10 microsievert (1 mrem), the collective dose would likewise be 10,000 person-Sv; however, it would be next to impossible to demonstrate any excess ill effects among the exposed group.

The following examples show how epidemiologic data and the concept of collective dose can be used to estimate the number of excess deaths (the fatal cancer probability coefficient) in a population group that has received a known equivalent dose to an individual organ or tissue.

Lung cancer. Studies of uranium miners and survivors of the nuclear bombings in Japan indicate that for each 10,000 person-Sv of collective dose to the lungs of a given population group, there will be an excess of about 90 lung cancers; that is, about 90 people will develop lung cancer who would not otherwise have done so. Medical experience has shown that about 95 percent of the lung cancers induced by ionizing radiation are fatal. This is referred to as the cancer lethality fraction (Table 14.4, column 2). On this basis, the risk of death from cancer as a result of radiation exposures to the lungs would be:

$$\left(\frac{\text{90 excess cancers}}{\text{10,000 person - Sv}}\right)(\text{0.95 fatality rate}) = 85 \times 10^{-4}\,/\,\text{Sv}.$$

This number has been designated by the ICRP as the fatal cancer probability coefficient for lung cancer (column 3).

Bone marrow (leukemia). Studies since World War II of survivors of the nuclear bombings in Japan indicate that if 10,000 people have received a dose of 1 Sv (a collective dose of 10,000 person-Sv) to their bone marrow, after a latency period about 50 excess cases of leukemia will develop in this group. Assuming that 99 percent of these cases are fatal (column 2), the risk of death (fatal cancer probability coefficient) from leukemia as a result of exposures to the bone marrow would be:

$$\left(\frac{\text{50 excess cancers}}{\text{10,000 person - Sv}}\right)(\text{0.99 fatality rate}) = 50 \times 10^{-4}\,/\,\text{Sv}.$$

Breast cancer. Epidemiological data indicate that radiation exposures of the breast produce in women an excess of about 80 breast cancers per 10,000 person-Sv. Assuming that breast cancer is fatal 50 percent of the time and

Table 14.4 Fatal cancer and adjusted probability coefficients, tissue-weighting factors, and annual occupational dose limits for individual tissues and organs

Tissue or organ	Cancer lethality fraction	Fatal cancer probability coefficient (10^{-4}/Sv)	Adjusted probability coefficient (10^{-4}/Sv)	Tissue-weighting factor(w_T)[a]	Annual equivalent dose limit (mSv)
Gonads	—	—	133		
Ovary	0.70	10	15	0.20[b]	100
Bone marrow	0.99	50	104	0.12	150
Colon	0.55	85	103	0.12	150
Lung	0.95	85	80	0.12	150
Stomach	0.90	110	100	0.12	150
Bladder	0.50	30	29	0.05	400
Breast	0.50	20	36	0.05	400
Liver	0.95	15	16	0.05	400
Esophagus	0.95	30	24	0.05	400
Thyroid	0.10	8	15	0.05	400
Bone surface	0.70	5	7	0.01	500[c]
Skin	0.002	2	4	0.01	500[c]
Remainder[d]	0.80	50	59	0.05	400
Total		500	725	1.00	

a. To avoid implications of accuracy beyond what the biological data will justify, the tissue-weighting factors for the various tissues and organs have been assigned one of four values: 0.01, 0.05, 0.12, or 0.20.

b. Total for the gonads (including cancer in the ovaries).

c. Based on deterministic effects.

d. The equivalent dose for the remaining body organs is the estimated mean equivalent dose over the whole body excluding the specified tissues and organs.

that the exposed population consists of 50 percent men and 50 percent women, the fatal cancer probability coefficient due to exposures of the female breasts would be:

$$\left(\frac{80 \text{ excess cancers}}{10{,}000 \text{ person-Sv}}\right)(0.5 \text{ fatality rate})(0.5 \text{ of population}) = 20 \times 10^{-4}/\text{Sv}.$$

Thyroid cancer. Radiation exposures of the thyroid produce an excess of about 80 thyroid cancers per 10,000 person-Sv. Assuming a fatality rate of 10 percent for thyroid cancer, the fatal cancer probability coefficient would be:

$$\left(\frac{80 \text{ excess cancers}}{10{,}000 \text{ person-Sv}}\right)(0.10 \text{ fatality rate}) = 8\times10^{-4}/\text{Sv}$$

Similar calculations can be made to estimate the excess deaths resulting from exposures of other body organs, including deaths due to cancers of the reproductive organs. These data are summarized in column 3. The sum of the fatal cancer probability coefficients for radiation-induced cancers in all body tissues and organs is 500×10^{-4}/Sv, the value currently used in radiation protection analyses and risk assessments.

CALCULATION OF ADJUSTED PROBABILITY COEFFICIENT

Recognizing that radiation-induced cancers have effects other than mortality, the ICRP has adjusted the fatal cancer probability coefficient to take into account not only the death itself but also the other mortality effects (as reflected by the years of productive life lost) among individuals who died, as well as the morbidity effects of the cancers that were not fatal (as reflected in the reduced quality of life for the victim, associated health-care costs, and emotional impact on family and friends) (ICRP, 1985, 1991).

In making this adjustment, the relative increase in the fatal cancer probability coefficient is less for a cancer that has a high cancer lethality fraction. Similarly, the relative increase is less for a cancer that results in fewer years of productive life lost. The techniques are shown in the following two examples, using lung cancer to illustrate a cancer that has a high lethality fraction and thyroid cancer to illustrate a cancer than has a low lethality fraction. In these two cases the years of life lost due to a cancer death are approximately the same (Table 14.5).

Lung cancer. Given that the percentage of lung cancers considered to be lethal—that is, the cancer lethality fraction—is 95 percent (Table 14.4, column 2), 5 percent are assumed not to be lethal. The fatal cancer probability coefficient for lung cancer (calculated earlier) is 85×10^{-4}/Sv. The adjusted probability coefficient is the product of the fatal cancer probability coefficient times one plus the percentage of lung cancers that are not fatal (0.05), multiplied by a factor of 0.9 to account for the years of life lost (Table 14.5, column 3):

$$(85\times10^{-4}/\text{Sv})(1.00+0.05)(0.9) = 80\times10^{-4}/\text{Sv}.$$

Table 14.5 Years of life lost and adjustment factors for various cancers induced by ionizing radiation

Type of cancer	Years of life lost	Adjustment factor
Bladder	9.8	0.65
Bone marrow	30.9	2.06
Bone surface	15.0	1.00
Breast	18.2	1.21
Colon	12.5	0.83
Esophagus	11.5	0.77
Gonads	20.0	1.33
Liver	15.0	1.00
Lung	13.5	0.90
Ovary	16.8	1.12
Skin	15.0	1.00
Stomach	12.4	0.83
Thyroid	15.0	1.00
Remainder	13.7	0.91

Here the adjustment in the fatal cancer probability coefficient is relatively small.

Thyroid cancer. Ninety percent of thyroid cancers induced by radiation are considered not to be lethal. The fatal cancer probability coefficient is 8×10^{-4}/Sv. The adjusted probability coefficient is therefore the product of the fatal cancer probability coefficient times one plus the percentage of thyroid cancers that are not fatal (0.90), multiplied by a factor of one to account for the years of life lost (Table 14.5, column 3):

$$(8 \times 10^{-4}/\text{Sv})(1.00 + 0.90)(1.00) = 15 \times 10^{-4}/\text{Sv}.$$

In this case the adjustment in the fatal cancer probability coefficient is relatively large.

Adjusted probability coefficients for cancers of the other organs can be calculated in the same way. These values are recorded in Table 14.4, column 4.

CALCULATION OF TISSUE-WEIGHTING FACTORS

As column 4 of Table 14.4 shows, the sum of the adjusted probability coefficients for all types of cancers in all organs of the body, plus the associated hereditary effects, is about 725×10^{-4}/Sv. This value can be used to calculate tissue-weighting factors, which in turn can be used to estimate the dose limits for individual body organs.

As calculated above, the adjusted probability coefficient for lung cancer is 80×10^{-4}/Sv. Therefore, cancer of the lungs represents 80/725, or about 12 percent, of the sum of all the adjusted probability coefficients. In other words, if the entire body were exposed to ionizing radiation, the risk and consequences of developing lung cancer would be about 12 percent of the total risk arising through the development of cancers in other organs and through the production of hereditary effects. This value, 0.12, is called the tissue-weighting factor (w_T) for the lung.

Similarly, the adjusted probability coefficient for thyroid cancer is 15×10^{-4}/Sv. Therefore, thyroid cancer represents 15/725, or 2 percent, of the sum of all the adjusted probability coefficients. Because the values calculated by this approach would imply far more accuracy than is justified by the biological data, the ICRP has recommended that tissue-weighting factors for the various organs be assigned one of four values: 0.01, 0.05, 0.12, and 0.20. The value assigned to the thyroid is 0.05. Tissue-weighting factors for all organs are listed in Table 14.4, column 5.

CALCULATION OF DOSE LIMITS FOR INDIVIDUAL ORGANS

Tissue-weighting factors can be used to calculate the effective dose (whole-body equivalent dose) for partial-body exposures. This process, in turn, can be used to set dose-rate limits for individual organs that are comparable (in terms of risk) to the dose-rate limits for the whole body. As noted above, the tissue-weighting factor for the lungs is 0.12. Since the longer-term ICRP dose-rate limit for the whole body is 20 mSv per year, the equivalent dose-rate limit for the lungs (assuming that this is the only portion of the body exposed) would be about eight times as high. That is, the dose rate for the lungs that is equivalent to 20 mSv per year to the whole body would be:

$$\frac{20 \text{ mSv}}{0.12} \sim 150 \text{ mSv per year.}$$

A similar annual limit would apply to the bone marrow, colon, and stomach.

Likewise, the dose rate for the thyroid that is comparable to 20 mSv per year to the whole body would be:

$$\frac{20 \text{ mSv}}{0.05} = 400 \text{ mSv per year.}$$

A similar limit would apply to the bladder, breast, liver, esophagus, and "remainder" organs.

By this same approach, the dose-rate limit for the skin and bone surface would be:

$$\frac{20 \text{ mSv}}{0.01} = 2{,}000 \text{ mSv per year.}$$

Because this is an unacceptably high dose rate, and because the technology for reducing the rate to values well below this number is readily available, the ICRP has based the limit for the skin and bone surface on deterministic or acute effects—that is, effects that have a dose threshold for occurrence. To avoid such effects, the ICRP recommends that the lifetime dose to any body organ for a radiation worker be limited to no more than 20 Sv. For a 40-year working lifetime, this would amount to 500 mSv per year. On this basis the ICRP has selected an annual dose limit of 500 mSv for the skin and bone surface.

Annual equivalent dose limits for a full range of body organs calculated on the basis of the tissue-weighting factors developed by the ICRP are listed in column 6 of Table 14.4. These limits are illustrative only. Although many radionuclides preferentially deposit in single organs (for example, radioactive iodine in the thyroid; radioactive strontium, radium, and plutonium in the bone; and airborne radioactive materials in the lungs), these materials may also cause radiation exposure to tissues adjacent to these organs. In addition, soluble radionuclides that are inhaled (and initially cause exposure solely to the lungs) may subsequently be taken up by the blood and deposited in other organs. Professional personnel responsible for calculating annual limits on intake for individual radionuclides (see below) must assess the doses to all portions of the body and keep the intake limits sufficiently low that the combined fractional doses to the several affected organs do not exceed the effective dose-rate limit (20 mSv per year).

Assessment of Internal Exposures

In the case of external radiation sources, particularly those that cause whole-body exposures, compliance with dose limits can be assessed through radiation surveys, personnel monitoring, or both. These evaluations are not possible in the case of doses resulting from the intake of radionuclides into the body. Instead, tertiary or derived limits must be developed that will assure that the basic dose limits are being observed. The derivation of two of these tertiary limits is described below.

The *annual limit on intake* (ALI) is the quantity of a given radionuclide that, if ingested or inhaled, will result in an uptake in the body that will yield a committed dose to the affected organ equivalent to the annual effective dose-rate limit (ICRP, 1991, 1994b). If a radionuclide causes exposure of the total body, the applicable limit on committed dose over the subsequent 50 years for a radiation worker will be 20 mSv. If a radionuclide causes exposure to a single organ only, the limit on committed dose for that organ over the next 50 years will be as indicated in Table 14.4, column 6. If the radionuclide causes exposure to several body organs, the sum of the fractions obtained by dividing the dose to each of the affected organs by the dose limit for that organ must not exceed one.

Estimates of ALIs require information about several factors. If the material is being inhaled, these include the deposition, retention, and uptake of the radionuclide in the lungs (which in turn depends on particle size and solubility) and the breathing patterns of the exposed worker (whether through the nose or mouth and the rate of breathing). If the material is being ingested, these factors include whether it is soluble and will be taken up by the blood and preferentially deposited in one or more organs. If the material is insoluble, the principal concern will be exposure of the gastrointestinal tract as the material moves through the body. Other considerations include the efficiency of transfer of the radionuclide to the bloodstream, the efficiency of the uptake of the radionuclide in the body, the mass(es) of the organ(s) in which the radionuclide deposits, the length of time it remains there, and the annual equivalent-dose limit(s) for the portion(s) of the body affected. It is also necessary to know the quantity of radionuclide ingested or inhaled, its physical half-life, the types and energies of the radiation it emits, and the associated radiation weighting factors. Combining this and other information, one can estimate the quantity of the given radionuclide that, if ingested or inhaled, would yield a committed dose that is equal to the respective annual dose limit for the portion(s) of the body affected. This quantity is the annual limit on intake. Table 14.6 lists ALIs for inhalation and ingestion of some of the more common radionuclides (ICRP, 1994a).

If the major avenue of intake is inhalation, an estimate can be made of the *derived air concentration* (DAC) of the given radionuclide, taking into account the hours of exposure per year and the breathing rate. For example, the DAC for occupational exposure to any radionuclide is the concentration in air that, if breathed by an adult for an assumed occupational exposure time of 2,000 hours per year, will result in the ALI via inhalation. Numerically, the DAC can be calculated as follows:

$$\mathrm{DAC} = \frac{ALI}{(2{,}000\ \mathrm{hr/yr})\,(60\ \mathrm{min/hr})\,(20{,}000\ \mathrm{cm}^3/\mathrm{min})}$$

THE COMMITTED DOSE

Once a radionuclide has been deposited in the body, the exposed person is "committed" to the dose resulting from the decay of that radionuclide as long as it is present in the body. For radionuclides with effective half-lives of only a few months (that is, those that have a short radioactive half-life or are rapidly excreted), the committed dose is delivered, in the main, during the year following intake. For radionuclides with long effective half-lives (that is, those that have a long radioactive half-life and are retained a long time in the body), only a fraction of the total dose will be delivered to the body during the year in which the radionuclide was initially inhaled or ingested. In fact, for radionuclides with very long effective half-lives, neither the total dose nor a full expression of the associated risk is likely to be manifested during the lifetime of the worker. Consequently, the committed dose from the lifelong intake of some radionuclides with long effective half-lives may overestimate, by a factor of two or more, the actual lifetime risk (NCRP, 1993). Care must be taken in using such data for epidemiological studies.

The NCRP recommends that use of the committed dose associated with the intake of a given radionuclide be restricted to radiation protection planning, as in the design of facilities and the development of manufacturing processes and research protocols; to the demonstration of compliance with those plans; and to the calculation of ALIs and DACs. In the opinion of the NCRP, the committed dose does not in itself constitute a sufficient basis

Table 14.6 Annual limits on intake for occupational exposure to radionuclides

	Most restrictive limit on annual intake[a]			
	Inhalation		Ingestion	
Radionuclide	Becquerels	Microcuries	Becquerels	Microcuries
Hydrogen-3[b,c]	1×10^9	3×10^4	5×10^8	1.5×10^4
Carbon-14[b]	3×10^7	8×10^2	3×10^7	8×10^2
Phosphorus-32	7×10^6	2×10^2	8×10^6	2×10^2
Sulfur-35[c]	2×10^7	5×10^2	3×10^7	8×10^2
Cobalt-60	1×10^6	25	6×10^6	1.5×10^2
Strontium-90	3×10^5	8	7×10^5	20
Iodine-131[b]	1×10^6	25	9×10^5	25
Cesium-137	3×10^6	80	2×10^6	50
Radium-226	2×10^3	5×10^{-2}	7×10^4	2
Plutonium-239	6×10^2	2×10^{-2}	8×10^4	2

a. Since inhaled radionuclides, particularly those that are relatively insoluble, predominantly expose the lungs, and since absorption through the lungs frequently differs from that through the GI tract, intake limits for inhalation often differ from those for ingestion. Intake limits also vary with the chemical compound in which the radionuclide is incorporated and particle size. The inhalation limits above are based on airborne particles assumed to have an activity median aerodynamic diameter of 5 micrometers.

b. Inhalation limit is based on intake as a vapor.

c. Ingestion limit is based on intake in an organically bound form.

for evaluation of the potential health effects of radiation exposures in individuals; rather, such evaluations should be based on estimates of the actual absorbed dose being delivered to the tissues and organs in question (NCRP, 1987).

The system that has been designed for recommending dose limits and standards for exposures to ionizing radiation has many desirable characteristics. The standards are based on associated risks to health; a procedure has been developed for expressing dose rates from internal and external exposures on the basis of their relative risks; there is a systematic relationship between limits for the general population and those for radiation workers; and, through the ICRP, the approach to these problems is

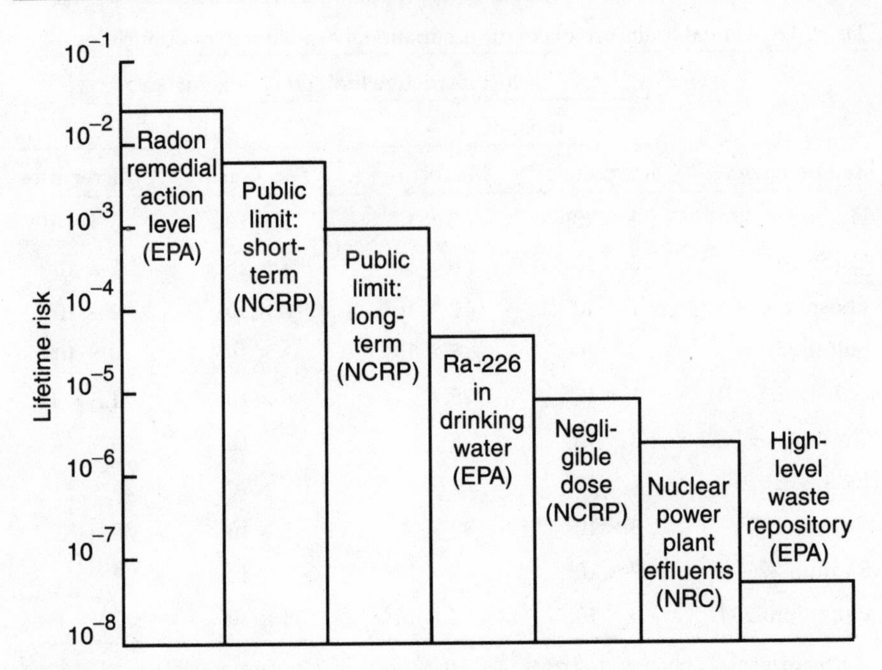

Figure 14.2 U.S. standards for exposures to various sources of ionizing radiation, 1995

uniform worldwide. Application of similar considerations and concepts might prove useful in the formulation of procedures for developing permissible limits for exposures to other occupational and environmental stresses.

Even this advanced system, however, has deficiencies. One of the most significant is that the given dose limits do not include exposures from natural radiation sources or the use of radiation in medicine and dentistry. In addition, the procedures for calculating annual limits on intake and the associated derived air concentrations are based on the assumption that the given radionuclide is the sole source of exposure. If an individual is exposed to several radionuclides (as many members of the public are), suitable reductions must be made in the allowable intake limits. Also, the existing system fails to take into account that radionuclides released into outdoor air and water undergo chemical and physical changes as they move through various environmental chains. The particle size distribution and chemical composition of individual airborne radionuclides will change. And seasonal variations will occur in the movement and behavior of individual radionuclides within the environment.

Another serious deficiency lies in the inconsistent manner in which the system is being implemented. There are orders-of-magnitude differences in the lifetime risks associated with the currently permissible limits in the United States for exposures to technologically enhanced, naturally occurring radiation sources (such as indoor radon), for short-term and long-term exposures of population groups to artificial sources, for exposures resulting from liquid and airborne releases to the environment from nuclear power plants, and for potential exposures from operation of the proposed deep geologic repository for disposal of high-level radioactive wastes (Kocher, 1988; see Figure 14.2).

The General Outlook

Clearly emerging from this review of U.S. environmental standards is the need to harmonize the methodologies for developing, and the procedures for applying, occupational and environmental standards. This is true with respect to the methods of calculating doses to the public and estimating the associated risks, as well as determining what level of risk is acceptable. One of the immediate benefits would be an enhancement in the cost-effectiveness of environmental controls. If all standards in this country had a common risk basis, trade-offs in the control of different sources of environmental contaminants could become routine. Such trade-offs are accepted industrial practice in meeting regulatory requirements for air pollution and water pollution. The approach has proved itself by enabling industry officials to apply controls to the contaminant sources that can be most cost-effectively reduced, regardless of whether they are located within the given industry or at a neighboring plant. Similar trade-offs could be applied to many types of contaminants, regardless of their nature or environmental pathway. One day it might even be possible to reduce the risks from chemical contaminants within the environment to compensate for risks associated with physical and biological stresses. The goal would be to reduce the total risk to an acceptable level, with controls being applied to the contaminant or stress that can be reduced or eliminated in the most cost-effective manner.

Although it is essential that such harmonization be accomplished among various federal, state, and local agencies within the United States, the effort should not stop there. It needs to be expanded so that ultimately all countries of the world will use the same approaches and methodologies in setting environmental standards. Only with such uniformity can one be sure that evaluations of environmental contaminants from country to country are conducted on a similar basis and that the resulting environmental standards are comparable. One immediate advantage internationally would be that the exportation of hazardous industries from one country to another to escape regulatory controls would no longer be a viable choice.

Also needing to be addressed is the common public misconception that compliance with environmental standards will assure risk-free conditions. Although many standards are expressed as single, numerical limits, organizations such as the ACGIH stress that, in the case of occupational exposures to airborne toxic chemicals, these limits do not represent "fine lines between safe and dangerous concentrations" nor do they represent "a relative index of toxicity" (ACGIH, 1996). Similar qualifying statements accompany the limits recommended by the ICRP and the NCRP.

One way to overcome some of these problems is to provide members of the public, particularly the stakeholders, an opportunity to be involved in the standards-setting process so that they will understand both the goals being sought and the means by which they are to be achieved. Experience has shown that although individuals voluntarily accept risks, they balk at accepting involuntary risks—particularly if they conclude that the risks are being imposed on them without full disclosure. The development and promulgation of standards must involve effective and continuing communication between the scientific community and the public at large (AAEE, 1983).

To increase public confidence in environmental standards, federal agencies need to avoid imposing limits that state and local agencies have insufficient funds to enforce. Because it is now possible to measure concentrations of environmental contaminants in the range of parts per billion or parts per quadrillion, and to assess the adequacy of cleanup in the same exacting manner, the costs of controls are increasing exponentially. Although the requirements for control are dictated by Congress and various federal agencies, most of the costs for cleanup are borne by local communities and business interests. As money for such activities becomes increasingly difficult to obtain, many of the standards are not being met. As a result, local environmental health officials have in some cases lost credibility, a situation that has tended to undermine the significance of the environmental standards involved (Pompili, 1995).

15

MONITORING

ENVIRONMENTAL monitoring programs were initially conducted on a local basis and had two basic objectives: (1) to estimate exposures of people resulting from certain physical stresses (such as noise and radiation) and from toxic materials that are being, or have been, released and are subsequently being ingested or inhaled, and (2) to determine whether the resulting exposures complied with the limits prescribed by regulations. Such programs were either "source" related or "person" related. *Source-related* monitoring programs were designed to determine the exposure or dose rates to a specific population group resulting from a defined source or practice. *Person-related* monitoring programs were designed to determine the total exposure from all sources to a specific population group. The latter was particularly useful in instances where several sources were causing the exposures (ICRP, 1985).

Although such programs continue to be important, it is increasingly recognized that assessing risks solely to human health or focusing on problems only on a local scale is inadequate. The purposes and goals of environmental monitoring programs today have expanded far beyond these earlier objectives. Significantly, it is now accepted that certain of these programs should have an *environment-related* component, and that conditions should be examined on a regional, local, and global basis. That is, they should be designed to assess the impact of various contaminants on selected segments of the environment, including ecosystems, and to evaluate factors that may have wide-scale, long-range effects. The types and purposes of current environmental monitoring programs are summarized in Table 15.1.

As a general rule, local environmental monitoring programs for indus-

Table 15.1 Types and purposes of environmental monitoring programs

Type of program	Purpose
Based on nature of the stress	
Physical stress	To assess the impact of environmental stresses such as noise and external radiation, where the evaluation is based primarily on exposure measurements made in the field, not on samples collected and returned to the laboratory for analysis
Chemical stress	To assess exposures resulting from the ingestion and inhalation of chemical and radioactive contaminants
Based on geographic (spatial) coverage	
Local	To evaluate the impact of a single facility on the neighboring area
Regional	To evaluate the combined impact of emissions from several facilities on a large area
Global	To determine worldwide impacts and trends, such as acidic deposition, depletion of the ozone layer, and potential for global warming
Based on temporal considerations	
Preoperational	To determine potential contamination levels in the environment prior to operation of a new industrial facility; to train staff; to confirm operation of laboratory and field equipment
Operational	To provide data on releases; to confirm adequacy of pollution controls
Postoperational	To assure proper site cleanup and restoration
Based on monitoring objectives	
Source related	To determine population exposures from a single source
Person related	To determine total exposure to people from all sources
Environment related	To determine impacts of several sources on features of the environment such as plants, trees, buildings, statues, soil, water, and ecosystems
Research related	To determine transfer of specific pollutants from one environmental medium to another and to assess their chemical and biological transformation as they move within the environment; to determine ecological indicators of pollution; to confirm that the critical population group has been correctly identified and that models being applied are accurate representations of the environment being monitored
Based on administrative and legal requirements	
Compliance related	To determine compliance with applicable regulations
Public information	To provide data and information for purposes of public relations

trial facilities are handled by plant personnel or environmental service contractors, whereas regional programs are handled by state and local environmental health and regulatory authorities, and the planning and coordination of national programs are handled by federal agencies. Close coordination between the facility operator and the local agencies is necessary if all objectives of the monitoring program are to be met. A well-planned program will usually involve some overlap in the activities of the several monitoring groups, including exchanges of samples and cross-checking of data.

Monitoring Physical Stresses and Toxic Materials

Because of differences in the nature of the exposures, the monitoring of physical stresses and toxic materials requires significantly different approaches. Monitoring physical stresses may simply involve identifying the sources and measuring the magnitude of the stresses. Data on the distribution of the energies of the physical stresses can be utilized to estimate the accompanying dose as a function of tissue depth and specific body organ. Monitoring toxic materials involves much more. For airborne or waterborne releases, the first step is commonly to measure discharges at the points of release. Additional steps include assessing the movement or transport of specific contaminants within given environmental media (air, water, soil), their transfer from one medium to another, and their chemical and biological transformation as they move within the environment. Data on the physical and chemical nature of toxic substances can be used to estimate their deposition and uptake by various body organs. These data, in turn, can be used to estimate the accompanying doses to people.

Because contaminants can cause exposures of people by so many avenues, most environmental monitoring specialists try to identify and trace the movement and behavior of key contaminants through several environmental pathways. They may supplement these measurements by analyzing the concentrations of selected contaminants in various ecological indicators, such as muds and biota from streams. As noted later, information on contaminants in such materials is useful in the conduct of environmental monitoring programs and in the evaluation of the resulting data.

Measurements to determine exposures from physical stresses, such as noise and ionizing and nonionizing radiation, must often be made on a real-time basis. Generally, instruments are placed near the people being exposed or in concentric rings at various distances from the source. It is important to recognize that the mere presence of people and monitoring equipment may alter the environment in such a way as to make accurate measurements (say, of electric or magnetic fields) difficult. In addition, the position and location of the people being exposed (for example, whether they are standing on the ground or near a tree, or sitting inside an automobile) can alter the resulting exposures (NCRP, 1994).

During sampling to collect data on the health effects of airborne contaminants, the presence of both particles and gases is noteworthy, for they behave differently. For example, the size of airborne particles will significantly affect whether and where they will be deposited in the respiratory tract, and their chemical composition will determine their movement within the body and their potential effects on health. It is also vital to know what other chemicals are associated with a given contaminant, since certain combinations are synergistic. Sulfur dioxide, a ubiquitous acidic gas that is highly soluble and is ordinarily taken up entirely in the throat and upper airways (where its effects on health are minimal), acutely impairs the functioning of the lungs when carried to the alveoli as an acid condensed on the surface of small particles (less than 5 micrometers in diameter).

Measuring Waterborne and Airborne Exposures

As noted above, one of the first steps in assessing potential exposures is to measure the concentrations of individual contaminants in samples of typical releases from the polluting facility. In most cases, air and water serve as the principal pathways for direct exposures (through inhalation and the consumption of drinking water) and as a vehicle for the transport of contaminants from the point of release to other environmental media (such as milk and food). Measurement of the airborne and waterborne contaminants leaving a plant can also provide advance information on impending problems in other environmental media. Since critical contaminants can be missed if only the obvious and easily measured effluents are monitored, or if monitoring ceases during key periods such as shutdowns for repairs and maintenance, sample collection and analysis should be conducted during all phases of plant operations.

ASSESSING WATERBORNE RELEASES

A range of samples can be collected to assess the impact of waterborne releases (Baker, 1992). These include:

Grab samples: collected on a one-time basis. They represent at best a snapshot in time of the characteristics of the waste. Unless the waste is uniform in composition, they will not provide useful information on the nature and characteristics of the waste.

Composite samples: a blending of a series of smaller samples. Composites are frequently prepared by combining a series of discrete samples, each collected in individual sample bottles and each representing, as do grab samples, a snapshot of the characteristics of the waste at the time of collection.

Timed-cycle samples: collected in equal volumes at regular intervals. Direct averaging of data on such samples is representative of the characteristics of the waste only if its flow rate is constant.

Flow-proportional samples: collected in relation to the volume of flow during the sampling period. To make up a composite that is representative of the waste, such samples can be proportioned manually, when flow records are available, or the sampling rate can be set on the basis of real-time measurements of the waste flow rate.

Indicator samples: contaminants biologically concentrated within various living organisms and plants, most especially in the aquatic environment. Data on contaminants biologically concentrated in biota can provide information on contaminants whose concentrations in the stream itself are below the limits of analytic sensitivity; data on contaminants as a function of depth in muds can provide information on the history of the release of contaminants into a stream.

One of the primary advantages of a composite sample is that it minimizes the expense of analyzing liquid-waste streams having a relatively uniform composition. For waste streams with a wide range of characteristics, oftentimes the collection and analysis of discrete samples is required to determine the temporal nature and concentration of various contaminants in the waste. As a general rule, the representativeness of the analytical data improves with sample-collection frequency. With highly variable conditions, samples should be collected as frequently as every five minutes to every hour. The prime consideration is the variability of the composition of the waste at the point of sampling.

If, after collection, the sample is placed in a bottle for transport to the laboratory for analysis, care must be taken to assure that ionic species or small particles suspended in the waste do not attach themselves to the walls of the bottle. This potential problem can be avoided by an appropriate choice of bottle or by adjusting the pH or adding stabilizing chemicals to the sample prior to placing it in the container. Similar steps, including refrigeration, will assure that the sample is properly protected against deterioration due to either chemical or biological processes. Such preservation is particularly important when there is a lag between collection and analysis (Baker, 1992).

The quantity of sample collected depends on the number and nature of the parameters being tested. The quantity should be sufficient to permit all desired analyses, allowing for possible errors, spillage, and sample splitting for purposes of quality control.

ASSESSING AIRBORNE RELEASES

Assessment of the impact of airborne releases will require the collection of samples from a variety of sites. As with liquid wastes, an initial step will be to sample the various release points at the polluting facility. In terms of evaluating the impact on the environment, samplers should be located in places where airborne concentrations and ground deposition of contaminants are estimated to be most likely to lead to human exposures. The selection of sites should be based on the best available meteorological information, coupled with data on local land use. Sites selected for monitoring the impact on various ecosystems will require a similar approach, the exposed entity in this case being environment, not people.

A frequently used sampling system employs a filter or electrostatic precipitator to collect airborne particles and an appropriate set of adsorbers to collect gaseous and volatile contaminants. Cascade impactors or other mechanical separation devices can be utilized to assess the size distribution of airborne particles. The choice of sampler depends on the desired sample volume, sampling rate, power requirements, servicing, and calibration. The minimum amount of air to be sampled is dictated by the sensitivity of the analytical procedure; the amount is often a balance between sensitivity and economy of time. As in any monitoring program, care must be taken to assure that the samples are representative (Sioutas, 1993).

As in the case of assessing occupational exposures, environmental monitoring specialists are becoming increasingly aware that it is difficult to relate the concentrations of airborne contaminants in the ambient environment to exposures to the public. Few people, for example, spend significant amounts of times at specific locations outdoors. For this reason, increasing efforts have been devoted to the development of methods to evaluate the concentrations of contaminants in the air actually being breathed by people.

One of the more significant advances has been the development of personal samplers that can be worn by individual members of the public and are designed to evaluate the quantity of various contaminants being inhaled (see Chapter 5). Such samplers are similar, in terms of the data they provide, to the personal monitoring devices available for assessing doses from external radiation sources. Personal samplers have been developed to collect airborne gases and particulates, or combinations of the two. For gaseous contaminants, the associated health effects generally are directly related to the quantity inhaled. For the reasons previously cited, assessment of the potential health effects of airborne particles requires information not only on the total quantity inhaled, but also on their size distribution, chemical composition, and solubility. Personal samplers have been developed to take all of these factors into consideration.

Important characteristics of personal samplers are that they have minimal power requirements and be relatively quiet and lightweight. Although operation and maintenance costs are relatively high, they are considered worthwhile because of the amount and direct applicability of the information they provide (Sioutas, 1993).

Table 15.2 summarizes the advantages and disadvantages of various sampling methods for the principal types of environmental contaminants and receptor media.

Designing a Monitoring Program

One of the first steps in designing an environmental monitoring program is to define its objective (what samples are to be collected and where and when) and how the data are to be analyzed (Table 15.3). Not only must the program be planned so that the right questions are asked at the right time, but also so that only the data necessary to answer these questions are collected. Other attributes of a successful environmental monitoring program are that (1) it is sufficiently inexpensive to survive unexpected reductions in supporting funds; (2) it is simple and verifiable so that it is not significantly affected by changes in personnel; and (3) it includes measurements that are highly sensitive to changes in the environment (Griffith and Hunsaker, 1994).

Most environmental monitoring programs have at least four stages: gathering background data, collecting and analyzing samples, establishing tem-

Table 15.2 Advantages and disadvantages of various environmental sampling methods

Type of sample	Advantages	Disadvantages
	Atmospheric environment	
Direct measurement		
Real-time field measurements of physical stresses such as noise and radiation	Monitors can be put in place to assess time-integrated exposures	Monitor often disturbs field being monitored; some monitoring equipment (e.g., for assessing electric and magnetic fields) expensive and complex
Airborne particulates		
Respirable fraction via air sampling	Direct-dose vector; provides data on potential effects on lungs	Omits larger particles that may be significant when deposited in nose, mouth, and throat
Total particulates via air sampling	Provides data for assessing doses to lungs as well as possible effects on skin and intake through ingestion	Not all measured contaminants respirable
Collection of settled particulates	Represents an integrated sample over known time and geographical area	Weathering may alter results; only large particles collected by sedimentation
Gases		
Integrated (concentrated) sample	Concentration of samples permits detection of lower concentrations in air	Samples must usually be analyzed in laboratory; chemical reactions may change nature of collected compounds
Direct measurement	Provides data on real-time basis	Lower limit of detection may not be adequate
	Terrestrial environment	
Milk	Direct-dose vector, especially for children; data easily interpreted	Milk samples not always available
Foodstuffs	Direct-dose vector; data easily interpreted	Samples not always available from areas of interest; weathering and processing may affect samples

Table 15.2 (continued)

Type of sample	Advantages	Disadvantages
Wildlife	Direct-dose vector	High mobility; not always available; data difficult to interpret
Vegetation	Samples readily available; multiple modes for accumulating contaminants (by direct deposition and leaf and root uptake)	Data difficult to interpret; weathering can cause loss of contaminants; not available in all seasons
Soil sampling	Good integrator of deposition over time	High analytical cost; data difficult to interpret in terms of population exposure and dose
	Aquatic environment	
Surface water (nondrinking)	Readily available; indicates possibility of contamination by aquatic plants and animals	Not directly dose related; difficult to interpret data
Groundwater (nondrinking)	Indicator of unsatisfactory waste-management practices	Not always available; data difficult to interpret because of possibility of multiple remote sources
Drinking water	Direct-dose vector; consumed by all population groups	Contaminant concentrations frequently very low
Aquatic plants	Sensitivity	Data difficult to interpret; not available in all seasons
Sediment	Sensitivity; good integrator of past contamination	Data difficult to interpret because of possibility of multiple remote sources
Fish and shellfish	Direct-dose vector; sensitive indicator of contamination	Frequently unavailable; high mobility
Waterfowl	Direct-dose vector	Frequently unavailable; high mobility; data difficult to interpret

Table 15.3 Questions to be answered when implementing an environmental monitoring program

Program stage or component	Question
Purpose	What is the goal or objective of the program?
Method	How can the goal or objective be achieved?
Analysis	How are the data to be handled and evaluated?
Interpretation	What might the data mean?
Fulfillment	When and how will attainment of the goal or objective be determined?

poral relationships, and measuring the validity of the results. Although the design depends largely on the purposes of the program, the following discussion (based primarily on assessment of exposures from nuclear facilities) applies in a general sense to the design of monitoring programs for pollution from all types of industrial facilities.

Although a nuclear facility can be a source of direct external exposure to nearby population groups, inhalation and ingestion of radioactive materials released into the environment are generally more important because they represent a greater contribution to dose, much the same as do toxic materials released by other types of industrial facilities. A key feature of environmental monitoring programs for such facilities is therefore identification of the potentially critical contaminants that might be released, their pathways through the environment, and the avenues and mechanisms through which they may cause population exposures. Figure 15.1 shows the major steps to be taken and the factors to be considered when designing an environmental monitoring program for a nuclear facility.

BACKGROUND DATA

Before monitoring begins, background information is needed on other facilities in the area, the distribution and activities of the potentially exposed population, patterns of local land and water use, and the local meteorology and hydrology. These data permit identification of potentially vulnerable groups, important contaminants, and likely environmental pathways whose media can be sampled.

People responsible for the background analysis must take into account the type of installation, the nature and quantities of toxic materials being

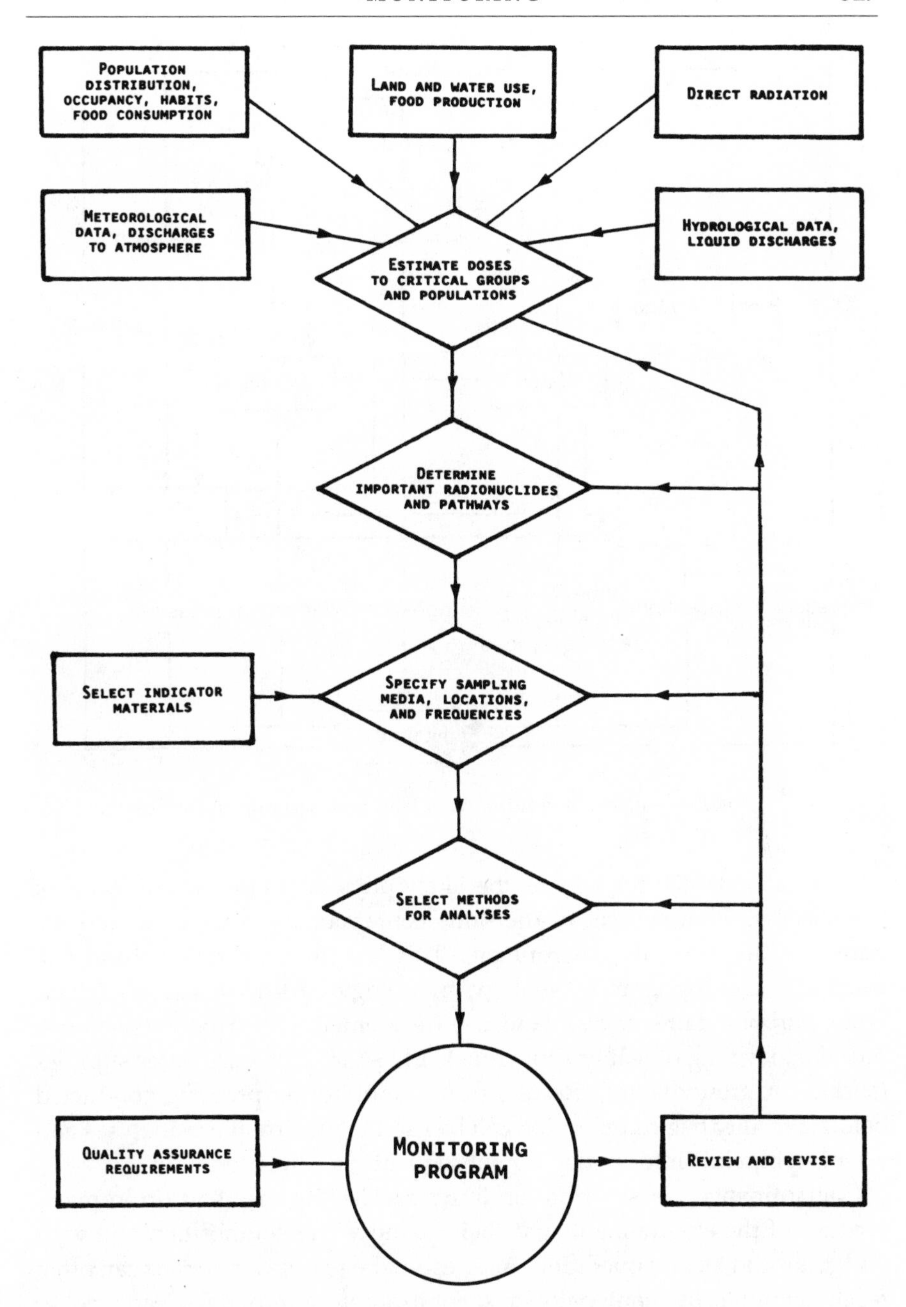

Figure 15.1 Example of the design of an environmental monitoring program and associated dose estimations for a major nuclear facility

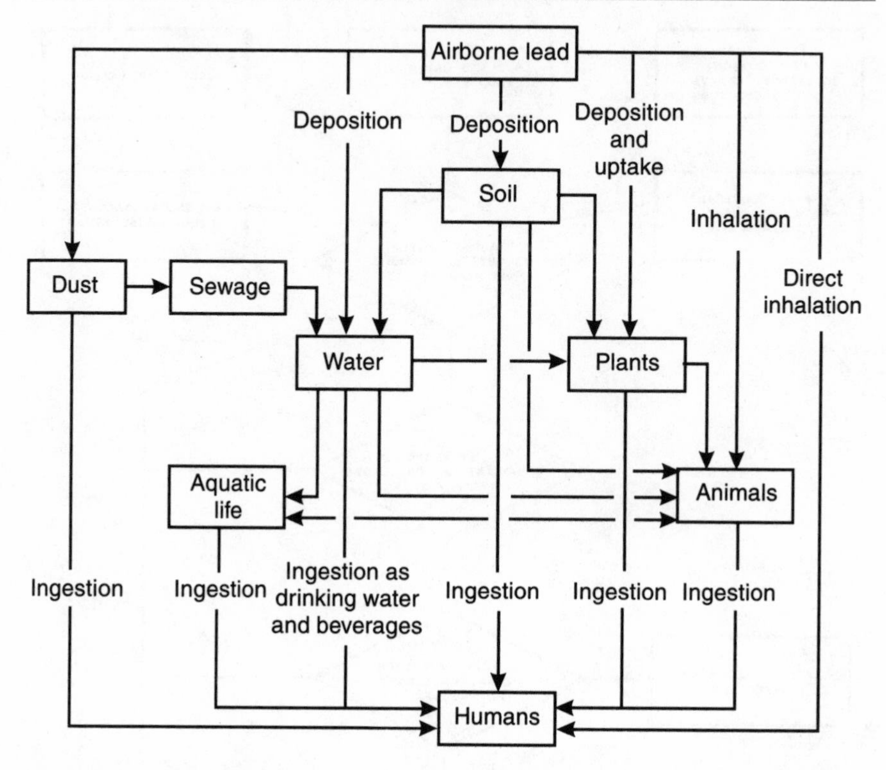

Figure 15.2 Possible pathways to humans of lead released into the atmosphere

used, their potential for release, the likely physical and chemical forms of the releases, other sources of the same contaminants in the area, and the nature of the receiving environment. This last item includes natural features (climate, topography, geology, hydrology), artificial features (reservoirs, harbors, dams, lakes), land use (residential, industrial, recreational, dairying, farming of leaf or root crops), and sources of local water supplies (surface or groundwater). Results from a monitoring program conducted before a facility begins operation can be used to confirm these analyses and establish baseline information for subsequent interpretation.

Contaminants released from an industrial facility may end up in many sections of the environment, and their quantity and composition will vary with time and facility operation. As a result, the released materials can often reach the public by many pathways. For example, a secondary lead smelter has the potential to release elemental lead and associated compounds into the atmosphere, whereupon they may become an inhalation hazard. The same facility can also release these contaminants to the soil, either directly or through the air, whereupon they may contaminate groundwater and subsequently be taken up by fish and agricultural products. Figure 15.2 illustrates the possibilities.

The milk from cows and the beef from cattle grazing on pastures adjacent to lead smelters can be expected to have a higher-than-normal lead content. Children playing on contaminated earth near such smelters have shown elevated lead concentrations in their blood. Arsenic emitted by copper smelters follows identical pathways of contamination and human exposure. In a similar manner, contaminants discharged to the liquid pathway can be a source of human exposure. Figure 15.3 outlines the principal pathways to humans for radioactive materials released in liquid form.

Tracing the movement of all contaminants through all potential pathways would be physically and economically impossible. Fortunately, in the

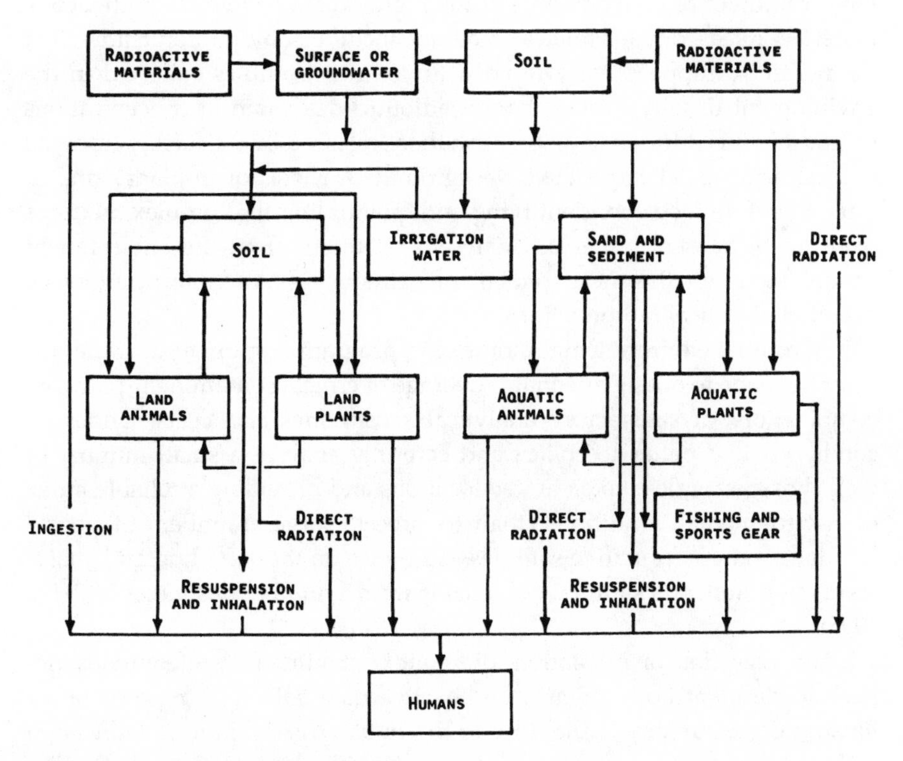

Figure 15.3 Possible pathways to humans of radioactive materials released to ground or surface waters (including oceans)

case of nuclear facilities, the primary contributors to population dose consist of no more than half a dozen radionuclides moving through no more than three or four pathways. Once these are identified, along with the habits of the people living or working in the vicinity, it should be possible to identify a "maximally exposed" group of individuals (the "critical group") whose activities and location would make them likely to receive the largest exposures. Since the regulatory requirements in most countries pertain to the dose to the critical group, environmental monitoring programs should provide this type of estimate. In addition, such programs should be designed to provide a means for estimating the total (collective) population dose, on the basis of which regulatory authorities can project the associated health impacts on the population as a whole.

SAMPLE COLLECTION AND ANALYSIS

The samples collected directly at an industrial facility usually contain the largest number of contaminants at the highest concentrations. Radioactive materials released from nuclear facilities include some radionuclides that decay (and disappear) very rapidly. In contrast, samples collected in the environment usually contain fewer radionuclides, often in concentrations so low that it is difficult to distinguish those released by the facility from the natural or artificial radioactive background. Analysis of in-plant samples must therefore focus on identifying and quantifying the complex mixtures likely to be present, whereas analysis of samples from the environment should focus on the measurement of extremely small concentrations of preselected critical radionuclides.

In a routine environmental monitoring program, a number of trade-offs must be made to obtain adequate coverage of critical contaminant-pathway combinations at satisfactory analytical sensitivities and costs. Under all conditions it is better to collect and carefully analyze a small number of well-chosen samples, taken at key locations and providing a reliable index of environmental conditions, than to process larger numbers of poorly selected, nonrepresentative samples. Guidance on the sampling and analysis of environmental samples is available from a number of sources (APHA, 1995).

Upon ingestion or inhalation, different contaminants preferentially deposit in different body organs. Radioactive materials are a representative illustration. Estimates of the dose to the public thus require identification and measurement of concentrations of individual contaminants. Besides collecting data to provide estimates of exposure, those in charge of a monitoring program may need to measure certain other contaminants because of the history of operations at the site or because of specific concerns of the local population. Such measurements may be necessary regardless of the levels of exposure involved (as in sampling and analysis for radioactive iodine in milk near commercial nuclear power plants, and for plutonium in the soil near nuclear facilities operated by the DOE). Such programs should also include collection and analysis of samples indicative of possibly higher-than-normal exposures to different plants and animals within various ecosystems. In general, however, if humans are being protected, aquatic and other organisms will also be protected (NCRP, 1991).

Careful analysis of specific radionuclides in various environmental media can also help determine the source of the contamination. A classic example is use of the ratio of cesium-134 (Cs-134) to cesium-137 (Cs-137) in a sample to determine whether the contamination has resulted from discharges from a nuclear power plant or from the detonation of a nuclear device. The energy source in both cases is nuclear fission, one of the products of which is radioactive Cs-137. Also produced during the fissioning process is Cs-133, a stable isotope of cesium. If the fissioning event results from the explosion of a nuclear device, most of the Cs-133 remains in that form. If the fissioning event occurs in a nuclear reactor, the Cs-133 remains in the reactor core for some time, is subjected to intense neutron bombardment from subsequent fissioning events, and a portion is converted into radioactive Cs-134. Therefore, when analysis shows that the radioactive cesium contamination in an environmental sample contains only very small quantities of Cs-134, the source of contamination was probably the detonation of a nuclear device (or perhaps a severe and unusual industrial accident) rather than releases from a nuclear power plant. Likewise, analysis for vanadium, a characteristic component of fuel oils, has been used to determine whether airborne contaminants in the environment come from the combustion of oils or of coal.

Accurate measurement of external radiation exposures from releases from a nuclear facility is particularly difficult because of the relatively high contribution of natural background radiation. Similarly, it is often difficult to measure the concentrations of man-made radionuclides in certain environmental media because of the relatively high concentrations of naturally occurring radionuclides. Under current USNRC regulations, the external dose-rate limit for the maximally exposed individual near a nuclear power plant is about 0.05 mSv per year. In contrast, the average external dose rate from naturally occurring cosmic and terrestrial radiation sources in the

United States is 0.6–0.7 mSv per year. The concentration limits in milk for key artificially produced radionuclides range from less than 0.5 to more than 5 becquerels (from a few tens to a few hundreds of picocuries) per liter; normal concentrations of such radionuclides are well below these values. Yet the concentration of naturally occurring radioactive potassium in milk exceeds 40 becquerels (1,000 picocuries) per liter.

Because they are faster and less expensive than specific radionuclide analyses for showing changes in environmental concentrations, gross measurements of radioactive materials in water and in air (in which the specific contaminating radionuclides are not identified) can be very useful as trend indicators. The process is similar to the measurement of "total suspended particulates" as a surrogate for, or indicator of, the health effects of particulate substances emitted into the atmosphere, or the use of the coliform organism as an indicator of the possible presence of disease organisms in drinking water. For media such as milk and food products, in which concentrations of naturally occurring radionuclides may be relatively high, gross measurements are of less value and are not recommended as part of a routine monitoring program. If the radionuclide composition of the releases from a specific facility is relatively stable, gross activity may be related in a simple manner to specific radionuclide concentrations of interest, and the gross activity measurements can then be related to estimates of population dose. Repetitive verification is advisable, however, to assure that conditions have not changed.

The lower limits imposed by revised standards and regulations (USNRC, 1991) have increased the need for more sensitive and more accurate measurement of radionuclide concentrations in environmental media. Increased accuracy (and thus improved exposure and dose estimates) can always be achieved through collecting larger samples or using longer counting times (to provide the required statistical sensitivity). Still, such strategies are costly. Although few would disagree that the sensitivity of the measurements should be adequate to detect concentrations below the applicable limits, there is no general agreement on what level of exposure to people is sufficiently small to be considered negligible. Although the NCRP has proposed that a dose rate of 0.01 mSv per year or less for a single source or practice be considered a negligible level of risk (NCRP, 1993), the concept has yet to receive widespread endorsement by regulatory agencies.

The installation of a network of air and water samplers and dosimeters around a site can provide continuous monitoring of exposures. Such a network may prove to be well worth the cost of years of maintenance and operation, should an accidental release occur. Where the need for measuring chemical and radioactive contaminants exists in the same site environment, the two programs should be coordinated to provide greater efficiency and lower overall monitoring cost. Since the behavior of certain radionuclides closely follows or is influenced by the behavior of chemically related stable elements, the use of common measurement or sampling locations may, if nothing else, permit the measurement of one constituent to serve as a tracer for or an indicator of the behavior of another. Examples of such combinations are strontium-90 and stable calcium, and cesium-137 and potassium.

TEMPORAL RELATIONSHIPS

Each type of release from a nuclear facility generates a characteristic pattern between the time of the release and the occurrence of exposures (Figure 15.4). Similar relationships exist with respect to certain aspects of the exposures to toxic chemicals. In the case of radioactive materials, releases to

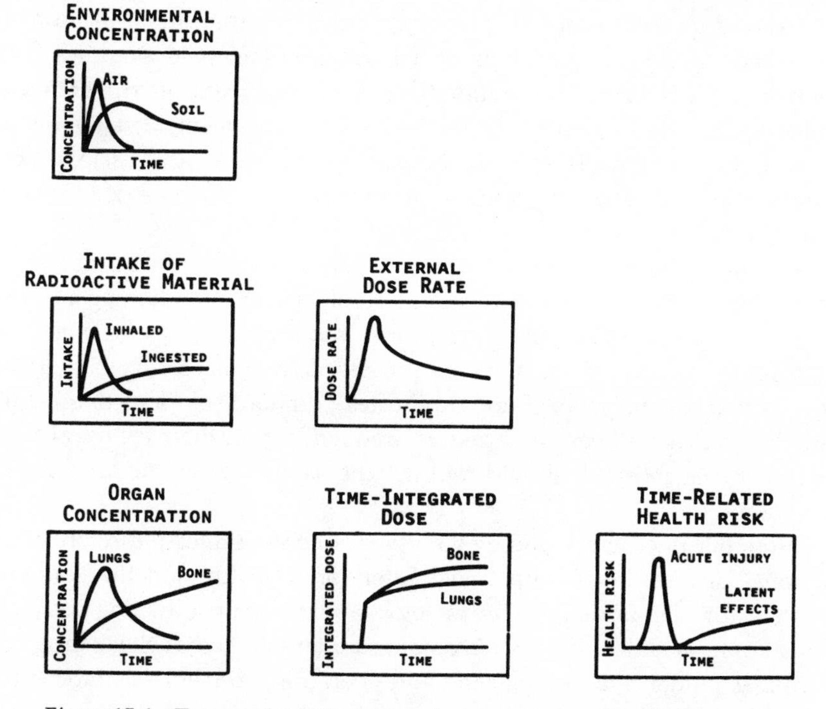

Figure 15.4 Temporal relationships of various types of contamination and accompanying human exposures resulting from environmental releases of radioactive materials

the atmosphere will subject people almost immediately to external exposures (direct radiation) from the cloud as well as exposures to the lungs as a result of the inhalation of radioactive material. Exposures resulting from the deposition of radioactive material from the cloud onto the soil will take longer, and uptake by agricultural crops and pasture grass longer still. Exposures to the bone and other organs such as the thyroid will be delayed, pending uptake and transfer of the material from the lungs. Likewise, exposures to specific organs (other than the stomach and gastrointestinal tract resulting from the ingestion of radioactive material in milk or food) will be delayed until the material is taken up by the blood and deposited in specific body organs. Acute effects from environmental exposures will appear within hours to weeks (depending on the dose); delayed effects (such as latent cancers) from lower-level exposures will not appear for some years.

QUALITY ASSURANCE REQUIREMENTS

To be effective, an environmental monitoring program must be supported by a sound quality assurance program. Such a program must include (1) acceptance testing or qualification of laboratory and field sampling and analytic devices; (2) routine calibration of all associated instrumentation, including flow measurements on field sampling equipment; (3) a laboratory cross-check program; (4) replicate sampling on a systematic basis; (5) procedural audits; and (6) documentation of laboratory and field procedures and quality assurance records.

Sampling validity and sample preservation also need to be addressed as part of the quality assurance program. Useful tools for maintaining analytic validity include duplicate sample analyses and control charts. As a general rule, 10–15 percent of the samples processed in a laboratory should be resubmitted for analysis as blind duplicates. Standard solutions (large bulk samples that have been analyzed so frequently that their chemical content is well established) should routinely be used to check the accuracy of new data.

Through services provided by the Environmental Protection Agency, the National Institute for Occupational Safety and Health, and the National Institute for Standards and Technology, laboratories conducting environmental radionuclide analyses can obtain standard and cross-check samples, as well as guidance in establishing and operating a quality assurance program. The USNRC requires all laboratories performing analyses of environmental samples from commercial nuclear power plants to participate in the

EPA program or its equivalent. Whenever discrepancies are noted, follow-up action is required to determine and eliminate the causes.

Computer and Screening Models

Computers offer an enormous capacity for collecting, storing, and organizing information that can assist in understanding the global environment and the effects of human activities. Computer monitoring systems are being used, for example, to study and maintain records on industrial and natural processes, including documentation of trends in carbon dioxide releases and increases in atmospheric temperatures. One form of industrial monitoring in the United States that has made rapid progress is pollution tracking, that is, collection of data on the identity and quantities of toxic substances being released and their sources. The Toxic Release Inventory, for example, includes annual data on chemicals released to land, air, and water, from about 24,000 industrial facilities in this country. Public release of these data has proved to be a powerful stimulus for the application of control measures, particularly when grassroots organizations become aware of local problems (Young, 1994).

Another major application of computers is in the development and use of models for evaluating the transport of environmental contaminants. The EPA has published guidelines for estimating doses from acutely toxic chemicals and carcinogens, and the Nuclear Regulatory Commission has published similar guidance for estimating doses to exposed individuals from radiation and radioactive materials (USNRC, 1977). Using this and related environmental transport information, programmers have developed models for estimating the doses to population groups from a variety of environmental contaminants. The initial phases of this effort aimed at development of models for estimating exposures due to inhalation of airborne releases. As regulatory requirements have become more stringent, and regulatory agencies have recognized the need to assess exposures through other avenues, these models have been expanded to incorporate terrestrial and aquatic food-chain pathways. The choice of model is normally based on the computer available, the type and complexity of the releases, the environmental pathways associated with the specific site or facility, and the living habits of the people being exposed.

Unfortunately, the ready availability of computers has led to models that are increasingly sophisticated and not readily applicable, particularly for operations involving small quantities of contaminants, as is the case in the release of certain radioactive materials. Recognizing the need for simpler models to assess compliance, the National Council on Radiation Protection and Measurements has developed a series of screening techniques (Till, 1988; NCRP, 1989, 1996) that can be used by the operators of any facility releasing radionuclides into the environment. Although the techniques incorporate all important transfer mechanisms, exposure pathways, and dosimetry parameters, they involve only a few calculations and require a minimum of site-specific data and decisions on the part of the user. Because of their general applicability, many of these models are being modified for assessment of nonradioactive environmental pollutants.

Comprehensive Exposure Assessment

Even if the monitoring procedures outlined above are followed, voids occur in the data—and large uncertainties in the accompanying exposure estimates—particularly in instances where it is desired to assess the total exposure of specific population groups from all possible sources. To meet these needs, environmental monitoring specialists have in recent years expanded and supplemented their procedures. One example is the National Human Exposure Assessment Survey (NHEXAS) currently being developed under the auspices of the EPA. It is a multiple-component program that includes: (1) the distribution of questionnaires to provide baseline information on the lifestyles, activities, and sociodemographics of population groups; (2) the collection of soil, house dust, indoor air, tap water, and diet samples; (3) the analysis of these samples for some 30 compounds, including airborne particulates in specific size ranges; and (4) the collection of samples of blood, urine, and hair as biological indicators of human uptake of individual contaminants (Newman, 1995b). Armed with the resulting information, it should be possible to conduct detailed assessments of the exposures of the monitored groups to each and every environmental contaminant and/or physical stress in their daily lives. Although such programs are more expensive to conduct than those directed at evaluations on a local basis, they are considered essential when it is important and/or necessary to make longer-range and fuller-scale assessments of the impact of environmental pollutants.

Further enhancing the capabilities of these more comprehensive monitoring programs are data being generated through ground-based remote-sensing and fast-response instruments. It is now possible to make real-time measurements of chemical species and atmospheric conditions in both the stratosphere and the troposphere. Extension of similar capabilities to other components of the environment is enabling scientists to develop programs for evaluation of the impacts of a wide range of contaminants on the capacity of the environment to sustain ecosystems. An example is the Environmental Monitoring and Assessment Program (EMAP) being developed by the EPA. It is designed to determine the extent (numbers, miles, acres) and geographic distribution of each ecosystem class of interest; to assess the proportions of each ecosystem class that is currently in good or acceptable condition; to evaluate what proportions are degrading or improving, in what regions, and at what rate; and to appraise the likely causes of harm and methods for seeking improvement (Bromberg, 1990; Thornton, Hyatt, and Chapman, 1993; EPA, 1993; Griffith and Hunsaker, 1994).

One of the basic premises of EMAP is that assessments of changes in the nation's ecosystems will require data collected over long periods of time (decades, at a minimum) and covering large geographical areas. Critical to the success of this program will be the selection, development, and use of ecological indicators that are relevant, robust, and reliable (Griffith and Hunsaker, 1994). There must be a clear tie between the indicator, the effect it is designed to measure, and the accompanying cause-and-effect connection, particularly as it relates to possible effects on human health (Newman, 1995a). If the indicators are properly selected, the resulting data will provide a broader understanding of ecosystems, help scientists anticipate emerging environmental problems before they reach crisis proportions, and assist legislators and environmentalists in addressing national and international monitoring and regulatory needs.

In essence, the data produced by EMAP will represent what might be called America's ecological report card (EPA, 1993). Some monitoring specialists have described EMAP's goal as determination of the health of an ecosystem, in much the same way as a doctor determines the health of a patient. Proponents believe that such a system, properly applied, would provide information on the point at which ecosystems begin to break down. Although analogies can readily be identified for making such a comparison (Table 15.4), some individuals oppose the analogy because ecosystems, unlike organisms, are not consistently structured, do not behave in a predictable manner, and do not have mechanisms such as the neural and hormonal systems of organisms to maintain homeostasis (Griffith and Hunsaker, 1994).

Another weakness of such an analogy is that it contributes little to identification of the available choices for control or to the decisionmaking process (Wolman, 1995). In spite of these difficulties, essentially everyone agrees

Table 15.4 Comparison of ecological health research and human health diagnosis

Ecologic research issue	Analogous human health area
Early warning of ecosystem transformation (e.g., localized fish kill in a river)	Early warning of disease, as PSA (prostate-specific antigen) test for prostate cancer
Exotic plant/animal/virus invasion or outbreak of native indigenous pathogens	Epidemiologic studies of disease outbreaks within a population group
Presence of "sensitive zones" in ecosystems	Study of certain body organs that are crucial to the functioning and well-being of the whole
Possible development of ecosystem immunity to particular classes or combinations of stress	Immune antibody responses to foreign antigens

that the program will prove to be an extremely valuable tool for understanding and anticipating future problems (Stone, 1995).

As currently planned, EMAP will serve a wide spectrum of users: decisionmakers who require information to set environmental policy, program managers who must assign priorities to research and monitoring projects, and scientists who desire a broader understanding of ecosystems. It is hoped that the program will promote the development of more cost-effective regulatory and remedial actions (Bromberg, 1990; EPA, 1993).

The General Outlook

An enormous amount of information is being generated through environmental monitoring programs. To assure that these data are properly analyzed and disseminated, the EPA annually reports on releases and concentrations of various pollutants, including toxic chemicals, in the ambient environment (EPA, 1989; 1994), and the USNRC publishes summaries of radionuclide releases and estimates of the associated population doses from the operation of commercial nuclear power plants in the United States (Baker, 1994). Estimates of radionuclide releases and population doses from government-operated nuclear facilities are published by the U.S. Navy (1995) and the DOE (Wilhelmsen et al., 1994). Similar reports on a wide range of contaminants are prepared and issued by a number of state environmental and regulatory agencies.

As a result of these and other efforts, applications of data obtained through environmental monitoring programs continue to expand. Increasing use is made of information obtained by earth-orbiting satellites. In accordance with a 1995 agreement, officials in Russia and the United States are exchanging data obtained through the operation of spy satellites during the cold war. Scientists believe that long-term records could provide valuable information on the effects of clouds on heating and cooling of the earth, as well as insight into the effects of pollution on cloud formation and better methods for modeling these effects. Other information provided by satellites may lead to methods for providing warnings about the occurrence of volcanic eruptions, earthquakes, and other types of natural disasters (Lawler, 1995).

With the recognition that people today spend most of their time indoors, emphasis on indoor monitoring, particularly of airborne pollutants, will increase. At the same time, recognition that some people will still spend considerable amounts of time outdoors and that outdoor pollutants often gain access indoors, the need for measurements in the ambient environment will continue. Analyses will also be necessary to assess the concentrations of key contaminants in water and food, and to evaluate the potential long-term effects of airborne contaminants such as carbon dioxide, chlorofluorocarbons, and oxides of sulfur. As more comprehensive databases are established, they will require better integration and more systematic methods of compilation and storage if they are to provide maximum benefit in assessing the long-term effects of various environmental contaminants. Computer models will also need to be refined not only to predict human exposures but also to determine the interrelationships among environmental factors, and to evaluate their potential impacts on ecosystems. New measures will need to be developed for monitoring transboundary air pollution flows, ultraviolet radiation, acid precipitation, desertification or land degradation, land conversion, deforestation, ocean productivity, biodiversity, and species destruction (Young, 1994).

Continuing attention should be directed to the support of environmental monitoring programs conducted by state and local regulatory and public health agencies. Such programs are essential as independent checks on the consistency of compliance by industrial and federal facilities. Incorporated as a mandatory component should be periodic reevaluations in response to quantitative and qualitative modifications in effluent releases, developments in analytic techniques and equipment, improved knowledge of the behavior of specific contaminants within the environment, new patterns of environmental use and population distribution, and changes in regulatory requirements.

16

RISK ASSESSMENT

IN A personal sense, risk can be defined as the probability that an individual will suffer injury, disease, or death under a specific set of circumstances. In terms of environmental health, the concept of risk must be expanded to include possible effects on other animals and plants, as well as on the environment itself. Knowing that a certain risk exists is not enough, however. People want to have some idea of how probable it is that they or their environment will suffer and, if they do, what the effects will be. Determination of the answers to these questions involves the science of risk assessment.

Risk assessment ranges from evaluation of the potential effects of toxic chemical releases known to be occurring, to evaluation of the potential effects of releases due to events whose probability of occurrence is uncertain. In the latter case, the risk is a combination of the likelihood that the event will occur and the likely consequences if it does. In essence, the process of risk assessment requires addressing three basic questions: What can go wrong? How likely is it? If it does happen, what are the consequences? (Kaplan and Garrick, 1981).

Once the risk has been assessed, it can be expressed in qualitative terms (such as "high," "low," or "trivial") or in quantitative terms, ranging in value from zero (certainty that harm will not occur) to one (certainty that harm will occur). At the same time, it must be recognized that a given risk assessment provides only a snapshot in time of the estimated risk of a given toxic agent at a particular phase of our understanding of the issues and problems. To be truly instructive and constructive, risk assessment should always be conducted on an iterative basis, being updated as new knowledge and information become available.

Once a risk has been quantified, the next step is to decide whether that risk is sufficiently high to represent a public health concern and, if so, to determine the appropriate means for control. *Risk management* may involve measures to prevent the occurrence of an event as well as appropriate remedial (protective or mitigative) actions to protect the public and/or the environment in case the event does occur. Each of these steps is accompanied by a multitude of related uncertainties. In fact, as many uncertainties are involved in deciding how to use risk assessment to make regulatory decisions as in conducting the risk assessments themselves.

Applications

As described in Chapter 2, two basic concepts are applied in assessing risks from toxic agents. These are exemplified by the two graphs shown in Figure 16.1. The graph on the left represents the linear nonthreshold dose-response curve that is generally assumed to apply to carcinogenic agents: any dose, regardless of how small, carries an associated risk. Through risk assessment, scientists seek to quantify the risk associated with a given dose and then to use that information to establish an acceptable level of exposure. The graph on the right represents the threshold type of response that is assumed to apply to many noncarcinogenic agents. Although a person is considered to be "safe" as long as the dose is below the threshold, there is still need for a quantitative estimate of the risk associated with any dose

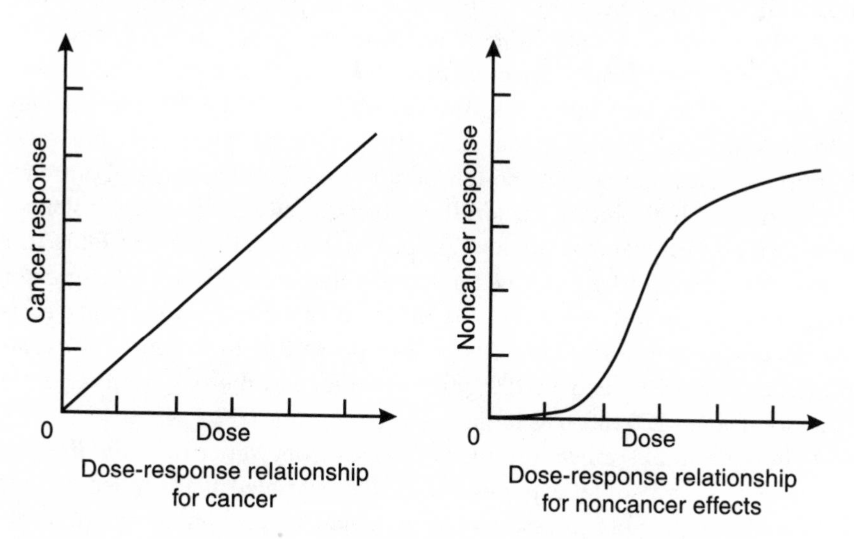

Figure 16.1 Dose-response curves for cancer and noncancer agents

above the threshold. It is in this range that scientists apply risk assessment procedures.

Among the earliest applications of the techniques of risk assessment were those of the National Aeronautics and Space Administration in assessing the safety of travel in space (Garrick, 1989), and those of the Nuclear Regulatory Commission in assessing the safety of nuclear power plants (USNRC, 1975). Although these applications were designed to provide quantitative estimates of the risk to the public in case of an accident, the same techniques have in subsequent years helped assess the risks associated with a wide array of industrial operations and products.

As a result of the issuance of Executive Order 12866 (Federal Register, 1993), such applications will undoubtedly continue to increase. The order requires that all federal agencies developing new regulations compare the risks each action is intended to address against other risks within that agency's jurisdiction, and provide cost/benefit analyses of the impacts of the proposed actions. Another impetus for the use of risk assessment is the severe budgetary constraints under which most regulatory agencies and the regulated industries are operating. Risk assessment appears to offer such organizations one of the best techniques for comparative evaluation of the risks associated with various toxic agents in the environment. Without such a tool, the monies needed for addressing critical issues would likely be spent on hazards of far less importance.

In spite of these apparent successes, applications of risk assessment are not without difficulties and controversies. The public, in particular, often finds the concept of risk difficult to understand. For many people, the concept of a "safe" dose is far more acceptable. Individuals also fail to realize that the risks associated with the hazards of everyday life vary widely and that the risks most feared are not necessarily those that are most important (Breyer, 1993). Although people may be concerned about contaminants (such as chloroform) in their drinking water or toxic substances in their food (for example, aflatoxin in peanut butter), the data in Table 16.1 show that many other risks in their daily lives should be of far greater concern: on a quantitative basis, the risk of death from motor vehicle accidents is over 40 times as large, the risk of death due to violence (as exemplified by homicide) is over 100 times as large, and the risk from cigarette smoking is over 250 times as large.

Although initial applications of risk assessments were primarily directed to the effects on people, experience has demonstrated that the techniques are equally applicable to assessment of ecological and environmental im-

Table 16.1 Estimates of comparative risks to individuals

Hazard	Probabilistic risk estimate	Degree of uncertainty
Death before age 85, all causes	0.70	Low
Death from cancer, lifetime risk	0.25	Low
Death from cigarette smoking, one pack per day for 40 years	0.13	Medium
Death from homicide (black male), lifetime risk	0.05	Low
Death from occupational exposure to benzene, 10 ppm concentration, for 30 years	0.05	Medium–High
Death from motor vehicle crash, lifetime risk	0.02	Low
Death of police officer on active duty, 30 years of service	0.007	Low
Death from eating 4 tablespoons of peanut butter per day, lifetime risk (aflatoxin)	<0.0005	Medium
Death from drinking water with EPA limit of chloroform, lifetime risk	<0.0005	High
Death from inhaling formaldehyde in urban air, 5 ppb concentration, lifetime risk	<0.000008	High

pacts. In line with this thinking, the Environmental Protection Agency is developing ecological risk assessment guidelines, the thought being that regulatory agencies will then be able to make better policy decisions on proposed developmental activities in geographical areas that might be particularly sensitive (Hamilton, 1992). Such guidelines might apply to the possible impact of the drainage of wetlands on the ecology of a region, or of discharges into the environment of chemicals that destroy the ozone layer and/or lead to global warming. A particular challenge is to identify the key species and to define the endpoints that can serve as indicators of the impacts of multiple stressors on entire ecosystems (Renner, 1996).

In evaluating the effects of toxic agents on people, researchers must determine whether the goal is to assess the impacts on individuals or on population groups. Although some may assume that it would be easier to estimate the risks to an individual, this is not necessarily the case. Most measurements of air pollution, for example, are designed to determine the average concentrations of specific contaminants in the ambient environ-

Table 16.2 Estimates of comparative risks to populations

Hazard	Annual total deaths in U.S.	Degree of uncertainty
Cigarette smoking	300,000	Low
Alcohol abuse	150,000	Low
Highway travel	45,000	Low
Homicide	20,000	Low
Indoor air toxins	6,000	Medium
Outdoor air toxins	3,000	High
Pesticide residues in foods	3,000	High
Airline travel	1,000	Low

ment. Consequently, these measurements are primarily limited to assessing the exposures of members of the public who spend large amounts of time outdoors. They are not indicative of the exposures to people who spend most of their time indoors, or to a specific individual who for various reasons may be subjected to exposures above the average. The accuracy of estimates of the risks either to individuals or to population groups depends on the nature of the toxic agent, the extent and availability of exposure measurements, the range and duration of the exposures, and other factors such as the physical characteristics and lifestyles of those exposed. The necessity of making assumptions concerning each of these factors adds to the uncertainties in the associated risk estimates, whether they are for individuals or for population groups.

The risk to an individual is generally expressed in terms of the likelihood of suffering a given detrimental effect as the result of exposure to a given agent. This is the type of data that were presented in Table 16.1. The risk to a population group, on the other hand, is generally expressed in terms of the estimated number of excess deaths that will occur as a result of exposures to a given agent. The goal of this type of risk assessment is to express the risk in terms of the impact on society as a whole, that is, in terms of the "societal risk." The concept of collective dose (Chapter 14) is one method of estimating this type of risk. Examples of the impacts of various risks, in terms of the estimated numbers of resulting annual deaths in the U.S. population, are shown in Table 16.2.

This table, like Table 16.1, includes an estimate of the degree of uncertainty for each number. It is important that this type of information be included; in fact, a "complete" risk characterization will include not only the uncertainties of the given estimates, but also the estimates that would have resulted had alternative assumptions and methods been utilized (Gray, 1994).

Qualitative Risk Assessment

Although most regulatory agencies attempt to develop "quantitative" risk assessments, the number of facilities having the potential for release of toxic chemicals makes universal quantitative assessment impossible. To provide some perspective, the EPA estimates that 33,000 uncontrolled toxic waste sites exist in the United States. If 10 percent of these, say 3,000, were to be assessed, and a community epidemiological study were to be implemented to quantify the harm to the neighboring population, the cost would be in excess of $6 billion (assuming a modest expenditure of $2 million per study). This sum is clearly beyond the economic capabilities of the country, much less within the budget and staffing capability of public health agencies.

For this reason the common approach is to apply as an initial step some type of "qualitative" or "semiquantitative" risk assessment. Possibilities include: (1) qualitative characterizations where health risks are identified but not quantified (for example, hazard evaluations and carcinogen classification schemes); (2) qualitative risk estimations where chemicals are ranked or classified by broad categories of risk (for example, chemical potency classification schemes); and (3) semiquantitative approaches where effect levels (for example, "no observable effect") are used in combination with uncertainty factors to establish "safe" exposure levels. Each of these approaches can and has been used to assess risks and, in a broad sense, constitutes a form of risk assessment (Smith, Kelsey, and Christiani, 1993).

One example of the qualitative approach is the procedure developed by the EPA for assigning each toxic agent to one of five categories, depending on its potential for causing cancer in humans: group A—carcinogenic to humans; group B—probably carcinogenic to humans; group C—possibly carcinogenic to humans; group D—not classifiable as to human carcinogenicity; and group E—evidence of noncarcinogenicity for humans (EPA, 1987). Simply knowing which of these categories applies to a given toxic agent can be very useful in assessing its risks.

Perhaps the best example of a qualitative approach to risk assessment is the "public health assessment" methodology that has been developed by the Agency for Toxic Substances and Disease Registry (ATSDR), primarily for evaluation of the potential health hazards of toxic waste (Superfund) sites. As employed by ATSDR, a public health assessment is "the evaluation of data and information on the release of hazardous substances into the environment in order to assess any current or future impacts on public health, develop health advisories or other recommendations, and identify studies or actions needed to evaluate and mitigate or prevent human health effects" (Johnson, 1992, p. 30). In this context, "evaluation of data and information" means employing professional judgment, according to specified guidelines, to characterize the nature and extent of the hazard to human health presented by releases from individual hazardous waste sites.

Public health assessments are also used to identify conditions of exposure to hazardous substances that, when reduced in severity, will prevent morbidity and mortality. In many ways public health assessments are designed to complement the quantitative risk assessments conducted by the EPA (see below), and in many ways they provide an independent review or evaluation of Superfund sites. They also furnish a mechanism for screening such sites to identify those that should be assigned priority for more detailed quantitative risk assessment.

Although public health assessments are basically qualitative in nature, ATSDR officials responsible for conducting them use a large amount of data. Obtained primarily through site visits, these data include information on environmental contamination (normally obtained from EPA and sometimes supplemented by data from state agencies), health problems (including community-specific cancer rates or adverse reproductive outcomes, normally obtained from state and local health agencies), and community health concerns of those residing near a site.

On the basis of their experience, ATSDR officials have identified ten key substances that are specifically considered in the evaluation of any hazardous waste site. These are: (1) lead, (2) arsenic, (3) mercury, (4) vinyl chloride, (5) benzene, (6) cadmium, (7) polychlorinated biphenyls, (8) chloroform, (9) benzo(b)fluoranthene, and (10) trichloroethylene. For each of these, and a host of other substances, ATSDR has developed a toxicologic profile that describes what is known about the related toxicity and human health effects.

In a similar manner, ATSDR has identified seven priority health conditions that receive specific attention in the evaluation of a waste site: (1) birth defects and reproductive disorders, (2) cancers, (3) immune function disorders, (4) kidney dysfunction, (5) liver dysfunction, (6) lung and respiratory diseases, and (7) neurotoxic disorders. This list was compiled on the basis of a comprehensive analysis of the literature on toxicologic and human health effects associated with known Superfund hazardous substances.

Based on the public health assessment, each site is placed in one of five categories in terms of its overall significance to public health and as a guide for follow-up action: (1) urgent public health hazard, (2) public health hazard, (3) indeterminate public health hazard, (4) no apparent public health hazard, and (5) no public health hazard (Johnson, 1992).

Quantitative Risk Assessment

In contrast to ATSDR, the EPA concentrates on quantitative risk assessments, its goal being to characterize in quantitative terms the potential adverse health effects of human exposure to toxic agents. Such assessments currently involve four primary steps (Rosenthal, Gray, and Graham, 1992; NRC, 1994). These are presented in schematic form in Figure 16.2 (Naugle and Pierson, 1991) and are described in more detail below.

Hazard identification. This is a qualitative determination of whether human exposure to an agent has the potential for adverse health effects. It generally requires certain types of information, including the identity of the toxic agent and its physical and chemical properties, and the outcomes of related mutagenesis and cell transformation studies, laboratory research using animals, and human epidemiological studies. Once the identity of an agent is known, databanks such as the EPA integrated risk information system (ICF, 1986) can be consulted for information on its potential toxicity. Knowledge of its chemical and physical properties is necessary to assess the degree to which it can become airborne, be inhaled, and be taken up by the body, as well as other factors such as its solubility in water and availability for transport through the food chain.

Dose-response evaluation. This is a quantitative estimate of the hazard potency (power to produce adverse effects) inherent in receiving a dose from a specific toxic agent. If available, dose-response estimates based on human data are preferred. In their absence, information from studies of other animal species that respond like humans must be used. As explained in Chapter 2, serious problems exist in applying animal data to humans and create a significant source of uncertainty in the accompanying risk estimates.

Exposure assessment. This is an estimate of the amount of exposure to the

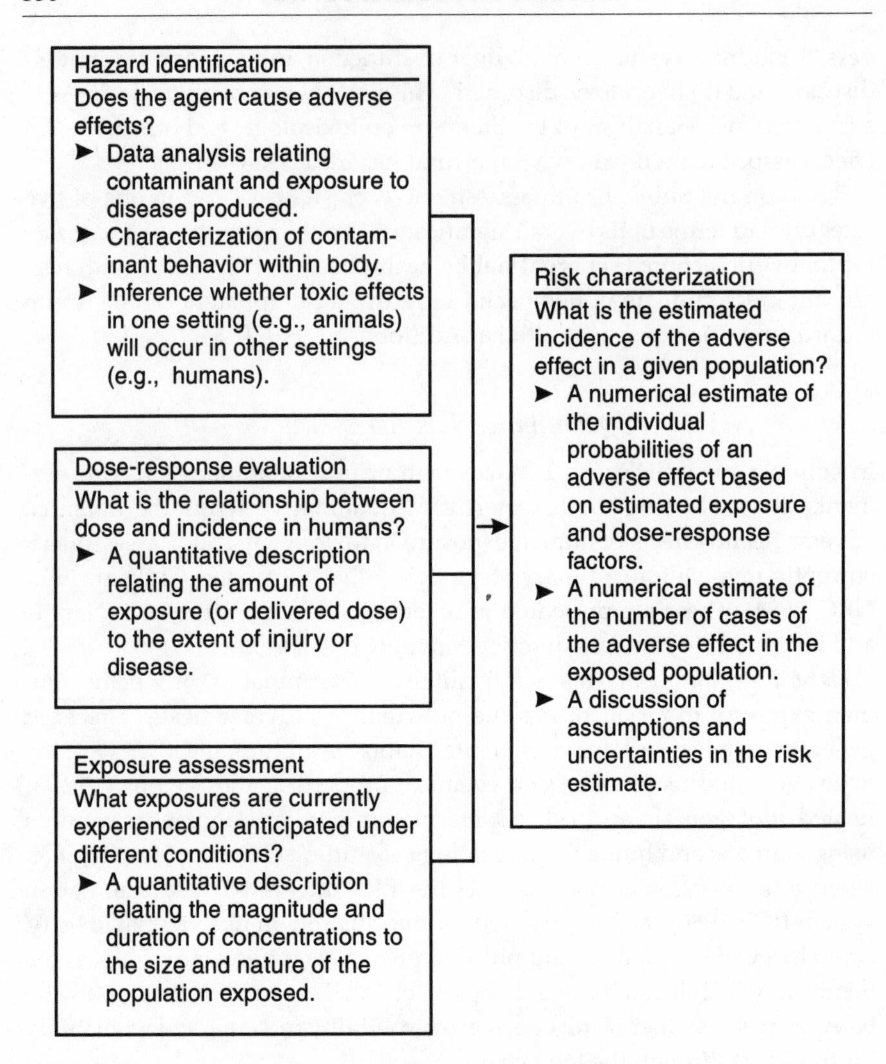

Figure 16.2 Components of a traditional risk assessment

agent and the resulting dose being experienced by people and the environment. It may be directed at routine or normal releases from a facility, or it may, in the case of accident assessment, involve estimating both the probability of the event and the magnitude of the accompanying releases. In making such assessments, certain factors must be considered.

1. The chemical and physical characteristics of the toxic agent, particularly those that will play a major role in its environmental transport

and behavior. These are similar to the data required for the hazard identification step. Key parameters include partition coefficients, retardation factors, bioaccumulation factors, and degradation rates.

2. Identification of the person to be protected. Often, regulations stipulate that the person to be protected is the "maximally exposed individual." At other times, care must be taken to identify and protect special groups within the population, such as children and pregnant women, who may be more vulnerable. This is one of the reasons for incorporating conservatisms into the assessment of risks (Caplan and Lynch, 1996).

3. Recognition of the difference in the exposure measured and the dose that will actually be received by the exposed individuals. The dose that a person receives depends on many factors. In the case of airborne materials, it depends on whether the individual breathes through the mouth or nose, the breathing rate, the person's age, variations in the deposition of the inhaled material within the lungs, and uptake of the agent in the blood and subsequent distribution within the body. Since it is not always possible to make direct measurements of the quantity of a toxic agent at the location within the body where it deposits and may cause harmful effects, exposure measurements are in essence used as a precursor and/or surrogate for dose (Herrick, 1992).

Risk characterization. This involves calculating or estimating the adverse consequences (risk) to people being exposed to a given amount of a specific agent. Such a process requires integrating the results of the above assessments to produce quantitative estimates the associated health and environmental risks. Because risk estimates have significant limitations, the EPA requires that, in addition to the estimate of risk (usually expressed as a number), the risk characterization contain a discussion of the "weight of the evidence" for human carcinogenicity (for instance, the EPA carcinogen classification); a summary of the various sources of uncertainty in the risk estimate, including those arising from hazard identification, dose-response evaluation, and exposure assessment; and a report on the range of risks, using EPA's risk estimate as the upper limit and zero as the lower limit (Rosenthal, Gray, and Graham, 1992; EPA, 1987).

As may be gathered from the preceding discussion, estimates of the exposures (doses) to the people potentially affected by a toxic agent carry with them large uncertainties. If those responsible for exposure assessments are too conservative in their assumptions, the associated risk estimates may be far in excess of what might be experienced in the real world. For this reason many public health officials urge that the outcomes of such assessments be as realistic as possible.

Accident Situations

The previous discussions have been directed primarily to the assessment of risks associated with routine releases from a facility or source. At times it may be equally important to assess the risks associated with accidents, or the performance of a system in a disrupted mode. Consider the accident in 1976 in an industrial facility in Seveso, Italy, in which a large amount of dioxin was released into the environment; a similar accident in Bhopal, India, in 1984, resulting in the release into the air of enormous quantities of methyl isocyanate (with over 2,000 deaths); the 1979 accident at the Three Mile Island nuclear power plant in Pennsylvania; and the 1986 accident at the Chernobyl nuclear power plant in the former USSR, which resulted in the release of huge amounts of radioactive material into the atmosphere. Evaluation of the risk of accidents of these types must include not only an estimate of the magnitude of the associated consequences but also the likelihood that an accident leading to such a release will occur. Assessments of this type require the following steps beyond those traditionally undertaken in the assessment of routine releases (Hunter, Cranwell, and Chu, 1986).

Accident scenario development and screening. The objective here is to postulate physically possible sequences of events and processes that could lead to a major disruption in a facility and the release and transport of toxic materials into the neighboring environment. Since industrial facilities are located in areas widely diverse in population distribution and land use, and since seismic and tectonic events frequently serve as initiators of disruptive events, the risk of an accident is generally site specific and facility specific. Human error can be a major source of disruptive events, so the magnitude of the risk may depend on the training and skill of the facility operators. For large industrial facilities, the risk assessment process involves identifying the events and processes that could initiate a release and then combining those events and processes into physically reasonable scenarios that can be analyzed and evaluated.

Consequence assessment. After the scenarios have been selected, a sequence of models is used to estimate the consequences of each. Typically, the first suite of models is designed to examine various mechanisms through which the toxic agent might be released. Once the release is assumed to have occurred, a follow-up analysis describes its transport from the source out into the environment. Such assessments have benefits far beyond the calculated risk estimate. One is the insight gained in identifying not only voids in the data needed to assess the accident potential of a facility, but also the need for research to better understand how the facility will behave in a variety of nonnormal situations. Often specific voids identified in the data—for example, estimates of the intensity and frequency of occurrence of natural events (such as earthquakes)—cannot be filled through laboratory or field experiments. Risk assessors then must depend on expert judgment to provide the missing information. The selection of appropriate experts, the solicitation of information from them, and methods of resolving conflicting opinions are often areas of controversy and sources of added uncertainty.

Uncertainty and sensitivity analysis. Because experimental data are often not available to confirm estimates of accidental releases, the acceptability of a facility must be based largely on its predicted behavior, as derived through the input of expert judgments and the utilization of computer models. Thus it is important to investigate the sources of uncertainty (scenario uncertainty, data or parameter uncertainty, model uncertainty) that affect the predictions of releases, and to share this information with those who will be taking action based on the risk assessments. Uncertainties also arise from a lack of understanding of the processes being modeled, a limited capability to represent these processes mathematically, or insufficient data describing the system or the processes acting on it. Another vital step is sensitivity analysis, which allows quantitative estimation of the amount of variation in model output (risk estimates) that will occur as a result of specified variations in the parameters fed into the model. Sensitivity analysis provides a means for identifying those parameters that are most significant in terms of the final product of the analysis—the risk estimate.

Regulatory-compliance assessment. Because accident experience is fortunately very limited, and associated risk assessments are generally performed prior to facility construction and operation, the goal of many risk assessments is to demonstrate whether the proposed facility will comply with the applicable regulations. In the case of nuclear power plants, the Nuclear Regulatory Commission has established safety goals that serve as a gauge of the adequacy of the protective measures; in the case of low-level radioactive waste disposal, the USNRC has established regulations that account for the potential long-term migration of radioactive materials through the environment as well as accidental human intrusion into the waste site after loss of institutional control.

Risk Management

The distinction between risk assessment and risk management is obvious in the matter of regulatory decisionmaking. Risk management is the process of integrating the risk assessment results with a variety of other information (engineering data, socioeconomic and political concerns), weighing the alternatives, and selecting the most appropriate public health action for reducing and/or eliminating the risk (Falk, 1992). In fact, these latter factors often play a more significant role in risk management decisions than the nature and magnitude of the risks themselves (Smith, Kelsey, and Christiani, 1993).

A successful program in risk assessment and risk management requires that certain aspects of the two processes be kept separated. Otherwise, it is difficult not to confuse the scientific conclusions about the nature of a risk with the social, political, and economic concerns over how the risk should be managed. This does not mean, however, that there should be no communication between the two groups (Wilson and Clark, 1991; Jasanoff, 1993). Once an assessment has been made, those responsible for management of the risk must be provided with both the risk estimate and the context under which it was developed. The primary areas of uncertainty must be defined, along with the degree to which they may influence the accuracy of the risk estimate (NRC, 1994). The techniques of risk management can then be used to aid in setting priorities for action and in analyzing alternative control strategies. Analyses must be conducted to determine the cost-effectiveness of a particular strategy and to weigh the relative costs, benefits, and risks of various control options (Douglas, 1985).

ACCEPTABLE LEVELS OF RISK

One of the primary considerations in risk management is the level of risk at which environmental or occupational controls should be applied. Such a level is required, for example, to determine when cleanup operations at a contaminated site can be terminated, or whether a product is sufficiently safe for consumption or use by the general public. The establishment of such levels has been an ongoing challenge among federal agencies for many years. If scientific evidence were the only basis for establishment of acceptable risk levels, decisions would be relatively easy. As will be noted below, public input and the need for flexibility have played a major role, the result being that acceptable levels are being set on a case-by-case basis.

One of the best examples of the application of risk levels is the Superfund toxic waste cleanup program being conducted by the Environmental Protection Agency. In its initial efforts, the EPA used a lifetime risk level of 1 in 1 million as justification for an affirmative decision about the acceptability of a site, that is, that no cleanup was required. In some situations this same risk level served as the goal for cleanup of a contaminated site. More recently, the EPA has defined acceptable excess cancer risk for the Superfund program as a range from 1 in 10,000 to 1 in 1 million. This approach was adopted to provide risk managers the flexibility to consider site-specific factors such as the number of people being exposed, and the feasibility and cost-effectiveness of cleanup. In providing guidance for state water-quality standards, the EPA has recommended a risk level ranging from 1 in 100,000 to 1 in 10 million. Meanwhile, in its air pollution control program the goal is to reduce the risk to as many people as possible to 1 in 1 million, while simultaneously assuring that the maximally exposed individual is protected against risks greater than 1 in 10,000 (Graham, 1993a).

Key decision factors such as the size of the exposed population, the resource costs of meeting risk targets, and the scientific quality of the associated risk assessments vary enormously from one context to another. Administrative discretion is necessary to weigh these factors; no magic risk number can substitute for informed and thoughtful consideration by accountable officials who work with the public to make balanced decisions.

RISK COMMUNICATION

Risk management actions can range from a program of public education to termination of the activity that is the source of the risk. Essential parts of the risk management process are to invite and encourage public participation, to communicate the nature and extent of the risk in understandable terms to both experts and laypersons, and to share background information fully with all affected groups to help make it possible to arrive at an informed estimate of the risk and a wise approach to its management (Falk, 1992; NRC, 1996).

Table 16.3 Activities with a one-in-a-million risk of death

Group 1. Living in the United States: time to accumulate a one-in-a-million risk of death

Activity	*Time*
Motor vehicle accident	1.5 days
Fall	6 days
Drowning	10 days
Fire	13 days
Firearms	36 days
Electrocution	2 months
Tornado	20 months
Flood	20 months
Lightning	2 years
Animal bite or sting	4 years

Group 2. Occupational risks: time to accumulate a one-in-a-million risk of death

Occupation	*Time*
Mining and quarrying	9 hours
Firefighting	11 hours
Coal mining	14 hours
Construction	14 hours
Agriculture	15 hours
Transport and public utilities	1 day
Police duty	1.5 days
Service and government	3.5 days
Manufacturing	4.5 days
Trade	7 days

Group 3. Personal lifestyle: activities required to represent a one-in-a-million risk of death

Eating and drinking
0.5 liter of wine
6 pounds of peanut butter (aflatoxin)
180 pints of milk (aflatoxin)
200 gallons of drinking water from Miami or New Orleans
90 pounds of broiled steak (cancer risk only)

Smoking
2 cigarettes

Other activities
Paddling a canoe for 6 minutes
Traveling 10 miles by bicycle
Traveling 30 miles by car

Risk communication can be a very difficult task. As previously noted, the public's perceptions of technological risk often bear very little relation to the risk assessments made by scientists (Davies, 1995). The gap between the lay and scientific communities is enormous. Many members of the public have difficulty dealing with risk estimates, particularly when they are expressed in terms of numbers with negative exponents. Individuals also have problems with the difference between numbers that are intended to represent upper bounds and those that are intended to represent best estimates of risk. Equally perplexing is the difference between a number quoted as an annual risk and one intended to represent a lifetime risk. Until recently, scientists and engineers were inclined to believe that the public was simply ignorant of the relevant facts or was irrational and could be disregarded. In reality, many (if not most) of these problems are directly attributable to the inadequate manner in which the scientific community has attempted to communicate with the public.

Although progress is being made, much work remains. One possible approach is to develop new ways of communicating the nature of risk. For years death was the principal effect considered, and risk was expressed almost exclusively in terms of its likelihood. In more recent years, regulatory and public health authorities have recognized the need to take other factors into account. Several examples will illustrate these points.

Because a high percentage of those killed in vehicular accidents are in the younger age groups, many public health authorities believe that "years of life lost" is a better measure of the associated impacts than the mere number of deaths. For similar reasons, people responsible for conducting risk assessments have begun to take into consideration the effects of occupational and environmental agents in terms of the resulting increased injuries and illnesses, and associated health care costs and emotional impacts. Alternative methods have therefore been developed for expressing the risks of various activities and events. These include the identification of various activities that carry with them a risk of death of one chance in a million (Table 16.3), the estimated amount by which various activities and hazards shorten one's life (Table 16.4), and the expenditures required to reduce a given risk sufficiently to save one year of life (Table 16.5).

Another lesson learned is that risk assessment must be conducted in the open and the public must be permitted to take part. To assure that this approach is followed, the EPA has published what it describes as the seven cardinal rules of risk communication (Table 16.6). These rules have been

Table 16.4 Days of life lost through various activities and events

Activity and/or event	Days lost
Natural hazard	
Earthquake	0.2
Hurricane	0.3
Flood	0.4
Tornado	0.8
Lightning	1.1
Sports activity	
Snowmobiling	2
Professional boxing	8
Hang gliding	25
Parachuting	25
Mountain climbing	110
Accident—general	
Electrocution	4.5
Fire	20
Drowning	24
In the home	74
Motor vehicle	207
Accident—occupational	
Trade and services	27
Manufacturing	40
Government	60
Mine quarrying	167
Construction	227
Agriculture	320

adopted by the Chemical Manufacturers Association. One of the essential steps is that the regulators be honest, frank, and open.

REPORTING RISK DISTRIBUTIONS

Closely intertwined with the problems of risk communication, as well as the lack of adequate coordination between the people responsible for risk assessment as opposed to management, is the need for a better system to reflect the uncertainties embedded within quantitative risk assessment (NCRP, 1996). Of utmost importance is that risk analysts avoid the "tyranny of a single number"; instead, they should report risk distributions. These may reflect the scientific uncertainties, or they may reflect unknown variabilities in the nature and extent of the exposures or in the sensitivities of

Table 16.5 Expenditures per year of life saved (in 1990 U.S. dollars)

Lifesaving program	Cost per year of life saved	Uncertainty
Childhood measles vaccinations	<0 (saves money)	Low
Phaseout of leaded gasoline	<0 (saves money)	Low
Safety rules at underground construction sites	52,234	Low
Hemodialysis at dialysis center	56,076	Low
Coronary artery bypass surgery	67,579	Low
Front-seat airbags	108,593	Medium
Dioxin effluent controls at paper mills	5,566,386	High
Control of routine radiation at nuclear power plants	164,875,379	High

Table 16.6 The EPA's seven cardinal rules of risk communication

1. Accept and involve the public as a legitimate partner.
2. Plan carefully and evaluate your efforts.
3. Listen to the public's specific concerns.
4. Be honest, frank, and open.
5. Coordinate and collaborate with other credible sources.
6. Meet the needs of the media.
7. Speak clearly and with compassion.

the exposed individuals or population groups. Another benefit of reporting uncertainties is that risk managers will be forced to make explicit policy judgments about what margins of safety are appropriate (Graham, 1991).

Obviously, a multitude of factors must be considered in developing risk estimates and conveying the results to the public. The Center for Risk Analysis at the Harvard School of Public Health has prepared a list of ten key principles of risk assessment and management. Because of their importance to the material covered in this chapter, these factors are reprised in Table 16.7 (Graham, 1993b).

Table 16.7 Key principles of risk assessment

1. Estimates of attributable health risk should make use of the best available science.
2. Since reputable scientists often do not agree about how to assess risk, scientific disputes should be acknowledged.
3. When hard data are lacking, risk assessments should be explicit about any assumptions and should indicate degree of sensitivity of results to plausible changes in assumptions.
4. Meaningful risk assessments usually develop a central estimate of risk, as well as upper and lower bounds on risk that acknowledge the extent of scientific uncertainty.
5. Public policy decisions about acceptable risk require public participation and application of democratic principles.
6. No quantitative level of risk exists that is universally acceptable or unacceptable; acceptable depends on the circumstances, the people affected, and the decision context.
7. Valid decisions about health risk require consideration of other cherished values such as quality of life, equity, ecological health, personal choice, and economic welfare.
8. Programs to reduce risk should be designed to avoid unintended side effects that may increase risk.
9. When risk reduction is desired, economic incentives and information should be considered in addition to conventional command-and-control regulation.
10. The context in which risk occurs (e.g., voluntary versus involuntary risk) may influence public reaction to risk as much as the magnitude of the risk in question.

The General Outlook

Ideally, risk assessment can provide structure to the process of setting standards and provide useful input to public debates on the acceptability of proposed industrial facilities. Quantifying risks and identifying the accompanying uncertainties can also help focus the political debate on how best to assign priorities to various categories of risk and how to decide what levels of risk are acceptable.

Through the risk assessment process, regulators as well as the public will be able to compare the risks of new technologies with familiar, socially accepted risks. This capability will help in deciding whether reduction of specific environmental risks is cost-effective. Risk assessments will also make it possible to avoid controls derived from technology-based standards that appear to be either unreasonably strict or lenient.

At the same time, regulators must be careful not to give disproportionate weight to the numbers produced by risk assessment. In particular, they must not let quantitative risk assessment relegate human values and perceptions to a secondary role. Individuals often have far different concepts of risk than regulators, sometimes overestimating and sometimes underestimating the judgments made by the experts. The sources of these differences are numerous, including how familiar they are with the risk, its magnitude, whether it can lead to fatal consequences, whether it was voluntarily accepted, and whether in their view it can be controlled. Another key factor is the public perception of the legitimacy of the regulatory process (Dwyer, 1990).

In spite of these potential drawbacks, it is anticipated that risk assessment will be increasingly applied in the United States and throughout the world. At the moment, the environmental regulations established by various federal agencies have no common risk basis. Depending on the agency, the nature of the toxic agent, and the number of people likely to be exposed, existing limits on the risk levels at which various environmental and occupational toxic agents are regulated cover a wide range. Similar differences exist from one country to another. Obviously, a strong effort is needed to resolve these differences. Without some type of agreement, chemicals viewed as unacceptable for application in one country might be deemed acceptable in another. The resulting confusion might dramatically affect international trade.

Also needing more attention are risk assessments of combinations of toxic agents. Few members of the public are exposed to a single chemical; most experience exposures to combinations of toxic agents, often accompanied by other biological and physical stresses. Although evaluation of such combinations through epidemiological or toxicological studies would be difficult, short-term cellular and molecular techniques do allow evaluation of a wide range of exposure conditions. Such techniques also permit examination of the mechanisms of interactions of combinations of agents (NCRP, 1993). The synergisms already observed in studies of combinations of risks, such as cigarette smoking and airborne asbestos, make such studies mandatory.

For optimal benefit from risk assessment, it would seem appropriate to maximize the number of toxic agents regulated rather than the stringency with which individual agents are controlled. It would likewise seem wise to concentrate on controlling the principal uses of such agents. A pesticide, for example, may have a dozen uses but only two of these may account for 98 percent of the production. In the face of limited resources, regulatory efforts should be concentrated on these major uses (Albert, 1980).

Finally, it is important to examine the total system to which a given risk assessment is being applied. Unless care is exercised and all interacting factors are considered, risk assessments directed at single issues, followed by ill-conceived management strategies, can create problems worse than those the management strategies were designed to correct. Experts have pointed out, for example, that the installation of devices on coal-fired electricity generating stations to control airborne effluents that lead to the production of acid rain makes the plants less efficient. The net result may be an increase of as much as 20 percent in the amount of carbon dioxide released into the atmosphere. This, in turn, can lead to a rise in global warming and associated greenhouse effects. The single-issue approach can also create public myopia by excluding the totality of alternatives and consequences needed for an informed public choice. A full evaluation includes not only presentation of the risks and consequences of the given hazard but also the risks and consequences of the alternative management strategies that might be applied (Starr, 1993).

17

ENERGY

MEETING energy needs and protecting the environment are inseparable goals. The mining of coal, for example, can lead to chronic diseases and injuries among miners and to degradation of the environment (as is often the case in strip mining); the drilling, acquisition, and transportation of oil can lead to spills that contaminate vast areas of land, water, or both; and the use of gasoline in cars leads to air pollution and smog. Internationally, the production and use of oil can lead to conflicts among nations and even to wars. The generation of electricity through either hydropower or the consumption of fossil or nuclear fuels leads to air pollution, problems of waste disposal, and other effects on the environment. Additional concerns in terms of the burning of fossil fuels include the long-range impacts, such as acidic deposition and global warming.

Conservation can play a substantial role in meeting energy needs. One of the most obvious places to practice conservation is the home, where one-sixth of the energy consumed in the United States is currently used. Newer generations of technology are leading to far more efficient ways to light, cool, and heat dwellings, and to more efficient home appliances. Conservation measures have been incorporated into the construction techniques for new buildings. Yet even actions such as these are not always without negative impacts; as an example, tighter, energy-efficient buildings have led to increased indoor radon concentrations and the "sick building syndrome" (ASME, 1989). Even the use of solar energy involves occupational health problems in the manufacture of photovoltaic cells, and injuries or deaths to home owners who fall while cleaning solar panels on the roofs of their houses.

After years of consuming energy resources as if they were unlimited, policymakers today recognize that continued health and safety will be possible only if these resources, particularly the supplies of nonrenewable fossil fuels, are carefully managed, conserved, and protected. The urgency of coping with this situation in the United States is illustrated by the fact that this country, with only about 6 percent of the world's population, consumes 30 percent of the world's energy. In fact, domestic sources of oil have been depleted to the extent that this nation now imports more than half of the oil it consumes. The seven largest countries of the Organization for Economic Cooperation and Development (OECD) consume over 40 percent of the world's production of fossil fuels (World Resources Institute, 1992).

Energy Uses in the United States

Making long-range plans for meeting energy needs requires detailed knowledge of current energy sources and how they are being utilized. The uses of energy in the three major sectors are summarized below (CEQ, 1995).

The *industrial sector* accounts for 38 percent of end-use energy consumption, relying on a mix of fuels. Of the energy it consumes, industry uses 70 percent to provide heat and power for manufacturing. In all, it uses 25 percent of the nation's petroleum, half of that as feedstocks.

This country devotes 36 percent of its end-use energy consumption to the *transportation* of people and goods. Virtually all of this energy consists of petroleum products that power automobiles, trucks, ships, airplanes, and trains. In fact, almost 200 million automobiles are in use, approximately three for every four members of the U.S. population. Over the past 44 years, the transportation sector's consumption of petroleum has more than tripled, but because of energy conservation and other measures, growth was slower in the 1980s and 1990s than in previous decades.

The *residential and commercial sector* accounts for 26 percent of U.S. end-use energy consumption. Of the total, almost 40 percent is used in the form of electricity. A typical house contains five basic energy-consuming items: central heating system, hot-water heater, cooking stove, refrigerator, and lighting system. Most of the energy consumption in commercial buildings is for space heating and cooling, lighting, and office equipment.

Although many uses of energy are beyond the control of the average person, individuals can contribute in many ways to conserving energy in the home, when commuting to work, and in the office.

Steps that individuals can take to minimize energy losses in the home include weather-stripping doors and caulking windows, closing the damper when the fireplace is not in use, and setting the thermostat at the lowest comfortable setting in winter and at the highest comfortable setting in summer (Carrier Corporation, 1994). Appliances such as television sets and lights can be turned off when not in use. Lighting accounts for 5–10 percent of the energy consumption in a typical residence. Incandescent bulbs can be replaced with compact fluorescent lights; the latter consume only 25 percent as much electricity.

Because a majority of people in the United States drive to work, energy consumption associated with transportation can be significantly reduced by either carpooling or using mass transportation. If every commuter carried one additional passenger, over a half-million gallons of gasoline would be saved each day (AT&T, 1994).

In a large office, electronic equipment such as personal computers (PCs), printers, copiers, and facsimile machines typically accounts for about 8 percent of direct electricity use and 7 percent of peak demand. If the additional heating, ventilation, and air conditioning required to compensate for the heat produced by electronic equipment is included, the combined load represents some 10–12 percent of the total electricity demand of a typical office. Computers and other desktop devices should be turned off at the end of the workday. The old canard that this practice wears out the components is no longer true. The energy-cost savings potential of turning off a PC and its associated printer during nights, weekends, and holidays ranges from $75 to $125 per year (Blatt, 1994). Even if the computer must be operated continuously because of a network connection, just turning off the monitor can save 30–60 percent of the computing unit's energy consumption.

Through its Energy Star Computers Program, the EPA is encouraging manufacturers to develop desktop PCs and monitors that go into a standby mode automatically after a period of inactivity. Because they demand less power and operate at cooler temperatures with lower thermal stresses, the reliability of the electronic components is improved. Desktop computers do not require cooling fans, so placing them in the standby mode further reduces energy requirements as well as accompanying noise.

Summarized in Table 17.1 are recommendations on equipment purchase and use for maximizing energy efficiency in the office.

Because of the widespread recognition of the need for reducing energy usage and conserving existing supplies, a variety of new energy-efficient products has been developed. Frequently, however, the public has been

Table 17.1 Recommendations on office equipment purchase and use for maximizing energy efficiency

Item	Guidance for purchasing	Guidance for operating
Personal computer	Buy a laptop computer Buy an energy-efficient unit	Turn off at night and weekends Activate power management features Turn off when not in use during the day
Computer monitor	Buy an energy-efficient unit Consider a monochrome monitor Consider an active-matrix color liquid crystal display Buy a monitor only as large as needed Buy only as much screen resolution as needed	Turn off at night and weekends Activate power management features Turn off when not in use during the day
Computer printer	Consider an inkjet printer Buy an energy-efficient printer Consider sharing a printer Consider a unit with double—sided printing	Turn off at night and weekends Activate power management features Reuse paper Use electronic mail
Copier	Choose a properly sized unit Consider a copier not based on heat and pressure fusing technology Compare ratings provided by the American Society for Testing and Materials Buy a unit with power management features Choose a unit offering convenient two-sided copying	Turn off at night and weekends Activate power management features Use two-sided copying whenever possible Batch copy jobs

slow to adopt such products. U.S. sales figures for 1993 show that only slightly over half (54 percent) of the refrigerators sold were energy-efficient models. The results for other products were even more disappointing, with energy-efficient models accounting for only 28 percent of the freezers, 26 percent of the water heaters, 24 percent of the heat pumps, and 17 percent of the central air-conditioning units sold. This resistance has occurred in spite of campaigns and rebate programs to encourage the purchase and use of such products.

The reasons for this response are varied. Studies conducted by the Electric Power Research Institute (Evans, 1994) indicate some of the contributing factors.

Skepticism. Consumers do not trust manufacturer claims relative to the energy savings that new technologies will provide. They are suspicious that the products incorporate unproven technology and may not be reliable, or that they will be complicated or difficult to operate or have other drawbacks. Some aspects of this skepticism can be overcome if the new energy-efficient product meets consumer needs better than existing technologies, as did the microwave oven.

First cost. Many energy-efficient products have a relatively high initial purchase price. Although over the life of the product the savings may be substantial, the customer does not always have that perspective. An example is the compact fluorescent lamp.

Savings difficult to measure. It is often difficult for customers to identify how and to what degree a new energy-saving appliance will reduce their electric bill. The typical bill reports only the total electricity used; there is no way to determine whether a new product has helped reduce that total.

Poor aesthetics. Energy-efficient appliances often are not pleasing to the eye. Consumers will opt for a product that looks significantly better and more modern over one that is simply more efficient. For example, a dishwasher with an electronic keypad is preferred to one with buttons and dials.

Requirements for change in behavior. New products that require consumers to change their behavioral patterns are often rejected in the marketplace, although there are exceptions. One is the microwave oven that required the use of special dishes, yet experienced rapid success in the marketplace.

Market-related issues. Consumers sometimes find that retailers do not stock adequate supplies of the newer energy-efficient models. If supply cannot keep up with demand, customer interest will wane. Customers also want to be assured that replacement parts and a reliable service organization will be available.

Energy Resources

The fuels available to meet the world's energy needs fall into two broad categories: (1) renewable sources, such as solar energy (including wind and waterpower) and geothermal energy, and (2) nonrenewable sources, such as fossil fuels (coal, oil, and natural gas) and nuclear fuels (uranium and plutonium) (ASME, 1989; CEQ, 1995). Fusion reactors, like solar energy, could present an almost limitless source of energy.

FOSSIL FUELS

Fossil fuels are the best examples of the nonrenewable sources. Experience shows that when such fuels are first discovered, they are utilized at a very modest level. Use then increases more or less continuously to a maximum and thereafter declines to zero when available supplies have been consumed (Hubbert, 1973). Worldwide, production of oil is expected to peak within a decade or two and will then decline. Production and consumption of coal and natural gas will ultimately meet a similar fate. The remaining worldwide and domestic supplies of these three fuels are estimated to be about as follows:

Natural gas—perhaps five or more decades. However, new supplies are being discovered, and the total recoverable reserves are probably not known at this time. Proved reserves in the United States are sufficient to meet domestic demand for about ten years (CEQ, 1995).

Oil (petroleum)—perhaps three to five decades. More than half of these resources are located in the Middle East. Again, proved reserves in this country are estimated to be sufficient to meet domestic demand for about ten years (CEQ, 1995).

Coal—perhaps two to four centuries. The United States has abundant reserves.

NUCLEAR FUELS

Nuclear fuels are another example of nonrenewable sources. The energy available from existing U.S. sources of uranium, if consumed exclusively in the current generation of boiling-water and pressurized-water nuclear power plants, is estimated to be about equal to that available through the combustion of the existing sources of natural gas or petroleum. With the

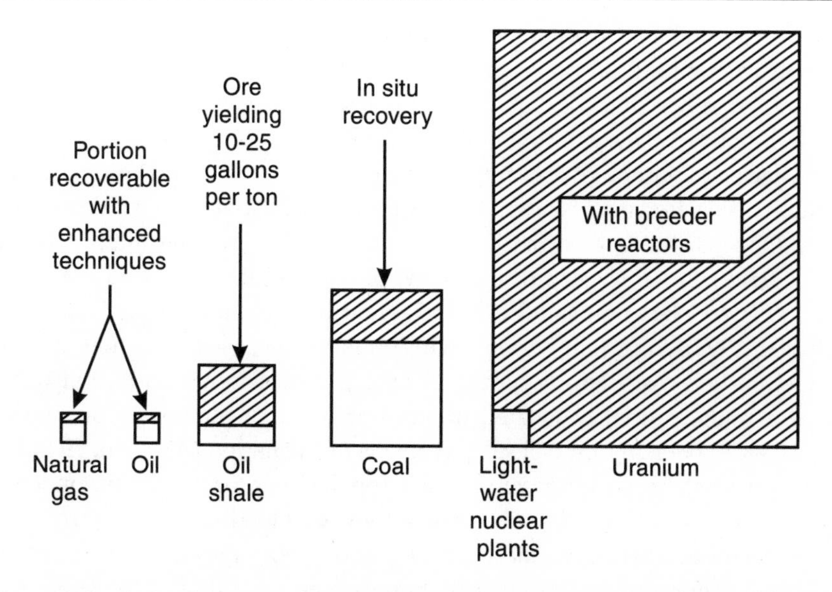

Figure 17.1 Comparison of available energy resources in the United States, 1990

development of breeder reactors (and the recycling of plutonium), nuclear fuels could become a far larger energy source than any of the fossil fuels (Figure 17.1). Even so, none of these sources begins to approach the amount of energy that would be made available by effective harnessing of the fusion process. This could provide essentially an unlimited supply of energy. Unfortunately, it may require another fifty years or more before a workable fusion reactor will be commercially available.

SOLAR ENERGY

Solar sources offer tremendous potential for meeting the world's energy needs. Since such sources are renewable, once the technology for a given use has been established it can be applied more or less indefinitely. Information on several sources of solar power is summarized below.

Hydropower. This is one of the most common sources of solar energy, a prominent example being the enormously successful series of hydroelectric power plants that were constructed by the Tennessee Valley Authority on major waterways in the south central United States during the mid-twentieth century. Still, hydroelectric plants can have severe impacts on environmental and public health.

Tidal power. Another source of solar energy, tidal power is manifested in the rise and fall of the ocean tides. Although a proposal in the early 1960s to construct such a tidal power plant in the Bay of Fundy in Maine was not approved, other countries are using this resource, an example being a large power station now operating in the United Kingdom.

Windpower. Although windmills have been used for centuries on an individual basis to pump water, few multiunit windpowered stations were developed in the United States until recently. The growing interest is due to several factors: larger turbines, new blade designs, advanced materials, and smarter electronics—all of which contribute to an increased efficiency (Stover, 1995). In certain regions, windpowered electricity can now be generated at less cost than by new coal-fired plants. Estimates are that windpower units, primarily located in rural areas, may ultimately provide as much as 20 percent of the electricity needs of the nation (Abelson, 1993).

Unfortunately, windpowered units too have impacts. Common complaints are that windmills can be noisy, they tend to kill birds, they are often considered aesthetically unacceptable, and their operation can interfere with television reception in nearby homes. The reliability of the generated power has also been questioned, for example, during periods of calm (lack of wind).

Heat pumps. Instead of burning natural gas or oil to provide heat, the heat pump uses electricity to collect and concentrate the latent heat in the ambient air or ground that has been warmed by solar radiation. Since the heat from the sun is essentially free, the heat pump can deliver significantly more energy for heating than it consumes in electric power (some three times more for the most advanced units). Heat pumps can cool buildings by reversing the process, collecting indoor heat and transferring it outside the structure (Jaret, 1992).

Photovoltaic systems. The use of solar collectors to provide power for small appliances, such as watches, typewriters, calculators, and answering machines, has also been very successful and is widespread (EPRI, 1989). Patio walkway lights that store the sun's energy during the day and release it at night have also proved popular. Today large, reliable, and efficient modules are being used in streetlights, highway call boxes, and microwave towers, and to serve remote installations such as offshore oil-drilling equipment, navigational lights, and signal buoys (Somerville, 1989). In fact, advances in this technology may soon lead to major electricity-generating units that can compete directly with conventional power plants (Service, 1996). For this and other reasons, electric utilities increasingly are exploring the opportunities for utilizing grid-connected photovoltaic systems.

Trees and agricultural crops. In many parts of the world, the primary source of fuel for cooking and heating is firewood. The growing use of wood as a source of energy, however, has led to the wholesale destruction of hardwood forests in many areas. One solution may be to encourage the planting and harvesting of rapidly growing trees such as mesquite (Ayres, 1995).

Fuel is being produced from various agricultural crops. Ethanol can be made from sugarcane, methanol can be made from wood materials, methane can be produced by the anaerobic digestion of animal and plant wastes, and biomass can be directly burned as a fuel. Recognizing these opportunities, many electric utilities are exploring the use of biomass feedstocks (wood, wood waste, and various herbaceous crops such as alfalfa) as renewable energy sources for power generation (Moore, 1996). Another source of biomass energy is the methane gas produced in municipal sanitary landfills. Estimates are that landfills in this country could provide almost 10 million metric tons of this gas annually. Over 100 such facilities are now equipped to take advantage of this energy source (Anonymous, 1995).

The potential applications of solar energy extend far beyond these examples. In the developing countries the sun is used for heating and distilling water, for drying crops, and for heat engines such as the sofretes pump, manufactured in France. In urban areas the trees planted around a house can promote not only comfort but also energy conservation. Deciduous trees, for example, shade a house in the summer and permit sunlight to warm it in the winter. And trees directly absorb greenhouse gases, such as carbon dioxide.

GEOTHERMAL ENERGY

Although geothermal energy is available in enormous quantities, it is often difficult to tap and is limited to certain geographic areas, most of which are remote from population and industry. The main potential of geothermal energy in the United States appears to lie in localized use as a source of inexpensive heat on a relatively small scale (Anonymous, 1981). Of the possible applications, heat pumps appear to be the most promising and could represent an energy-efficient way to heat and cool buildings in many areas of the country (GAO, 1994).

Environmental Impacts

In any discussion of energy use, one topic that inevitably arises is the environmental impact of electricity-generating power plants. In plants us-

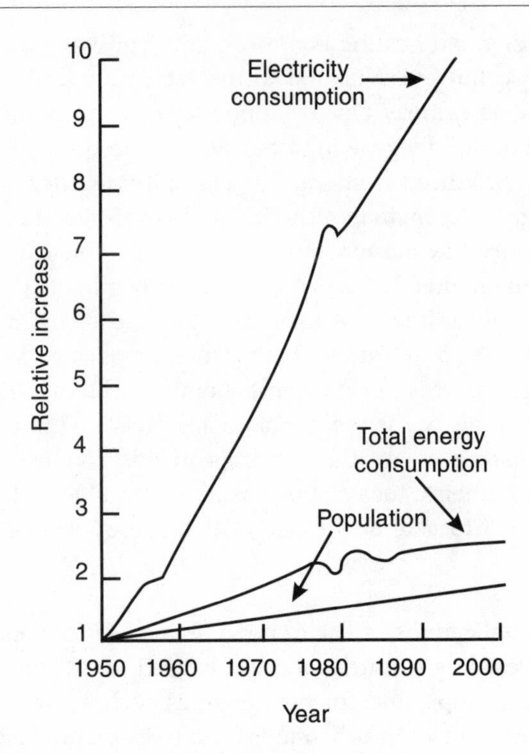

Figure 17.2 Relative increase in U.S. population, total energy use, and electricity consumption, 1950–2000 (data normalized to 1950)

ing coal and oil, smoke from the tall stacks is clearly visible and thus readily subject to public comment. Nuclear power, the main alternative to coal and oil, remains a highly emotional and controversial subject. Yet demand for electricity continues to grow. Figure 17.2 shows the relative increase in population, total energy use, and electricity consumption in the United States since 1950. As in many other countries, the use of electricity has far outstripped population growth or total energy use: in 1973–1988 alone, the use of electricity in the United States increased by over 50 percent.

The following sections assess and compare the environmental impacts of various methods of generating electricity: hydroelectric power, geothermal energy, fossil-fueled power plants, and nuclear-powered plants.

HYDROELECTRIC POWER

Harnessing waterpower on a major scale generally involves the construction of a dam on a river or stream. A lake is formed thereby, and the potential energy in the water is used to turn turbines and generate electricity. Experience has shown, however, that large-scale projects can alter the environment and people's lives drastically, bringing as many risks as benefits. The Aswan High Dam on the Nile River in Egypt exemplifies the problems (White, 1988). As first planned, this project had two objectives: to produce electric power and to irrigate the nearby desert. But many unforeseen results ensued.

In its original state, the Nile served as a mechanism for the transport of tremendous quantities of silt and organic matter. During the annual spring floods some of this material was deposited on the banks of the river, where it served as fertilizer for agricultural crops. Material remaining in the river ultimately reached the Mediterranean Sea, where it served as food for large numbers of fish. After construction of the dam, the river no longer flooded its banks and the fertility of the farming areas declined. Without the discharge of nutrients into the Mediterranean, the fishing industry was essentially destroyed. In addition, Lake Nassar, created by the dam, raised the water table, dissolved salts in the ground, and brought them up into the desert topsoil, making it less usable for agriculture than anticipated. Control of the flow rate in the river led to increased growths of algae and phytoplankton, which adversely affected water quality. Quiescence also promoted the growth of the snail population, which dramatically increased the incidence of schistosomiasis among people living nearby. It is for reasons such as these that scientists in recent years have advocated the deliberate opening of dams on a periodic basis to flood downstream areas. Experiments conducted by the U.S. Geological Survey show that such releases, through lifting and churning river-bottom sediments, rejuvenate the beach habitats for certain endangered species and various forms of wildlife (Wuethrich, 1996).

Experience in the United States has shown that dams can also have serious effects on ocean fish, such as salmon, sturgeon, steelhead, and striped bass, that spawn in freshwater streams. This is especially true of the dams located in the Northwest. Even with special ladders installed to assist the fish in continuing their upstream journey, dams block the passage of many fish. They also slow the flow rate of the streams and warm the waters. All of these changes lead to reductions in fish populations, notably of salmon.

Also to be considered are the consequences of sudden failure of a dam. More than 400 people were killed in the failure of the Malpasset Dam, in southern France, during a flood in 1959. One can only hope that steps are being taken to avoid or ameliorate these types of problems in ongoing projects such as the hydropower facility being constructed across the Yangtze River in the People's Republic of China.

GEOTHERMAL ENERGY

The use of geothermal energy has a variety of environmental impacts. The pressure in pressurized hot-water reservoirs is often a result of the weight of the overlying land; withdrawal of the water can lead to subsidence. Accidental spills of the withdrawn water, which often has a high mineral content, can lead to soil salination and water pollution. Where water is injected to be converted into steam using the heat from dry rocks, induced seismicity is a risk (NRC, 1979). In addition, significant problems may arise from the release of radon and volatile gases such as hydrogen sulfide, which accompany the steam. The quantities of radon being released from such operations in northern California equal or exceed the airborne releases of other types of radionuclides from nuclear-powered plants of comparable generating capacity (Hellums, 1990).

NUCLEAR OR FOSSIL FUELS

Facilities that generate electric power with nuclear fuels use similar processes as those that employ fossil fuels. In both situations the fuel is used to produce heat, which converts water to steam. The steam is sent to a turbine, which is connected to a generator for producing electricity. Cooling water, usually from a nearby river or lake, condenses the steam leaving the discharge side of the turbine, and the hot water is returned to the heat source for reheating. Condensation of the steam leaving the turbine produces a vacuum so that the incoming steam will have the necessary pressure to turn the turbines. In this process, twice as much energy—that is, twice as much heat—is discharged to the environment with the cooling water as is converted into electricity, and power plants of this type are increasingly being challenged as sources of thermal pollution. These plants also release various gases and particles into the atmosphere, along with smaller quantities of liquid wastes.

To avoid thermal pollution of water sources, many electric utilities have equipped their power plants with cooling towers, which release the excess heat into the atmosphere instead of into a lake or river. During cold weather the steam leaving the towers may condense and freeze on nearby roads, causing accidents. Some scientists have speculated that a large number of cooling towers, concentrated in so-called power parks, could cause local changes in the weather.

Fossil-fueled and nuclear-fueled power plants have environmental impacts at five stages:

Fuel acquisition

Fuel transportation

Power plant releases

Processing and disposal of spent fuel or ashes

Power transmission

Fuel acquisition. The acquisition of fossil fuels has considerable environmental impact. In the case of coal and oil it is largely related to the volumes required. For example, a standard 1,000-megawatt electric (MWe) power plant requires more than 2 million tons of coal or more than 500 million gallons of oil per year.

Strip mining of coal pollutes the air with dust and defaces the earth's surface if the land is not restored to its original state after the mining is completed. Underground mining of coal frequently produces "acid mine drainage"—sulfuric acid and iron salts that drain out of or seep from the mine during operation and for some years thereafter. When these materials flow into surface streams, they are toxic to most forms of aquatic life. Underground coal miners experience an array of occupational health problems. On average, the mining of sufficient coal to provide fuel for one 1,000-MWe power station results in two to four accidental deaths and two to eight cases of black lung disease and other respiratory ailments among coal miners each year (Wrenn, 1979).

Underground mining of uranium can also have serious effects on health. Of the 6,000 people who have worked or are now working in uranium mines in the United States, an estimated 15–20 percent, or about 1,000, have already died or will die from lung cancer. In addition, the processing of uranium ore produces large quantities of tailings that can be sources of liquid and gaseous releases to the environment.

Drilling for oil has environmental impacts both on land and offshore. Oil refineries release airborne wastes, and leaks from offshore drilling operations can contaminate the marine environment—as did the spills some years ago near Santa Barbara, California, and more recently in the Gulf of Mexico and the North Sea.

Fuel transportation. A standard 1,000-MWe coal-fired electric power plant requires about 8,000 tons of fuel per day, enough to fill at least 100 railroad cars. The transportation of this amount of coal is estimated to result in 2–4 deaths and 25–40 injuries each year, primarily as a result of accidents at railroad crossings (Gotchy, 1987). For this reason and for economic considerations, the concept of the "mine-mouth power plant" is being implemented for coal-fired plants in several Western states: the power plant is constructed near the source of the coal, and although the electricity must subsequently be transmitted over longer distances, that strategy is considered preferable to transporting the coal.

As noted previously, much of the oil consumed in the United States is obtained overseas. Tanker accidents are common. The accident involving the *Exxon Valdez* in Alaska in the spring of 1989 released more than 10 million gallons of oil into Prince William Sound and onto nearby beaches. Most tanker accidents occur within 10 miles of shore, usually affect a recreational area, and spill an average of more than 25,000 barrels of oil, or more than 1 million gallons per accident. The cost of cleanup is substantial, in the case of the *Exxon Valdez* exceeding $2 billion.

Ocean shipment of natural gas poses similar problems (Martino, 1980). A modern tanker can transport about 125,000 cubic meters of liquefied natural gas at a temperature of $-160°$ C. When the tanker reaches its destination, the liquified gas is allowed to warm and regasify as it is transferred to storage tanks for later distribution to consumers. During this process the volume of the gas increases by more than 600 percent. An accident while a tanker is unloading its cargo could release millions of cubic meters of natural gas (in expanded cloud form) that would blanket a city. If the cloud were subsequently ignited, widespread death and destruction would result.

A 1,000-MWe plant powered by nuclear fuel requires only about 30–50 tons of fuel per year. Because the original uranium fuel is sealed within fuel rods and is not a significant radiation source, its transportation to the power plant does not present any unusual occupational or environmental health problems.

Power plant releases. Fossil-fueled plants release sulfur oxides, nitrogen oxides, carbon monoxide, and some naturally occurring radioactive material originally present in the fuel (Wilson et al., 1981). In fact, the United Nations Scientific Committee on the Effects of Atomic Radiation estimates that the collective dose from a coal-fired electricity-generating plant is equivalent to 25 percent of the local and regional population dose from a nuclear-powered station of the same generating capacity (Mettler et al., 1990). Fossil-fueled plants also release significant amounts of carbon dioxide into the atmosphere.

Table 17.2 Airborne emissions from fossil-fueled electric power plants (assumed capacity 1,000 MWe)

	Emissions (thousands of tons per year)[a]		
Pollutant	Coal	Oil	Natural gas
Sulfur dioxide	69	31	—
Nitrogen oxide	25	13	15
Carbon monoxide	0.7	1.0	1.1
Particulates	120	0.6	<0.1

a. Emissions will vary depending on nature of fuel, plant design, and operating parameters.

Table 17.2 summarizes the estimated quantities of major airborne pollutants released by coal-fired, oil-fired, and gas-fired 1,000-MWe plants. The numbers, which vary depending on a range of plant conditions and are presented primarily for purposes of illustration, indicate that coal-fired plants discharge more than twice as much sulfur dioxide per unit of electricity generated as do oil-fired plants; gas-fired plants emit essentially none. Coal-fired plants also release almost twice as much nitrogen oxide as plants fueled by either oil or natural gas. In terms of carbon monoxide, the discharges by the three types of plants are not significantly different. However, coal-fired plants emit several hundred times as much particulate matter as oil-fired plants and several thousand times as much as plants fueled by natural gas.

Clearly, a plant fueled by natural gas has far less impact on the environment than a plant fueled by either coal or oil. Why, then, do public utilities not use natural gas as their sole source of fuel for generating electricity? The answer is twofold: apparently not enough natural gas is available to meet such a demand, and the fuel can be used more effectively for other purposes, such as home heating. Even so, many power plants have installed natural-gas turbine systems to handle periods of peak demand. Although facilities for converting coal into gas have been considered, such conversion has not yet proved to be cost-effective on a large-scale commercial basis.

The kind and quantity of releases from nuclear-powered plants depend on the type of reactor. Two basic types are in use in the United States today: boiling-water reactors (BWRs) and pressurized-water reactors (PWRs) (Figure 17.3).

In a BWR, the water is heated by the fuel and converted into steam, and

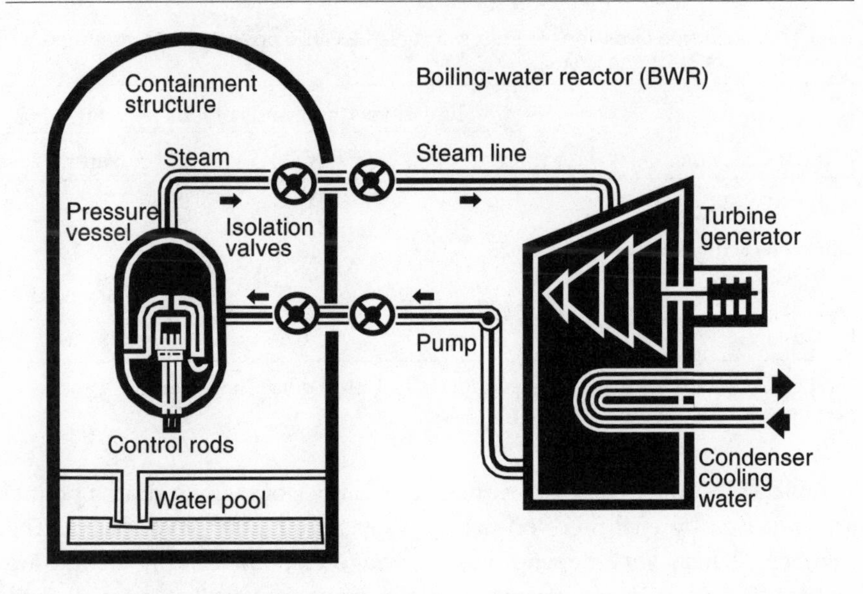

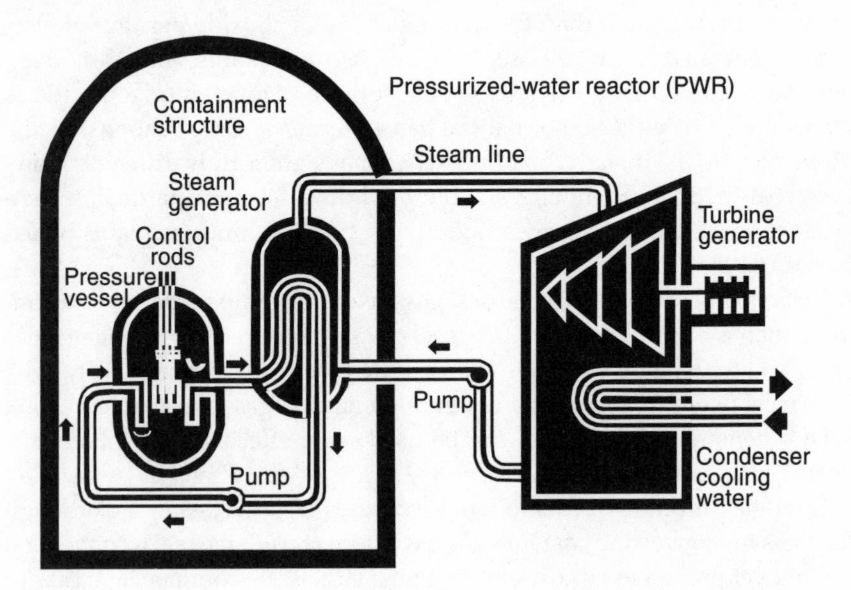

Figure 17.3 Schematic diagrams of boiling-water and pressurized-water nuclear power plants

Table 17.3 Annual airborne emissions from nuclear power plants (assumed capacity 1,000 MWe)

	Pressurized-water reactor		Boiling-water reactor	
Pollutant	Curies	Becquerels	Curies	Becquerels
Krypton-85	20	7×10^{11}	—	—
Xenon-133	2,000	7×10^{13}	100	4×10^{12}
Carbon-14	10	4×10^{11}	10	4×10^{11}
Iodine-131	0.002	7×10^{7}	0.003	1×10^{8}
Tritium	150	6×10^{12}	100	4×10^{12}

this steam turns the turbine. Neutron irradiation of the cooling water as it passes through the reactor core converts stable oxygen into radioactive nitrogen, which is carried out to the turbine by the steam. As a result, personnel cannot work in the vicinity of the turbine during plant operation. If leaks occur in the fuel cladding, the water and steam will also contain radioactive fission products.

In a PWR, the water heated by the reactor is kept under sufficient pressure that it is not converted into steam. Through use of an intermediate heat exchanger (steam generator), this water in turn transfers heat to water in a secondary system, which converts the water to steam and turns the turbine. Under normal conditions the water in the secondary system will be clean, and any leakage will not release radioactive material. However, the tubing in the steam generators often fails, releasing radioactive water and gases from the primary to the secondary system. Although the liquid releases can be readily controlled, some gases are released to the environment.

Table 17.3 summarizes the estimated airborne releases from representative 1,000-MWe PWRs and BWRs brought into operation since 1979. The two types of plants release about equal amounts of carbon-14, iodine-131, and tritium; in general, PWRs release more krypton-85 and xenon-133.

Although many factors must be considered in evaluating the environmental and public health impacts of the operation of fossil-fueled and nuclear-fueled electric power plants, some perspective is achieved by comparing the various plants on the basis of the quantities of air needed to dilute their most critical releases to concentrations that will comply with federal standards for the ambient environment. These data, presented in Table 17.4, are based on the assumption that all the sulfur is released as

Table 17.4 Annual dilution requirements for a 1,000 MWe power plant

Type of plant	Limiting pollutant	Required dilution[a]	
		Cubic meters	Cubic miles
Coal	Particulates	2×10^{15}	500,000
Oil	SO_2	3.5×10^{14}	85,000
Natural gas	NO_2	1.5×10^{14}	30,000
Nuclear (pressurized-water reactor)	Xe-133	4×10^{9}	1

a. Approximate volume of air required to dilute the most critical pollutant from each type of plant to the permissible concentration as prescribed by federal standards for the ambient environment.

sulfur dioxide, that all the nitrogen is released as nitrogen dioxide, and that all the particulate emissions have diameters less than 10 micrometers.

In the case of a coal-fired plant, the releases requiring the largest dilution are the particulates, for which the EPA standard is 50 $\mu g/m^3$; for an oil-fired plant, the controlling pollutant is sulfur dioxide (SO_2), for which the EPA standard is 80 $\mu g/m^3$; for a gas-fired plant, the controlling pollutant is nitrogen dioxide (NO_2), for which the EPA limit is 100 $\mu g/m^3$ (EPA, 1994). For a nuclear-powered plant, in this case the PWR, the controlling airborne pollutant is Xe-133, whose permissible ambient air concentration is 2×10^4 becquerels (5×10^{-1} microcuries) per cubic meter (USNRC, 1991). On this basis, the volume of air required to dilute the most critical airborne release from a nuclear-powered plant to the acceptable concentration in the ambient environment is less than 0.004 percent of that needed for the most critical pollutant from a comparable-sized facility fueled by coal, oil, or natural gas. For a BWR the volume of diluting air would be even less.

On the basis of these data, one might conclude that a nuclear plant is far safer in terms of airborne releases than any other type. However, these calculations do not include the associated dilution requirements for the airborne gases and dusts that would be released from the tailing piles produced in conjunction with milling the uranium ore. Also not included is any consideration of the airborne releases in the course of enriching the fuel used in these plants or in chemically processing the spent fuel. Although chemical processing is not currently being done in the United States, it is common practice in many other countries and should be taken into account in those cases. Even if these other emissions are considered, on a comparative basis the associated emissions for a nuclear-powered plant are small.

Other factors need to be considered in making comparisons of this type (Wilson et al., 1981; Gotchy, 1987). For example, the quantity of SO_2 released from fossil-fueled plants depends on the concentration of sulfur in the fuel. The results also depend on the applicable environmental and public health standards. Calculations of the volume of air required to dilute a given contaminant to an acceptable level assume the existence of a single universally accepted concentration to which a population group can be exposed. If a higher acceptable concentration is chosen, the calculated volumes of diluting air will be lower. There is also no assurance that the standards for the permissible airborne concentrations of each of the critical pollutants are based on equivalent risks.

Another factor is that natural processes remove pollutants such as SO_2, NO_2, and particulates from the atmosphere. Were these processes taken into account, the quantities of diluting air required for the coal-fired, oil-fired, and gas-fired plants would be less. Similarly, the half-life of Xe-133, the critical airborne release from a nuclear power plant, is only about 5.3 days; it too will quickly dissipate (decay) in the ambient environment.

Finally, the conclusions reached depend on whether the potential health impacts of the airborne contaminants are evaluated individually or collectively. In a coal-fired plant, particulates and gases are released simultaneously. If the combination of pollutants has the potential to produce synergistic effects, then the pollutants should be evaluated as a totality, not simply on the basis of the most critical individual contaminant.

Many people would want to have still other factors taken into account in an overall evaluation of the environmental and public health impact of electric power plants. In the case of fossil-fueled facilities, these include acidic deposition through the release of sulfur and nitrogen compounds, and the potential impact of global warming as a result of the release of carbon dioxide. In the case of nuclear-powered plants, additional considerations include long-term disposal of the associated high-level radioactive wastes and the potential for serious accidents that might release large amounts of radionuclides into the atmosphere. As illustrated by the Chernobyl accident in 1986, such events can have far-reaching effects.

When all these factors are taken into account (including the atmospheric pollution associated with the processing of spent fuel), it appears that a nuclear-powered plant has far less environmental impact than a plant fueled by oil or natural gas. All three types of facilities appear to be superior to a plant fueled with coal. If adsorption systems are used to delay the release of radioactive gases from spent-fuel chemical reprocessing facilities or if, as in the United States, the spent fuel is not reprocessed, then nuclear plants remain far superior in terms of airborne releases.

Power plants fueled by natural gas or oil have no spent-fuel disposal problems because these fuels burn cleanly and produce no ash. In coal-fired plants, however, 12–25 percent of the fuel ends up as ash. Thus, a 1,000-MWe plant would require 12–25 railroad cars for the daily removal of ash. Where and how this ash is disposed of is also important, since it contains many toxic compounds. As indicated above, a nuclear-powered plant produces some 30–50 tons of spent fuel each year. Because it is intensely radioactive, spent fuel poses significant problems from the standpoint of radiation protection and waste disposal. In this country such fuel is currently being stored at the power-plant sites either in water pools or in above-ground dry casks. It will subsequently be transported offsite for disposal in a geologic repository.

Power transmission. The type of power plant has no effect on the efficiency with which electricity is distributed to consumers. An estimated 12–14 percent, perhaps even 20 percent, of the electricity generated in the United States is lost during transmission. Development of more efficient transmission systems has long been discussed, one promising system being the use of superconducting cables installed underground. The power losses in such a system would be less than those currently experienced in high-tension overhead power lines, and perhaps some of the possible health effects of the associated electric and magnetic fields would be avoided (Abelson, 1989). In addition, an underground system would eliminate much of the environmental degradation caused by overhead lines. The United States has more than 325,000 miles of such lines, 10,000 miles of which operate at 750,000 volts or higher. Each mile of transmission line requires up to 100 acres of land as right-of-way (Abrahamson, 1970).

The General Outlook

Energy policies that are sensitive to environmental concerns can reduce the impact of energy production and the consumption of air, land, and water resources (CEQ, 1995). Unfortunately, many actions in past years have not recognized that fact. Subsidies to the airlines, for example, have encouraged the growth of less efficient forms of transportation; lack of support of the railroad industry has discouraged the use of rail transportation; and federal support for the development of a nationwide highway system, though making road travel far safer, has led to greater use of cars and increased gasoline consumption. The government has also not addressed the problem of rate structures for natural gas and electricity, which promote higher consumption by offering large-volume industrial users a lower price per unit.

Appropriate energy policies should also encourage conservation (Galvin, 1994). One useful step would be tax credits for the installation of insulation and solar units in dwellings. More comprehensive fuel efficiency requirements could be promoted in products such as home appliances. Congress has passed legislation that sets energy-efficiency requirements for refrigerators, hot-water heaters, and furnaces; similar requirements could be established for lighting. With about one-quarter of all the electricity in this country being used for illumination, the potential for savings is enormous. In fact, utility companies throughout the United States are actually encouraging customers to adopt more efficient lighting (Lamarre, 1989); they see conservation as preferable to the alternative—developing new electricity-generating capacity.

Household appliances should also be regulated in terms of their use of water, another limited environmental resource. Conserving water will in turn reduce the energy used to purify and heat it. The use of low-flow showerheads could be encouraged, also aerators on sink faucets, toilets that flush with less water, and more efficient methods of irrigating lawns, gardens, and agricultural crops.

The development of smaller, passively safe nuclear-powered electric generating units (for example, ones that are designed to shut down automatically, without the aid of external control devices, in case of a power excursion) could help make this source of energy more acceptable to the public. Such plants are in advanced stages of design and appear to hold promise for avoiding, or at least greatly reducing, both the probability and the consequences of accidents. The latest epidemiological data indicate that, during normal operation, nuclear plants have little impact on the nearby population. On the basis of a detailed review of mortality from 16 types of cancer among people living in 107 counties near 62 major nuclear facilities in the United States, scientists at the National Cancer Institute concluded that there was no convincing evidence of increased risk of death from living near these facilities (NCI, 1990).

Although progress related to harnessing fusion energy continues to be promising, it is generally acknowledged that it will be decades, at a minimum, before commercial utilization of this process will be feasible (Lawler, 1995). The potential advantages of this source of energy, however, make pursuit of the process a desirable goal. Such advantages include the provision of a virtually inexhaustible source of energy; no airborne emissions that will contribute to acid rain or global warming; the impossibility of a large uncontrolled release of energy and therefore the impossibility of a major accident; and radiological hazards whose magnitudes are thousands of times less than those associated with nuclear fission (DOE, 1993).

Even with nuclear power and increased emphasis on energy conservation and the use of solar power, environmental pollution from electricity-generating plants will pose problems for generations to come. Many of the existing coal-fired plants are expected to have lifetimes of almost 60 years. If stringent backfitting requirements are not enforced to reduce their releases of NO_2 and SO_2, these pollutants will continue to be a problem well into the next century.

However undesirable the environmental impact of electricity-generating facilities, the fact remains that electricity is essential to the quality of modern life. Better lighting reduces accidents on highways and crime in cities. Electricity is necessary to clean the air, to operate water purification facilities and sewage treatment plants, to dispose of old automobiles, and to recycle other types of solid waste. It powers radio, television, microwave ovens, computers and office equipment, and labor-saving home appliances. Although conservation can help reduce the overall demand for energy, the need for electricity, and thus for more power plants, will almost certainly continue to grow. The basic challenge is to educate people to use energy more efficiently and more conservatively and to encourage the commercial sector to design, construct, and operate generating stations that function at maximum efficiency with minimal impact on environmental health.

18

DISASTER RESPONSE

THROUGHOUT the world, natural disasters—floods, tornadoes, hurricanes, earthquakes, volcanic eruptions—and accidents involving industrial and technological facilities—oil spills and accidents at chemical and nuclear power plants—are having a significant impact on both people and the environment. Natural disasters alone have caused an estimated 3 million deaths worldwide over the past two decades (NAE, 1988). Table 18.1 summarizes the impacts of some of these events. The economic and human costs vary widely, depending on the concentration of population, the existence of emergency response capabilities, the area's accessibility to outside assistance, the efficiency of rescue operations, building design and construction practices, and soil conditions (Ward, 1989).

Types of Disasters

In a broad sense, disasters can be classified as natural and man-made. Upon close examination, however, few disasters are either totally "natural" or totally "man-made." For example, the consequences of geological and climatological disasters are often exacerbated by inadequacies on the part of humans—witness the failure to design a building to resist an earthquake or a ferry to resist heavy seas. Even events such as droughts may be influenced by human destruction of forests.

NATURAL DISASTERS

In terms of the numbers of people affected, the natural disasters having the largest impact are hurricanes and floods (that is, events related to clima-

Table 18.1 Examples of major natural and man-made disasters, 1985 and later

Event	Location	Date of occurrence	Impact
Natural disasters			
Volcanic eruption	Nevado del Ruiz, Colombia	13 November 1985	25,000 deaths, thousands homeless
	Mount Pinatubo, Philippines	June-July 1991	700 deaths, 300,000 homeless
Gas release	Lake Nyos, Cameroon	21 August 1986	1700 deaths
Earthquakes	Mexico City, Mexico	19 September 1985	9,000 deaths, 30,000 injured, 95,000 homeless
	Armenia, former USSR	7 December 1988	>55,000 deaths 250,000 homeless
	Erizincan, eastern Turkey	13 March 1992	4,000 deaths, 180,000 homeless
	Northwestern Iran	21 June 1990	>40,000 deaths
	Latur District, India	30 September 1993	11,000 deaths, 80 villages destroyed
	Los Angeles, California	17 January 1994	60 deaths, 8,500 injured, $40 billion damage
	Kobe, Japan	17 January 1995	>5,000 deaths, 25,000 injured, >300,000 homeless
	Sakhalin Island, Russia	June 1995	>3,000 deaths
Hurricane	Gilbert: Jamaica, Yucatan Peninsula, northeast Mexico	September 1988	300 deaths, widespread destruction
	Andrew: Florida, Louisiana, Bahamas	August 1992	51 deaths, 250,000 homeless, $25 billion damage
	Bertha: Caribbean, coastal North Carolina	July 1996	>$300 million in crop and property damage

Table 18.1 (continued)

Event	Location	Date of occurrence	Impact
Cyclone	Bangladesh Southeast India	30 April 1991 6 November 1996	138,000 deaths, 9 million homeless 2,000 deaths, 500,000 homeless
Landslide	Leyte and Negros Island, Philippines	6 November 1991	3,000 deaths
Flood	Pakistan and India	September 1991	2,000 deaths, thousands missing, millions homeless
	Midwestern United States	July 1993	46 deaths, 2 million acres affected, $25 billion in crop loss and property damage
Sinking of passenger ferry	Baltic Sea Tanzania	28 September 1994 21 May 1996	>900 deaths 615 deaths
Man-made disasters			
Nuclear power plant accident	Chernobyl, former USSR	26 April 1986	42 immediate deaths, many latent cancers
Fuel storage fire	Dronka, Egypt	2 November 1994	>470 deaths, thousands of homes destroyed
Forest fire	Oakland and Berkeley, California	October 1991	16 deaths, 49 missing, 3,000 homes destroyed, $5 billion damage
Gasoline pipeline explosion	Guadalajara, Mexico	22 April 1992	190 deaths, 1400 injured, thousands homeless
Oil tanker spill	*Exxon Valdez,* Prince William Sound, Alaska	March 1989	11 million gallons, heavy contamination
	Braer, off Garths Ness, near Scotland	January 1993	26 million gallons
	Maersk Navigator, Indian Ocean off Indonesia	January 1993	78 million gallons

Table 18.1 (continued)

Event	Location	Date of occurrence	Impact
	Sea Empress, Milford Haven, Wales	February 1996	19 million gallons
Oil spill (not tanker related)	Three major spills into Persian Gulf from refineries in Kuwait	January 1991	250–350 million gallons, heavy contamination
Oil spill (from pipeline)	Komi Republic in Arctic	October 1994	100 million gallons spilled, with widespread contamination and fires
Oil well fires	Kuwait	February–November 1991	>700 burning wells, initially widespread air pollution
Rioting	Los Angeles, CA	April 1992	>60 deaths, 2,400 injuries, billions of dollars damage
Terrorism	Oklahoma City, Oklahoma	19 April 1995	169 deaths, federal building destroyed

tological factors), as contrasted to earthquakes, volcanic eruptions, and tsunamis (events related to geological factors). Climatological disasters also occur more frequently than geological disasters. In terms of deaths, however, geological disasters, particularly earthquakes, have the largest impacts. Because their homes are less sturdy and their resources more limited, peoples in the less-developed countries suffer far more devastation from natural disasters than those in the developed countries. Ninety-five percent of deaths from natural disasters occur in the developing countries of the world, and the economic losses, computed as a percentage of gross national product, are almost 20 times higher than in the developed countries (Platt, 1994). On a geographical basis, Asia is most prone to natural disasters; Latin America and Africa are intermediate; North America, Europe, and Australia are least prone. For each major natural disaster in Europe and Australia, 10 occur in Latin America and Africa and 15 in Asia (Sidel, et al., 1992).

On an individual basis, floods are the most common natural disaster, accounting for about 40 percent of all such events worldwide. Water levels of rivers can rise gradually or very rapidly, in the latter case leading to so-called flash floods due to rapid snowmelt or heavy or repeated rains. Flash flooding is the leading cause of weather-related mortality in the United States, accounting for about 200 deaths annually. This compares to some 100,000 lives lost annually from such events in the People's Republic of China. The public health impacts of floods also include damage to or destruction of homes and displacement of the occupants, which in turn can lead to the spread of infectious diseases because of crowded living conditions and compromised personal hygiene (CDC, 1993).

Common effects of floods are contamination of drinking-water sources, disruption of sewer systems, release of dangerous chemicals from storage tanks, and interruption of solid waste collection and disposal. Flooding also enhances the opportunities for mosquito breeding as well as for snake and dog bites. As a result, floods are frequently followed by unsubstantiated reports of epidemics such as typhoid, cholera, or rabies (CDC, 1993). Further problems from hurricanes and tropical storms, as well as floods, are caused by the migration of people into disaster-prone areas, especially along coasts and rivers. Today an estimated 60 percent of the world's population lives within 100 kilometers (62 miles) of a coastline (Platt, 1994).

Recognizing the need for improved capabilities for coping with natural disasters, the United Nations General Assembly in 1989 designated the 1990s as the International Decade for Natural Disaster Reduction (UN, 1989). Summarized in Table 18.2 are the probabilities of occurrence of a range of health effects caused by various types of natural disasters.

MAN-MADE DISASTERS

Man-made disasters also cover a wide range. Examples of those that can cause immediate and widespread effects are the release of toxic chemicals from an industrial plant, as occurred in Bhopal, India, in 1984, and the widespread distribution of radioactive materials as the result of a nuclear power plant accident, as occurred at Chernobyl in 1986. Estimates of the number of people killed in the Bhopal accident range from 2,500 to 7,000. Although immediate deaths due to acute exposures from the Chernobyl accident were less than 50, the long-term effects, particularly the development of latent cancers due to chronic exposures, may lead to many more deaths. In terms of potential direct effects on the environment, few disasters are more dramatic than those involving major oil spills, which unfortunately continue to occur on a regular basis (see Table 18.1). Other man-

Table 18.2 Health effects of natural disasters

Health effect	Earthquake	Hurricane, high wind	Volcanic eruption	Flood	Tidal wave, flash flood
Deaths	Many	Few	Varies	Few	Many
Severe injuries (requiring extensive medical care)	Overwhelming	Moderate	Variable	Few	Few
Increased risk of infectious disease	A potential problem in all major disasters; probability increases with overcrowding and deteriorating sanitation				
Food scarcity	Rare (may occur as a result of factors other than food shortages)	Rare	Common	Common	Common
Major population movements	Rare (may occur in heavily damaged urban areas)	Rare	Common	Common	Common

made disasters that may have equal or even more far-reaching effects are not so readily apparent. If the predictions of global warming due to the release of carbon dioxide into the atmosphere prove true, the worldwide impacts could be catastrophic. Consequences of comparable magnitude could occur as a result of the depletion of the ozone layer due to the use and release of chlorofluorocarbons (Sidel et al., 1992).

Compounding the problems of man-made disasters is that they are more difficult to plan for, especially in developing countries. The effects of unanticipated system failures in the chemical plant in Bhopal were exacerbated by several factors: public housing located next to the plant; the occurrence of the release at night; delays in alerting the public; little public knowledge of the potential toxicity of the emissions; and the fact that the only medical facility in the area was a small community hospital unable to cope with the thousands of affected people (Merchant, 1986). Few countries even today provide the public with adequate information about the location of chemical manufacturing plants or the nature and quantity of chemicals being manufactured (Wasserman, 1985). Few have regulations governing transportation routes for hazardous chemicals or systems for registering their importation, distribution, and storage. No centralized and coordinated international system exists for reporting either chemical accidents or their long-term consequences. Recognizing these problems, the U.S. chemical industry has initiated programs both to reduce the types of accidents that can be caused by chemical releases and to limit their consequences (Holder and Munson, 1996). In some instances, the amounts of toxic intermediates have been reduced by modification of the processes. In other cases, the quantities of dangerous substances in storage or present within plant systems at any given time have been minimized (Abelson, 1992).

The importance of such steps is illustrated by the fact that more than 25,000 hazardous material incidents were reported in the United States in 1991. These included 893 reports involving injuries, 352 reports involving evacuations, and 97 reports involving deaths. Over 7,000 of the incidents were transportation related. Specific incidents that have occurred over the past few years include a release to the atmosphere of aromatic hydrocarbon vapor in 1992 that required the evacuation of nearly 50,000 residents in Minnesota and Wisconsin, and a railroad accident in northern California in 1991 that resulted in the release of 19,000 gallons of methyl dithiocarbamate, which destroyed all aquatic life within a 45-mile stretch of a river, caused huge fish kills in a downstream lake, and affected residents in a nearby town (ATSDR, 1992, 1994).

General Emergency Response

A well-designed and well-executed emergency plan is essential to coping with any type of disaster. Such a plan can ensure quick and effective mobilization to respond to the immediate health-care needs of the people affected and to restore disrupted services. The plan should be clear, concise, and complete. It should also be dynamic, flexible, and subject to frequent evaluation and update. It should designate precisely who does what and when, and everyone involved should be familiar with it. Its top priority should be to provide an immediate response to the event, by locating and providing emergency medical services to the victims, controlling fires, removing downed power lines, and controlling leaks of natural gas (Waeckerle, 1991). On a longer-range basis, the goals should be to provide health care and shelter for victims, and to restore important services such as a safe water supply and basic sanitation. Next in importance are arrangements to provide a safe food supply and to meet needs for personal hygiene.

In general, there are two types of emergency plans. One is national or regional in scope and defines the responsibilities and mobilization procedures of personnel in key public and environmental health departments and emergency preparedness agencies. Planning at this level frequently includes coordination of civil defense and military services. The other level, which is local in scope, is much more definitive and includes detailed listings of the personnel involved, their individual responsibilities during an emergency, and the range of countermeasures available for implementation. The local plan should be closely coordinated with the national or regional effort. Together the two can provide a cadre of well-trained personnel to cope with natural disasters or industrial accidents of almost any size.

Most disaster plans have four phases: the years before the event (the pre-event phase); the warning or alerting period, just before the occurrence of events that can be predicted; the response phase, immediately following the event; and the recovery (rehabilitation) phase.

Pre-event phase. The objectives during this phase are to anticipate that accidents and disasters will occur and to plan for responding to them. Specific steps should be taken to identify all available organizational resources; inventory the types and locations of available supplies and equipment, including hardware and medical supplies; identify private-sector contractors and distributors who can provide otherwise-scarce specialized personnel and equipment; review essential community and industrial facilities to identify those that may be vulnerable to a disaster; and define the responsibilities of each agency or group and establish lines of communication. In support of these activities, an emergency operations center should be designated and properly equipped.

Warning or alerting period. For certain types of disasters, particularly those caused by natural forces (hurricanes, tornadoes, floods), advance warning may be possible. If so, there will be an opportunity to alert emergency planning personnel and to have them move, where appropriate, to the emergency operations center. Timely and accurate information should be furnished to the media and the public about what to expect (Waeckerle, 1991), including specific details on what preparations should be made.

Response phase. Usually fire, emergency medical, and police personnel are the first to arrive with help at the site of a major disaster. The laypeople already present will inevitably include well-meaning volunteers. Properly managed, volunteers can be helpful; otherwise they can hinder the response. Their sheer numbers may create a logistical problem. It is important to provide security to the affected area to assure the safety of both victims and workers. The most experienced senior person should take charge, immediately surveying the area and carefully assessing the scene, the number of victims, and their injuries. This person should relay information to the emergency operations center and make recommendations for action. Officials at the center must then determine whether the police, fire, and emergency medical personnel on-site can adequately meet the needs (Waeckerle, 1991).

Recovery phase. During this phase substantial numbers of injured people may need follow-up care. All survivors will require food, water, shelter, clothing, and sanitation facilities (Waeckerle, 1991). Sometimes conditions will favor rapid increases in insect and rodent populations (PAHO, 1985). Floods in particular promote unsanitary conditions not only through the buildup of debris and blockage of sewer systems, but also through the creation of breeding habitats for insects, such as mosquitoes, in rain and floodwaters remaining on the soil, in empty receptacles, and elsewhere. Planners should maintain up-to-date information on the distribution of vector-borne diseases in a given area and nearby (PAHO, 1982).

A key ingredient of a sound emergency plan is to be essentially independent of which members of the emergency response staff are on duty when the event occurs. Emergency responders need to be able to readily communicate with one another and provide information to the public and the media. The communication system must be independent of the landline telephone network, which is subject to failure in almost any type of disaster, or of cellular telephone systems, which quickly become overburdened. Experience has shown that radio communication systems overcome these problems.

Also essential is an adequate staff of environmental health professionals who can monitor water quality and food, and assure proper facilities for the disposal of human excreta, supplemented by coworkers qualified to assess various types of dangers, such as downed electrical power lines, leaking natural-gas lines, and buildings subject to collapse. On a longer-term basis, attention must be directed to the exacerbation of the homeless problem, the lingering disruption of the health care system, and the emotional and psychological effects of the disaster (APHA, 1994).

Assessment and Mitigation of Natural-Disaster Impacts

Although the ability to respond to a natural disaster is important, equal attention should be directed to advance efforts to minimize its impacts. Some of these measures are discussed below.

EARTHQUAKES

The value of prior efforts to minimize the impact of earthquakes is exemplified by the wide variation in the human costs of five recent earthquakes: one in Mexico City in September 1985 left 9,000 people dead, 30,000 injured, and more than 95,000 homeless; one in northwestern Iran in June 1990 caused more than 40,000 deaths and left hundreds of thousands of people homeless; one in Kobe, Japan, in January 1995 caused over 5,000 deaths, in excess of 16,000 injured, and left 250,000 people homeless; and one that struck Sakhalin Island in Russia's far east in June 1995 is estimated to have killed in excess of 3,000 people. In contrast, an earthquake in southern California in January 1994 claimed only 60 lives and injured only 8,500.

One of the primary reasons for the high tolls in the first four earthquakes was the density of the population, coupled with the magnitude of the event. Exacerbating the effects in Mexico City, Iran, and Russia were inadequate building codes, lack of planning for disasters, insufficient rescue and debris-removal equipment, and either an inadequate number of local medical facilities or the fact they were damaged by the earthquake and no longer in operation. The relatively low level of damage in California was primarily the result of strict building codes, sound construction techniques, well-organized and well-rehearsed emergency response capabilities, ample communication facilities, and the increasing reliability of forecasting techniques. Research has clearly demonstrated that engineering techniques can increase the resistance of buildings to earthquakes (Housner, 1991). In Mexico City the most heavily damaged buildings were those constructed on certain types of subsoil, which under earthquake conditions becomes like quicksand and provides essentially no support (Canby, 1990). Response in northwestern Iran was delayed by the mountainous terrain and landslides that blocked the major roads.

The impacts of earthquakes on transportation systems and sources of electrical power pose significant problems in themselves, in terms of delayed rescue and medical care and food shortages. Earthquakes can also threaten human health on a massive scale by disrupting water supplies and the safe disposal of wastes. Just as it is possible to design and construct buildings that are resistant to earthquakes, it is possible to design, build, and maintain water-supply and sewage-disposal facilities that will withstand such events.

HURRICANES

Owing to the tremendous destruction caused by several recent U.S. hurricanes, increasing attention is being given to both preventive and mitigative measures. Nonetheless, such efforts still do not receive an appropriate share of the funds being directed to disaster planning. For example, the Federal Emergency Management Agency spends $50 on earthquake-related activities for every $1 spent on hurricane preparedness programs (Sheets, 1995). Experience shows that hurricanes cause three primary losses: loss of life, direct property destruction, and associated loss of commerce. The following mitigative measures can alleviate these problems.

Improved forecasting. Computer models to predict the paths and rates of movement of hurricanes have improved significantly during the past decade. Another advance has been the use of satellite-based tracking systems. However, the accuracy of the forecasts has not improved rapidly enough to offset the lengthier evacuation times now needed to accommodate the larger number of people living in the coastal areas most subject to the damaging effects of hurricanes. Additional research is needed to develop models that will provide longer-range forecasts.

Reduced evacuation time. What is needed are improved road systems, controlled residential and commercial development, better building practices, and safe in-place shelters for people who might otherwise have to leave the impacted area. Evacuation efforts are more likely to be successful if people are asked to evacuate 10 miles rather than 100.

Provision of refuge. Shelters providing a safe haven, if used properly, would minimize potential loss of life when complete evacuations are not feasible. Ironically, the location of such facilities should not be publicized in advance; otherwise people will delay evacuating, knowing that such shelters exist.

Other measures that can be used to mitigate the effects of hurricanes include restricting development and redevelopment in high-risk areas; enforcing hurricane-resistant building codes; and educating the public on successful implementation of mitigative measures, particularly steps to reduce loss of life.

Technological forecasting has also proved effective in reducing the impact of other types of natural events, such as volcanic eruptions. The eruption of Mount Saint Helens in Washington State in 1980, for example, took few lives—not only because it occurred on a Sunday morning, when few loggers were in the area, and because the primary blast was toward the more sparsely populated north and northeast, but also because the National Geological Survey had forecast the approximate time and the area likely to be affected by the release (Buist and Bernstein, 1986; Merchant, 1986).

FLOODS

During the summer of 1993, the once-in-a-hundred year rains that fell on the midwestern sections of the United States led to both gradual and flash flooding (CDC, 1993). Prompt response by public health and disaster-control agencies prevented the usual impact of the historic causes of death (drownings, infectious diseases, lack of accessible medical care). Nonetheless, an estimated 46 people died and an estimated $25 billion was lost because of crop and property damage. The floods carried away some entire towns and all the possessions of the inhabitants. That the loss of lives was as low is a tribute to the effective emergency response of public health and relief agencies, and the cooperative efforts of many groups, including the military services. Advanced meteorological and communication equipment permitted the National Weather Service to deliver timely information to response agencies. This allowed time for evacuation, reduced the risk of entrapment by flash floods, and enabled planners to predict where dams and levees might fail (Orme, 1994).

Although coping with the direct effects of a disaster such as a flood is of immediate importance, many problems need attention in the days and months following the event. Posttraumatic stress disorder among adolescents is common after many disasters, especially hurricanes and floods (Sidel et al., 1992), as is the increased potential for injuries during the cleanup phase. All too often, these sorts of problems are overlooked. Injuries and deaths result from electrical hazards, carbon monoxide exposures, musculoskeletal hazards, thermal stresses, and the use of heavy equipment by unqualified operators (NIOSH, 1994).

Assessment and Mitigation of Disaster Impacts at Industrial Facilities

As noted above, coping with the impacts of any type of disaster requires a careful, well-thought-out emergency preparedness plan. Essential to the development of such a plan, in the case of industrial facilities, is an assessment of the potential internal, external, and natural phenomena that can cause accidental releases of toxic materials.

The first steps are to consider the full spectrum of accidents that can occur due to facility operations; estimate their individual likelihood of occurrence; analyze the potential consequences; assess the systems that have been incorporated into the facility to prevent such releases; identify significant structures, systems, and components that are present to mitigate the consequences; and identify a selected subset of accidents and related scenarios that will need to be formally considered. Applications of these principles to industrial facilities are discussed below.

PRE-EVENT ASSESSMENTS

The initial conduct of what is called a hazard identification and evaluation is largely a qualitative exercise. It is designed to provide a comprehensive evaluation of natural phenomena and external hazards that can affect the public, the workers, and the environment owing to single or multiple failures within the facility. An inventory of all hazardous materials and energy sources within the facility is prepared. The next step is an evaluation of the consequences of an unmitigated release from the facility, and use of the resulting information to classify the facility and/or its operations into one of three hazard categories (DOE, 1994):

Category 1—having the potential for significant *offsite* consequences;

Category 2—having the potential for significant *onsite* consequences;

Category 3—having the potential for significant consequences on a localized basis only.

Since this ranking is based on the assumption that none of the safety systems mitigate the release, it represents what might be called a bounding calculation—which is one of the requirements of the Clean Air Act Amendments of 1990 (Mukerjee, 1995b). If the analysis indicates that a facility and its operations fall in Category 3, they do not present much of a problem. More detailed evaluations must be conducted for those facilities that pose more significant problems. For these facilities, it is necessary to identify the possible accidents, the nature of which depends on the facility and the processes considered. As a general rule, the range of accidents considered includes those that might be caused by internal as well as external initiators (DOE, 1994). The latter include naturally occurring events such as earthquakes and tornadoes; and man-made events such as nearby explosions, fires, and aircraft crashes that might negatively impact operations within the facility.

In such an evaluation, attention should be given to all modes of facility operation, including startup, shutdown, and abnormal testing or maintenance configurations. It is vital to develop a series of scenarios that link the initiating events with the responses of preventive and mitigative systems and other contributing phenomena. These, in turn, can be used as the basis for the sequence of events to be analyzed. Many designers and operators of industrial facilities employ the "defense-in-depth" philosophy to limit releases in case of an accident: they incorporate successive barriers to prevent the release of hazardous materials into the environment. No one layer by itself, regardless of how adequate it appears to be, is considered completely reliable. The intent of the multiple-barrier concept is to avert damage to the plant and to the barriers themselves, as well as to protect the public, the workers, and the environment from harm in case any one of the barriers proves not to be fully effective. Although many barriers are physical in nature, others may be administrative. The latter include procedural restrictions on operators; monitors designed to alert plant personnel to failures within critical systems; and procedural guides for actions to limit abnormal conditions, accident progression, or potential exposures (DOE, 1994).

Estimates of population doses in case of an accident can be made on the basis of "realistic" or "conservative" assumptions, or they may involve the bounding calculations described above. Any such estimates are accompanied by large uncertainties. For this reason, field surveys should be conducted as soon as possible following any unanticipated release to confirm that the theoretical estimates of exposures are reasonable.

ACCIDENT RESPONSE

Although it was an unfortunate experience, the accident at the Three Mile Island nuclear plant in 1979 stimulated increased attention to disaster response in the United States. Similarly, the accident in Bhopal in 1984 and a subsequent release at a similar plant in Institute, West Virginia, led to the passage by Congress of the Emergency Planning and Community Right-to-Know Act (Mukerjee, 1995b). Under this act, operators of industrial facilities in which certain chemicals are used are required to report the amounts and locations to local emergency planning committees, which in turn can use this information in developing emergency preparedness plans.

Some years back, the plans and organizations developed in the United States for responding to accidents in nuclear power plants were separate from those developed to cope with other types of emergencies. Recognizing that countermeasures for the control of exposures from nuclear power plant accidents are, for the most part, equally applicable to other types of industrial accidents, emergency planners at the state and local level are now combining all such activities into one organization. One of the primary benefits of this consolidation is preparedness and response capabilities that are better organized, staffed, and focused than would otherwise have been possible.

COUNTERMEASURES

According to the Public Health Service (USPHS, 1962), an acceptable countermeasure must be *effective* (substantially reduce population exposures below those that would otherwise have occurred), *safe* (introduce no health risks with potentials worse than those presented by the releases), *practical* (capable of being administered at a reasonable cost and without creating legal problems), and *defined* (with no jurisdictional confusion about responsibility and authority for applying the measure). Almost any countermeasure will carry with it health risks and social and economic disruption, depending on when and where it is applied. The following are countermeasures that have proved useful.

Evacuation. This is generally one of the first protective actions considered in the event of an industrial accident, especially one that releases some type of toxic material into the atmosphere. The feasibility of evacuation depends on a wide range of factors: magnitude and likely duration of the release, weather, time of day, potential for vehicle accidents and personal injuries, availability of transportation, availability of suitable shelter in a "safe" area, time interval between the accident and the order to evacuate, movement

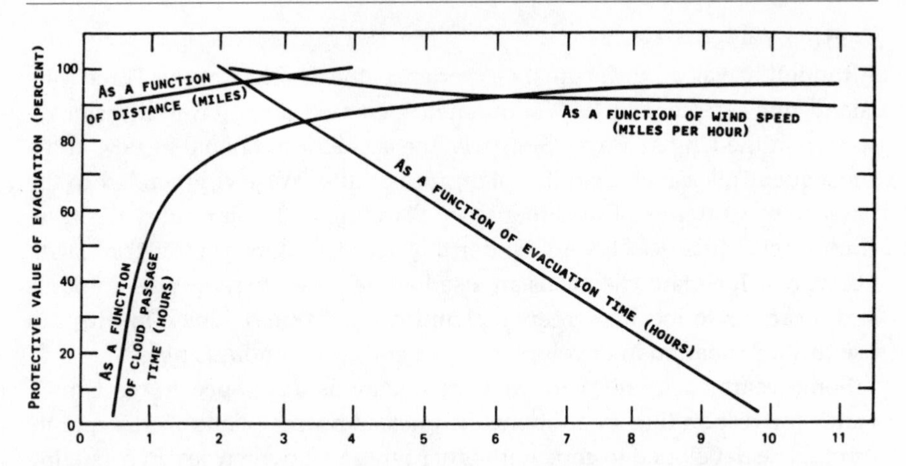

Figure 18.1 Protective value of evacuation as a function of time required for evacuation, wind speed, distance, and cloud passage

of children from schools, removal of patients from hospitals and nursing homes, and potential for increased uptake of releases as a result of the exertions involved (NCRP, 1977). Figure 18.1 presents data on the protective value of evacuation as a function of various conditions. As might be expected, the protective value increases with the rapidity with which evacuation is carried out and the distance between the airborne release and the evacuated population. Similarly, the longer the airborne cloud remains in the area, the greater the value of evacuation.

Sheltering and respiratory protection. The simplest and least disruptive of all proposed countermeasures is instructing people to remain indoors, as they would for other weather-related emergencies. The value of sheltering can be enhanced by encouraging the exposed population to use common household materials, such as handkerchiefs and bath towels, for respiratory protection. In the Bhopal accident, those exposed could have protected themselves simply by placing a wet cloth over their face to protect their eyes and to prevent inhalation of methyl isocyanate into their lungs. Unfortunately, this information was not made available to them (Mukerjee, 1995a). Sheltering and respiratory protection, because of their simplicity, are often proposed as viable alternatives to mass or partial evacuation, especially for people located several miles from the accident site, and when data concerning the magnitude, duration, and direction of an airborne release are unavailable or unreliable.

Protective prophylaxis. If the toxic chemical that has been released is identifiable and a known antidote is available, it is possible that the antidote can be administered to counteract or negate the exposures. If humans, for example, have a sufficient intake of stable iodine, the uptake of radioactive iodine into their thyroid will be minimized, depending on how soon the stable iodine is administered (Figure 18.2). The decision to give stable iodine should be based on a preplanned estimate of the probable degree of contamination from an accident. Even so, its administration to the general populace is not without problems. If stable iodine is to be distributed after an accident has occurred, health personnel may have difficulty locating all members of the population. If it is distributed before an accident, the measure may later prove to have been unnecessary. In addition, the intake of stable iodine may have adverse health effects on a small percentage of people (NCRP, 1977).

Other countermeasures. During later phases of an industrial accident, it will frequently be necessary to utilize other types of countermeasures. Airborne toxic materials can be transported many miles from the accident site, leading to the contamination of milk and agricultural products over large geographical areas. Since these foods are widely distributed, those produced in the more highly contaminated areas may be consumed by people some distance away. Ingestion frequently represents the primary

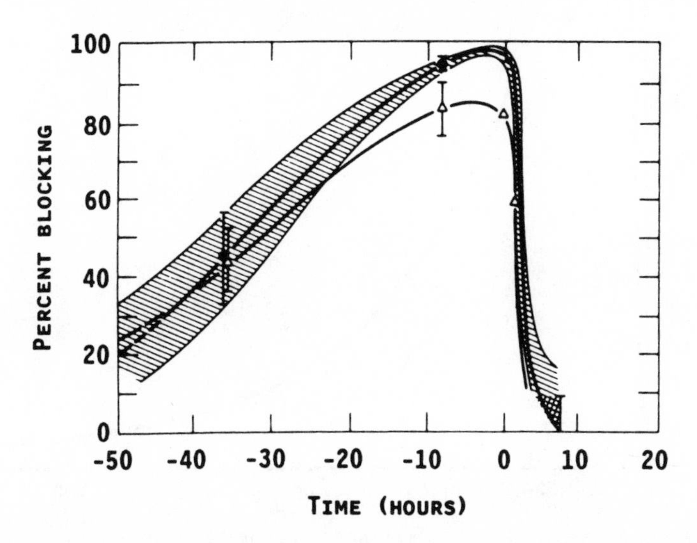

Figure 18.2 Percentage of thyroid blocking afforded by the administration of 100 milligrams of stable iodine as a function of time before and after an assumed intake of 1 microcurie of I-131

pathway of exposure, often exceeding inhalation by a factor of several hundred. Adding to these concerns is the fact that the movement of contaminants through the food pathway can be extremely rapid. Radioactive iodine, for example, can appear in milk within hours of being deposited on pasture grass (Figure 18.3). As a result, protective action needs to be taken promptly. In the case of milk, one of the most effective countermeasures is to shift cows from outdoor feed to stored (noncontaminated) feed. The effectiveness of this countermeasure as a function of time is shown in Table 18.3.

The major agricultural crops of immediate concern after an accidental airborne release will be those that grow above ground (such as lettuce and cabbage), since they may have been directly contaminated. Crops that grow underground, such as carrots, beets, and potatoes, should pose no problem

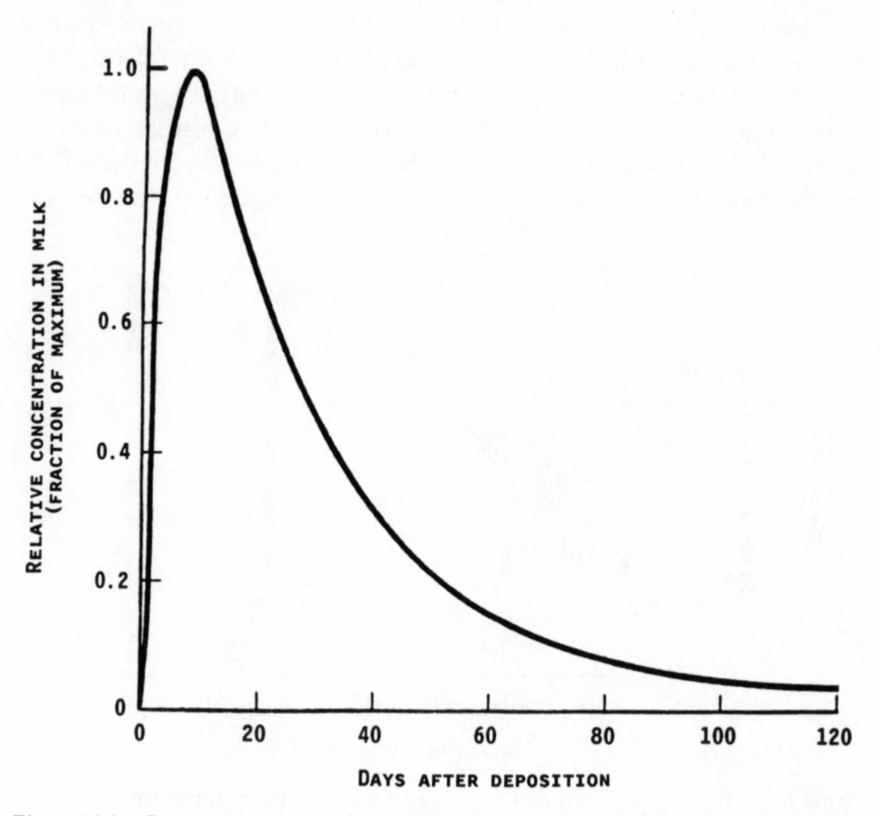

Figure 18.3 Concentration of radionuclides in milk as a function of time following a single deposition on pasture grass

Table 18.3 Intake avoided via milk pathway as a function of time of initiation of protective actions following an accidental single airborne release of radioactive material[a]

Projected intake avoided (%)	Iodine-131 (days)	Strontium-89 (days)	Strontium-90 (days)	Cesium-137 (days)
50	5–7	13	17	14
75	2–4	6	7	5
90	<1	2	2	2

a. Days after initial contamination of pasture when cows are assumed to have been shifted to uncontaminated feed.

until enough time has passed for the contaminant to gain access to them via root uptake through the soil. Washing with clean water may decontaminate previously harvested vegetables that were subjected to airborne deposition. Similarly, it may be possible to make fruit such as oranges and bananas acceptable for human consumption simply by removing the outer peel.

On farmland that has suffered extensive exposure, the soil must be treated so that contamination is reduced to safe levels. Idling, or nonuse, for a specific period is one strategy. Another is deep-plowing to move the contaminant below the root level, to prevent the plants from taking up contaminated nutrients, and to allow the contaminant to dissipate or be converted into a nontoxic form. Highly contaminated soil may have to be removed and disposed of elsewhere. In some circumstances it may be possible to grow alternative crops, such as cotton and flax, which would contribute no contaminant to the human diet.

Covered wells and underground sources of drinking water will probably not be contaminated. Although contaminants deposited on the soil will take a long time to travel to underground sources, care should be taken to assure that groundwater supplies are acceptable for consumption. Effective measures for treating contaminated surface water will depend on the nature of the contaminants (see Chapter 7).

For certain types of disasters, the need for countermeasures and for care of the victims may continue for years after the event. For example, follow-up medical studies are continuing for the victims of both the release at Bhopal and the Chernobyl accident. Such studies include not only medical care of the victims, but also epidemiologic research to learn more about the long-term effects of the accompanying exposures. In the case of the Bhopal accident, it is estimated that 50,000 people continue to suffer health effects; the reported mortality rate for those who were severely exposed is 15–20 percent higher than for those who were not (Mukerjee, 1995a). The number of people affected as a result of the accident at Chernobyl is even larger. In addition to the evaluation and care of those members of the exposed population who have developed cancer, work continues to resolve the emotional and economic impacts on the nearly one million people who were moved out of the contaminated areas subsequent to the accident (Filyushkin, 1996).

In recent years, the scope of planning for natural and man-made disasters has been expanded to include events resulting from unusual cases of prolonged hot weather and from acts of terrorism, riots, and wars. Examples include the heat wave in July 1995 that resulted in over 500 deaths in Chicago; the riots that occurred in Los Angeles in 1992 and in Miami in 1995; the bombing of the Murrah Federal Building in Oklahoma City in April 1995 (with an accompanying loss of about 169 lives); and wars such as the one in Bosnia, which have led to thousands of deaths and made untold numbers of people homeless. The deaths in Chicago demonstrated the importance of air conditioning in modern society (and of the continuing availability of electrical power), as well as the need for food and shelter, especially for the elderly. The riots in Los Angeles and Miami were a grim reminder that acts of violence can involve an entire community, not just a single family within a home.

As a follow-up to the Los Angeles disaster, a concerted effort has been made to promote dialogue among community groups, health-care providers, and other leaders to determine what factors led to the riots and what can be done to solve the underlying problems. Among the specific actions being taken is the development of violence prevention programs for children and adults. Far too often, children view violence as a way of solving problems at home. Another factor being addressed is the role of poverty in such disasters and what can be done to correct the inequities that exist. This is an area of increasing public health concern (Davidson, 1996).

Coordinating Emergency Responses

Ideally, attempts to mitigate the effects of a disaster will not add to its negative consequences. Often this is not the case, especially when governments and organizations outside the affected area attempt to assist. Physicians and nurses frequently are sent into the region in numbers far exceeding the need. Yet for most disasters it is unlikely that medical personnel from outside the affected country will be required. Another typical response is an influx of volunteers who neither speak the local language nor have had disaster relief experience. These people only exacerbate the problem. Usually any such response is too late to provide help; the need for search and rescue, life-saving first aid, and other immediate medical procedures in most emergencies is short-lived. A somewhat reverse situation applies to the supplies needed by medical and paramedical personnel in the disaster area: such supplies often are not brought in for 48 hours or more after the disaster has occurred. By that time victims with traumatic injuries will have already been cared for or will have died (Sidel et al., 1992).

In some ways, the problems enumerated above highlight what some have called the secondary disaster, one that all too frequently occurs in the days, weeks, and months following the initial event. The Pan American Health Organization (PAHO), in cooperation with the World Health Organization, accordingly has established a program on emergency preparedness and disaster relief coordination. One of its foundations is the recognition that individual countries take pride in relying on national resources, and prefer to limit their requests for external assistance to a few specific items or specialized skills. One of the primary activities undertaken by PAHO, in support of this effort, is the *su*pply *ma*nagement (SUMA) program. It is designed to provide a standardized methodology for managing relief supplies and equipment efficiently at the ports of entry into disaster-affected countries. Although this program is an attempt to provide a systematic approach under conditions that at the time of implementation will be very unfavorable, its principal objective is to provide a mechanism for sorting, classifying, and preparing an inventory of relief supplies sent to the disaster-affected country. It is hoped that the resulting data will enable local emergency managers to control the incoming supplies, and reconcile the offers of goods and services with the demands (de Goyet, 1993).

The General Outlook

The planning involved in coping with natural and man-made disasters is an enormous undertaking. Government officials at all levels must work together to devise a comprehensive strategy to meet the multitude of needs created by casualties, severe property damage, and disruption of services. Because the same principles of response apply to a wide range of natural disasters and man-made accidents, the most successful plans assign responsibility for responding to all such emergencies to a single group.

As experience has been gained, the trend has been to focus more on reducing risk than on coping with a disaster after it has occurred, and to provide information more rapidly to those responsible for making decisions after an event has occurred. Careful inspections can warn of impending accidents in industrial plants, and proper follow-up maintenance and repairs can avoid future accidents. Engineering technology can be used to design plants for containment of damage in the event of accidents. More rapid methods of exchanging information can be made feasible by perfecting reliable communication systems; improving the models used to forecast consequences, particularly of airborne releases; and developing models that can assimilate monitoring data and use it to update predictions (Lakey, 1993). Another step has been to factor into accident prevention and mitigation the possible contributions from malevolent acts such as terrorism.

Informing the public about what to expect and what measures to take to avoid harm can considerably reduce the impact of a natural disaster or industrial accident. Public officials need to provide advance information about the possible effects of chemicals and about appropriate countermeasures, as well as timely warnings so that people can seek shelter or evacuate. One technique is for emergency planning officials to involve the public in developing the plans for response. Not only does such an approach help assure cooperation and support, but it also leads to plans that are superior to those that would otherwise result. Table 18.4 lists a range of documents that provide guidance on emergency planning and response. These include the Federal Response Plan, developed by the Federal Emergency Management Agency in cooperation with a host of other federal departments and agencies.

Most natural disasters, such as earthquakes, floods, tornadoes, and hurricanes, leave buildings, bridges, and roads heavily damaged. Repair and replacement are costly, but techniques for restoration do exist. For disasters such as oil spills and major chemical or radioactive material releases, which may involve extensive contamination of water, land, or facilities, no simple techniques are yet available for removing or stabilizing the contamination and restoring such areas to a usable state. These problems represent significant challenges for the future.

The attention being given to disaster response worldwide was highlighted by the World Conference on Natural Disaster Reduction held in Japan in 1994. Among the outcomes was agreement that disaster reduction

Table 18.4 Sources of information on emergency planning and response

Agency	Document
ATSDR	*Managing Hazardous Materials Incidents* (Washington, D.C., 1992) Vol. 1, *Emergency Medical Services: A Planning Guide for the Management of Contaminated Patients* Vol. 2, *Hospital Emergency Departments: A Planning Guide for the Management of Contaminated Patients* Vol. 3, *Medical Management Guidelines for Acute Chemical Exposure*
DOE	*Planning and Preparedness for Operational Emergencies,* DOE Order 5500.3A, and supplemental *Emergency Management Guide: Interim Guidance for Protective Actions* (Washington, D.C., 1993)
DOT	*1993 Emergency Response Guidebook,* Report RSPA P 5800.6 (Washington, D.C.: Government Printing Office, 1993)
EPA	*Manual of Protective Action Guides and Protective Actions for Nuclear Incidents,* Report EPA 400-R-92-001 (Washington, D.C., 1991)
	Implementing Protective Actions For Radiological Incidents at Other Than Nuclear Power Reactors, Report EPA-R-92-001 (Washington, D.C., 1992)
FEMA	*Guidance on Offsite Emergency Radiation Measurement Systems, Phase 2—The Milk Pathway,* Report FEMA REP-12 (Washington, D.C., 1987); *Phase 3—Water and Non-Dairy Food Pathway,* Report FEMA REP-13 (Washington, D.C., 1989)
	Post-Emergency Response Resources Guide, Report FEMA-REP-17 (Washington, D.C., 1991)
	The Federal Response Plan, Report FEMA-229 (Washington, D.C., April 1992) (periodic updates issued thereafter)
HHS	*Accidental Radioactive Contamination of Human Food and Animal Feeds: Recommendations for State and Local Agencies,* draft report (Washington, DC, 1994)
IAEA	*Medical Handling of Accidentally Exposed Individuals,* Safety Series no. 88 (Vienna, 1988)
	Guidelines for Agricultural Countermeasures Following an Accidental Release of Radionuclides, Technical Reports Series no. 363 (Vienna, 1994)
	Intervention Criteria in a Nuclear or Radiation Emergency, Safety Series no. 109 (Vienna, 1994)
ICRP	*Principles for Intervention for Protection of the Public in a Radiological Emergency,* Publication 63, Annals of the ICRP (New York, 1993)

Table 18.4 (continued)

Agency	Document
NCRP	*Developing Radiation Emergency Plans for Academic, Medical or Industrial Facilities,* Report No. 111 (Bethesda, Md., 1991)
	Advising the Public about Radiation Emergencies: A Document for Public Comment, Commentary no. 10 (Bethesda, MD, 1994)
NRT	*Hazardous Materials Emergency Planning Guide,* Report NRT-1 (Washington, D.C., 1987)
	Criteria for Review of Hazardous Materials Emergency Plans, Report NRT-1A (Washington, D.C., 1988)
	Developing a Hazardous Materials Exercise Program: A Handbook for State and Local Officials, Report NRT-2 (Washington, D.C., 1990)
USNRC	"Standard Format and Content for Emergency Plans for Fuel Cycle and Materials Facilities," Regulatory Guide 3.67 (Washington, DC, 1992)
	"Emergency Planning and Preparedness for Nuclear Power Reactors," Regulatory Guide 1.101, Revision 3 (Washington, DC, 1992)
	Health Effects Models for Nuclear Power Plant Accident Consequence Analysis, Part II: Scientific Bases for Health Effects Models, Report NUREG/CR-4214, revision 1, part 2, addendum 2 (Washington, D.C., 1993)
WHO	*Nuclear Power: Accidental Releases—Practical Guidance for Public Health Action,* WHO Regional Publications, European Series no. 21 (1987)
	Guidelines for Iodine Prophylaxis Following Nuclear Incidents, Publication no. 35, Environmental Health Series (1989)

ATSDR Agency for Toxic Substances and Disease Registry
DOE U.S. Department of Transportation
DOT U.S. Department of Transportation
EPA Environmental Protection Agency
FEMA Federal Emergency Management Agency
HHS U.S. Department of Health and Human Services, Food and Drug Adminstration
IAEA International Atomic Energy Agency
ICRP International Commission on Radiological Protection
NCRP National Council on Radiation Protection and Measurements
NRT National Response Team (representing 14 Federal agencies)
USNRC U.S. Nuclear Regulatory Commission
WHO World Health Organization

activities can be extremely cost-effective; that better communication among the many interest groups, particularly at the national level, is crucial; and that the lessons learned and management techniques developed for coping with natural disasters can be usefully applied to man-made emergencies (Douglas, 1994). Other conclusions were that in most cases the first people to respond in a disaster will be those in the affected community; and that although the development of building codes, early-warning systems, and approaches to training and community preparedness require international support, they are best carried out at the regional and national level. For example, codes that will assure that buildings can withstand earthquakes in California will not necessarily lead to the design and construction of buildings that can withstand wind damage from hurricanes in the Caribbean (PAHO, 1994).

19

A MACROSCOPIC VIEW

MANY of our environmental problems—air and water pollution, solid waste, food contamination—are consequences of large-scale cultural patterns. Some are the net result of millions of people making individual decisions; others are triggered by a small number of people with key decisionmaking powers in industry, government, and academia. Although many of the problems are local, ozone depletion, global warming, acidic deposition, and the resulting climatic and environmental changes have global implications. Solutions will require cutting across national jurisdictions, and a shift in focus from protection and restoration to planning and prevention.

Our global problems appear to reflect three major trends (Speth, 1989). First, as the result of a threefold increase in the world's population and a twentyfold increase in the values of goods produced since 1900, the quantity of pollutants being generated has significantly increased. Second, there has been a shift from the use of natural products to the production and use of synthetic chemicals. For example, a billion pounds of synthetic pesticides are used every year in the United States. Many have proved to be highly toxic, and some persist and accumulate in biological systems and in the atmosphere. Third, as a result of expanded technological capabilities, and in some cases the export of hazardous technologies, the developing countries have become as polluted as the developed nations.

Although there is no consensus on how to solve these increasingly difficult environmental problems, serious international efforts are under way and significant progress is being made. An outstanding example is the Montreal Protocol on Substances That Deplete the Ozone Layer, an international treaty developed in 1987 that established ways to reduce the use of chlorofluorocarbons and other chemicals that destroy stratospheric ozone. The protocol was a forerunner of the Earth Summit, sponsored by the United Nations and held in Brazil in 1992, which led to the world's first treaty concerning climate. A key part of the treaty was the establishment of a groundwork for stabilizing the atmospheric concentrations of the so-called greenhouse gases. In a follow-up meeting held in Germany in 1995, delegates from countries around the world, including the 116 that had ratified the original treaty, concurred that the commitments made in Brazil were not adequate. New goals are being negotiated. The delegates have essentially agreed that the crux of the matter is not whether releases of gases such as carbon dioxide and methane will cause a change in the climate of the world. Rather the questions are How much? How fast? Where? and What must be done to prevent the accompanying effects? (Koenig, 1995).

Ozone Depletion

Chlorofluorocarbons (CFCs) have been widely used over the past 60 years as refrigerants in household appliances and air conditioners, as industrial solvents, as blowing agents in manufacturing foam products, and as propellants for aerosol sprays. As their name implies, these chemical compounds consist of chlorine, fluorine, and carbon atoms. Some also include hydrogen atoms. When released into the air, CFCs mix with other compounds and rise slowly into the stratosphere, where they may remain for years. In the stratosphere, ultraviolet (UV) radiation destroys the CFC molecules, releasing highly reactive chlorine atoms. These react with ozone, converting it into normal oxygen. A single CFC molecule can destroy tens of thousands of molecules of ozone.

Although ozone is considered a pollutant when it is near the ground, in the stratosphere it shields the earth's surface from UV radiation. The harmful effects of excess UV radiation include increased skin cancers and cataracts, lower crop yields, and damage to materials such as vinyl plastics. Phasing out production of CFCs, however, has not been without its complications. The CFCs are nontoxic and have significantly better refrigeration properties than any of the alternative chemicals currently being considered. They also require little energy to produce. Although the processes required to manufacture alternatives have proved to be more complicated than existing CFC processes, a range of substitutes have been developed and are now being commercialized. It is encouraging that several CFC producers have formed consortia to share costs and expedite the toxicity testing of these new compounds and to evaluate their effects and those of their degradation products on the environment (Manzer, 1990). It is also encouraging that efforts to control CFC releases are producing results. Data show that the concentrations of chlorine in the atmosphere have been decreasing in recent years (Kerr, 1996).

Acidic Deposition

Some airborne pollutants, after being discharged by electric power plants, industrial installations, and automobiles, are chemically transformed into acid compounds. Although some of these compounds remain in solid form, the nitrogen oxides are transformed into nitric acid and certain of the sulfur oxides are transformed into sulfuric acid. As a result, the concentrations of acid compounds today in the atmosphere of the northeastern United States and eastern Canada is significantly higher than it was 30 years ago. The deposition of these compounds, either dry or as nitric or sulfuric acid in rain or snow, is imposing an unprecedented and alarming burden on forests, streams, and lakes in these areas. Prevention of further ecosystem damage will require a substantial reduction in the discharge of these pollutants, especially SO_2.

The principal measures for controlling acidic deposition and its effects are (1) reduction of the discharges of sulfur and nitrogen oxides to the atmosphere and (2) treatment of sensitive ecosystems to make them less susceptible to damage. Experts believe that SO_2 accounts for about two-thirds of the total acidic deposition in the northeastern United States and eastern Canada; nitrogen oxides account for the rest. In acknowledgment of this fact, the Clean Air Act Amendments of 1990 mandated a 50 percent reduction by the year 2000 in the releases of SO_2 from coal-fired plants in the Midwest. To control acidic deposition in the southwestern United States, including cities such as Los Angeles where nitrogen oxide emissions from automobiles are a principal source, the Clean Air Act Amendments require stringent controls on emissions from automobiles. Since some of the mandated emission standards may be difficult to meet with current designs, automobile manufacturers are actively exploring alternative fuels and electric units. No remedial actions have been developed for treating terrestrial ecosystems damaged by acidic deposition, but it is common practice to add lime to lakes to increase their ability to absorb the resulting acidity. Many uncertainties remain about the cost-effectiveness and environmental consequences of this procedure (Harrington, 1988).

Global Warming

When present in the atmosphere, chemical compounds such as carbon dioxide, methane, and CFCs, are transparent to shortwave electromagnetic radiation reaching the earth from the sun, but they absorb this radiation when it is reflected back from the earth at reduced energy. In theory, increasing the concentration of these gases in the atmosphere is causing the temperature near the surface of the earth to increase, as it does in a greenhouse. Although there is still controversy about the degree of warming that is actually occurring, what is now known appears to be more than sufficient to warrant serious attention (Kerr, 1995).

Atmospheric concentrations of CO_2 are increasing by about 0.4 percent per year, and overall concentrations have increased by 25 percent in the last century. In fact, the concentrations today are higher than at any time during the past 160,000 years (Warrick and Jones, 1988). Once released into the atmosphere, CO_2 can be removed by several processes. Some is absorbed into the oceans; some is absorbed by trees and other vegetation. However, the exact nature of these processes is not known. Further complicating the situation are the complexities and uncertainties of the mathematical models necessary for predicting the trends and consequences of the greenhouse effect (Douglas, 1994). Progress is being made on improving these models, and our ability to predict the future impact of increasing concentrations of carbon dioxide in the atmosphere has improved considerably (Nierenberg, 1995).

If the concentrations of greenhouse gases continue to increase, it has been projected that temperatures at the surface of the earth will be rise by 1.5–4.5 degrees centigrade (3–9 degrees Fahrenheit) by the end of the next century (Koenig, 1995). The changes are not expected to increase temperatures uniformly, but rather to change climate in ways that will vary both regionally and seasonally. The winters at high latitudes will probably be warmer and wetter, with less change in the summer and in the tropics. Sulfate emissions may offset greenhouse warming over industrialized regions, which could potentially alter global atmospheric circulation.

Overall global warming could lead to a rise in sea level both from the expansion of warming seawaters and the added contribution from melting glaciers. Current estimates are that the overall rise in the level of the seas could approach two-thirds of a meter (about 2 feet) over the next century (Hammitt, 1995). Such a change would make the low-lying and relatively flat Atlantic and Gulf coasts of the United States particularly vulnerable to inundation (Hanson and Lindh, 1996). Similar problems would exist for the Netherlands and Bangladesh. In Central Europe higher temperatures, the accompanying changes in rainfall patterns, and the potential for extended droughts could turn fertile farmlands into deserts. Some experts estimate that even if discharges of greenhouse gases were to be terminated within the next several decades, sea levels would continue to rise for perhaps another thousand years.

The potential impact of global warming, however, ranges far beyond the dangers posed by melting ice caps and coastal flooding. Threats to wildlife may be even more far-reaching than those to humans. Scientists estimate that populations of zooplankton in the Pacific Ocean off the coast of California have decreased 80 percent in the last 40 years. This has had a ripple effect in terms of declines in fish and seabird populations (Hill, 1995). Terrestrial animal species can migrate because of changing climatic conditions, but a variety of topographical obstacles pose formidable barriers. Largely unknown is whether the changes will be gradual or abrupt. An abrupt shift in ocean circulation, for instance, could lead to a discontinuation of the Gulf Stream and an accompanying cooling of Western Europe (Hammitt, 1995).

Associated impacts could include the disappearance of 50–90 percent of the boreal forests, consisting largely of conifers and representing the world's largest ecosystem (Taubes, 1995). Instead of encouraging exuberant growth, the warming appears to be stressing these forests by speeding up moisture loss and subjecting them to more frequent insect attacks. The spread of insects due to global warming is projected to have other negative effects, one being an increase in a host of human diseases transmitted by various insect and other vectors whose habitat will expand beyond the tropics. Diseases most likely to increase include malaria and dengue, as well as schistosomiasis; others include filariasis, onchocerciasis, African trypanosomiasis, and yellow fever (Stone, 1995a). Global warming can also cause increased frequencies of natural disasters, including hurricanes.

Over the past 200 years the cumulative addition of CO_2 to the atmosphere has come in approximately equal proportions from deforestation and the combustion of fossil fuel. Today combustion accounts for about 80 percent of annual CO_2 emissions. Thus, CO_2 production is closely related to energy usage. Although the industrialized nations, particularly the United States and countries constituting the former Soviet Union, have been the primary contributors of CO_2 in the past, Brazil, China, and India currently also rank as major global warmers. Projections are that by the year 2025 the developing nations will be emitting four times as much CO_2 as the developed world now produces (World Resources Institute, 1990).

During this same 200-year period, atmospheric concentrations of methane have approximately doubled as the result of a surge in worldwide population and in rice farming and animal husbandry. Methane is produced by the decomposition of plant matter in the stomachs of cattle and is released as flatulence; decaying vegetation in wet rice paddies also produces large quantities. Another source is biomass burning related to agricultural expansion. India, through its production of methane, ranks among the top five producers of greenhouse gases. Methane concentrations in the atmosphere currently are increasing at the rate of about 1 percent per year (Warrick and Jones, 1988).

From a policy perspective, the ideal solution would be to slow the rate of increase of atmospheric concentrations of greenhouse gases. However, it will be extremely difficult to reconcile such control measures with the needs of the world's burgeoning population. One obvious near-term step would be to promote the use of energy sources such as solar and nuclear power, which do not produce CO_2. Another option is to develop methods making it easier to adapt to the change (through improved climate forecasting and the development of more resilient agricultural plants.) It will also be necessary to stimulate research to better understand the causes and consequences of climate change and to encourage continuing international discussion of the issues and how best to manage them (Hammitt, 1995).

Deforestation

At one time forests covered 70 percent of the land area of the world. Today this is no longer the case. India has lost two-thirds of its forests since 1600 (CEQ, 1995). The contiguous United States has lost 30 percent of its forests since 1620. Worldwide, satellite sensing shows that tropical rain forests are being destroyed at a rate of 40–50 million acres (60,000–75,000 square miles) per year (CEQ, 1993). If current trends continue, most of the forests of the Amazon River basin will be lost by the year 2000 (Table 19.1).

Forests are a source of ecological services and economic goods, ranging

Table 19.1 Loss of tropical rain forests, 1991

Forest	Original area (thousands of acres)	Remaining area (thousands of acres)	% loss
Tropical Andes	25,000	8,600	66
Atlantic forest, Brazil	247,000	4,940	98
Madagascar	15,314	2,470	84
Indonesia	301,340	130,910	57
Philippines	61,750	1,976	97
Total	650,404	148,896	77

from soil and watershed protection to timber and fuelwood to carbon sequestration, wildlife habitat, and recreation. Forests protect soils, and they play a major role in hydrological cycles. Forests also exert a gyroscopic effect in atmospheric processes and other elements of global climate, having an influence second only to that of oceans. As much as half of the carbon in the world's biomass may be stored in forests. As a result, deforestation reduces the earth's capacity to absorb carbon dioxide from the atmosphere. Forests are also critical to the energy budget and albedo (reflectivity) of the earth, and they harbor a majority of the species found on land (Myers, 1995). The ecology and economics of forests cannot be separated, since forest values are at once local, national, and global in scope.

The cutting and hauling of trees from a forest is equivalent to removing the essential nutrients and topsoil. Water and nutrient cycles are destabilized, and the soil itself is left without protection from flooding and erosion. The resulting sharp decline in soil fertility reduces agricultural productivity. Without the moisture retention capability provided by trees, floods occur. Without the trees themselves, less moisture transpires into the air and less rain falls in neighboring areas; the slow release of water from forests to streams during dry periods ceases. Droughts result, then cycles of drought and floods, and—finally—deserts. Constructing dams to control the water can cause a multitude of other public and environmental health problems.

A serious impediment to the survival of forests is a practice known as clear-cutting, which essentially destroys the entire forest ecosystem. Not only live trees are removed, but dead ones as well. Had they been permitted to remain, the dead trees would have provided soil for the future. Also removed are bark beetles and fungi that transport nitrogen-fixing bacteria into the soil; decaying wood that serves as the main purifier of water; whole communities of plants and animals; and all the other elements necessary for the complex cycles of regeneration. Although in many cases regulations require that new trees be planted to replace those that have been removed, the net result is the replacement of complex ecosystems with monocultures (Ayres, 1994).

With 60 percent of the people in the world still using wood for cooking, and with businesses relying on forest resources for timber products, tobacco curing, and other industrial applications, additional deforestation is inevitable. Intensified management of the remaining forests to raise yields and an increase in the areas devoted to timber plantations will have both positive and negative effects on the environment. Well-managed plantations and production forests can reduce the harvesting of trees from natural forests, but plantation forests are biologically less diverse, are poorer habitats for native animals, and in some instances may be inferior in retaining the water and in conserving the soil. In addition, successful operation may require the use of pesticides, herbicides, and energy-intensive fertilizers.

Forests can be destroyed by causes other than human actions. Two that are sometimes overlooked are fires and insects. Fires can be devastating in areas where dead and dying vegetation has accumulated, and it is not unusual for fires in the western states to destroy millions of acres of trees annually. Insects can be equally devastating. In 1991 the European gypsy moth defoliated 4 million acres of trees in the Northeast. Especially vulnerable are oaks and maples. Augmenting the problem was the introduction in 1993 of the Asian gypsy moth in foreign cargo unloaded at a port in North Carolina. Concurrently, populations of the southern pine beetle reached epidemic levels in seven states and threatened timber, wilderness values, and habitats for certain birds (CEQ, 1993, 1995).

Steps that can be taken to counteract the loss of forests include programs for reforestation, such as the one in Brazil that offers tax incentives to industry to encourage the planting of trees; restrictions on land clearing, based on soil capability studies; better management of existing rangelands, including controls on grazing and improvement of pasture; more thoughtful management of existing forest resources; use of more efficient wood cooking stoves; and development of biogas and solar stoves to replace wood burners.

Loss of Biodiversity

According to current estimates, 3–10 million species of flora and fauna exist on earth. Their diversity provides an index of the ecological health of the planet. Yet in the United States at least 500 species and subspecies of plants and animals have become extinct since the 1500s. Worldwide, some 75 mammals and birds became extinct between 1600 and 1900; between 1900 and 1980, another 75 mammals and birds met a similar fate (CEQ, 1995). Although in isolated cases these losses have been due to natural causes, in the main they have occurred as a result of human actions. As many as three species are disappearing worldwide every day, and the rate of loss appears to be accelerating. In fact, extinction rates are estimated to be 100 to 1,000 times their prehuman levels. Global patterns of extinction are dominated by those regions in which the threatened species are unique (Pimm et al., 1995). If current trends continue, by the year 2050 as many as 25 percent of the species now on earth could be extinct (Wilson, 1992; Chadwick, 1995).

Extinction is an irreversible process. Lost with these plants and animals will be critical genetic information and biotic resources, including natural chemical compounds that could have useful medical applications. Benefits already derived from such sources include a drug to treat high blood pressure, which uses venom from the pit viper in Brazil; extracts from the root and flower of the kudzu plant that may serve as a cure for alcohol abuse; and extracts from lobsters that may provide medicines for the treatment of tumors and the AIDS virus. Indications are that a protein in the saliva of the sand fly may prevent blood clots, modify the immune system, and serve as a vaccine against leishmaniasis (Cromie, 1996). According to the National Cancer Institute, one of the most promising anticancer agents currently being evaluated is taxol, a substance that concentrates in the bark of Pacific yew trees (CEQ, 1993).

Several pharmaceutical companies, recognizing that forests are possible sources of new medicines, have established agreements through which they provide funds to preserve forests in various countries of the world in exchange for access to the associated plants, microbes, and insects (French, 1995). Other benefits being derived from exotic plants include the use of genes from tropical tomatoes to increase the density of U.S. tomatoes, with added profits to catsup manufacturers; the potential use of natural protein from wild Mexican beans to repel insects without poisoning soil or water; and the use of high-yield African plants, rather than trees, as a source of pulp for paper (Linden, 1989). One scientist has likened the current reduction of biodiversity to "burning a library before the books have been read" (Bertrand, 1990).

Recognizing this as a problem, Congress in 1973 passed the Endangered Species Act, which was based on the assumptions that each life-form may prove valuable in ways that cannot yet be measured or understood and that each species is entitled to exist for its own sake. In essence, this act was a "bill of rights" for nonhumans. The original list prepared at that time contained the names of 109 species considered to be endangered; by 1995 the total had reached over 1,400, including foreign species. Waiting in line are almost 4,000 additional candidates that have been officially recognized (Chadwick, 1995).

In attempting to protect various animal species, governmental agencies have set aside certain areas of the United States as national parks or forests. But experience has shown that biotic diversity cannot be preserved simply by fencing off a few protected areas within an overall landscape of exploitation. This approach will fail for a number of reasons. Within many countries of the world, a large percentage of the existing empty space has already been occupied; little space is left whose use can be restricted. Many animal species do not remain within the boundaries of the restricted areas. While certain terrestrial animals may have a limited range, others do not. Some common U.S. birds migrate across the North American continent; others migrate all the way to the tip of South America. Having an especially serious impact on the survival of such birds is the destruction of habitats that they formerly used on a temporary basis during their migrations (Babbitt, 1995).

One of the newer approaches being considered for protecting endangered species is to provide financial incentives, ranging from tax breaks to future development rights, to owners who choose to provide such protection, as opposed to concentrating on threatening them with fines or jail (the approach currently used to enforce the Endangered Species Act). The incentive approach has received endorsement from a coalition of scientists, environmentalists, and businesspeople and may indeed serve as the basis for future legislation (Stone, 1995b). It coincides with the recommendations of the World Resources Institute (1995), which concluded on the basis of an extensive study that the key to protecting biodiversity is to ensure that species do not *become* endangered, not waiting until they are endangered to protect them.

Loss of Topsoil

Accelerated erosion occurs where human activity disrupts the plant cover that usually protects the soil. Poor farming techniques such as deep-plowing, followed by wind and water erosion, are major causes of topsoil losses. As a consequence of such practices, more than 1.5 billion tons of topsoil were lost in the United States in 1989, 300 million of which ended up being deposited in the Gulf of Mexico (AMA, 1989). Urban and suburban sprawl also contributes: farmlands with acres of the world's richest topsoil are being covered by asphalt parking lots, housing developments, and shopping malls. On a global basis, estimates are that over 40 percent of the earth's terrestrial vegetated surface has a diminished capacity to supply benefits to humanity because of recent direct impacts of land use (Daily, 1995).

Soil loss by erosion involves more than just depletion of the topsoil. Soil washing off cropland, pasture, or forest land ends up in surface water, in the air, or on other lands. Sediment in water bodies causes turbidity, silting, and deterioration of aquatic habitats, decreases storage capacity in lakes and reservoirs, and interferes with water distribution systems. Fertilizers, pesticides, and salts in eroded sediment reduce water quality. Soil particles blown by wind cause dust storms that physically damage crops and buildings. Windblown particles also contribute to air pollution, exacerbating respiratory ailments and impairing vision (CEQ, 1989). The accompanying direct and indirect economic costs in the United States, through losses in agricultural production and associated pollution, damage from sediments, and associated increased flooding, are estimated to exceed $40 billion annually (Glanz, 1995). Concerns about these problems have led some scientists to propose the creation of a "soil health index" that could be used by policymakers in evaluating and maintaining the quality of key agricultural areas (Haberern, 1995).

Once removed, topsoil requires hundreds to thousands of years for regeneration. Without appropriate conservation practices, farming, livestock grazing, logging, and other activities will continue to cause soil erosion. Among the most successful techniques for erosion control are various forms of conservation tillage, such as no-till farming, in which the residue from a previous crop is left in the field, and new seeds and/or plants are placed in the ground without prior plowing. Other soil conservation techniques include contour plowing, maintenance of vegetative buffer strips between fields and along waterways, planting highly erodible soils with permanent trees or grass cover, and keeping a vegetative cover on idle land. It is estimated that implementation of such practices will reduce soil erosion in the United States by over 50 percent within the next few years (CEQ, 1989, 1995).

Destruction of Wetlands

Wetlands include tidal marshes, swamp forests, peat bogs, prairie potholes, and wet meadows. Biologically, they are the most productive ecosystems in the world. Wetlands serve as nurseries and feeding grounds for a range of commercial fish species, as nesting and feeding grounds for waterfowl and migratory birds, and as a habitat for many other forms of animal life (otters, turtles, frogs, snakes, insects). In addition, wetlands trap nutrients and sediments; purify water by removing coliform bacteria, heavy metals, and toxic chemicals; provide flood protection by slowing and storing water; and anchor shorelines and provide erosion protection.

Some 80 percent of coastal fisheries and one-third of the endangered species in this country depend on wetlands for spawning, nursery areas, and food sources. Wetlands are home to millions of waterfowl and other birds, plants, mammals, and reptiles. They also perform hydrologic functions: serving as recharge areas, they help protect the quantity and quality of the groundwater. Unaltered wetlands in a floodplain can reduce flood peaks by 80 percent. Wetlands are also vital to commercial and recreational sectors of the economy, such as sports fishing and waterfowl hunting. In addition, the diversity of plant and animal life in wetlands make them a valuable resource for nonconsumptive fish- and wildlife-related recreation. Wetlands also provide educational and research opportunities and a variety of historical and archaeological values (CEQ, 1993).

In essence, the protection of wetlands is essential to preservation of the trove of biodiversity beneath the sea. Dredging, filling, and other development activities have altered many coastal habitats, including coral reefs, bays, marshes, rocky shores, and beaches. This has led to reductions in popular edible fish and shellfish species, and reduction or loss of species that have strong potential for biomedical products. Unfortunately, the ability of scientists to evaluate the scale and consequences of these changes has been seriously compromised by inadequate knowledge of the basic processes that control the diversity of life in the sea (NRC, 1995).

More than half of the wetlands that existed in the contiguous 48 states have been converted to other uses; several states have lost as much as 80 percent of their wetlands. Estimated annual wetland losses between the mid-1950s and the mid-1970s totaled almost a half-million acres, leading to a total loss of almost 10 million acres over these two decades. Estimated annual losses between the mid-1970s and the mid-1980s were between 120,000 and 290,000 acres (CEQ, 1993). As of 1980, more than 100 million acres of wetlands had been drained, cleared, or filled (Canby, 1980). Although most of these lands were converted to agricultural use, urban development accounted for approximately 10 percent of the loss; another 5 percent was lost to suburban residential developments, highways, airports, industrial facilities, and marinas (CEQ, 1989).

Section 404 of the Federal Water Pollution Control Act of 1972 assigned the Army Corps of Engineers responsibility for defining and supervising wetlands. Initially, its regulations applied only to navigable waterways and their adjacent wetlands. In 1976 the final regulations expanded the definition of wetlands to include lands adjacent not only to navigable waterways and their tributaries but also to other rivers and streams and to natural lakes of at least 5 acres (O'Connor, 1977). As a result of these and other efforts, 104 million acres (some 5 percent of the contiguous 48 states) are now officially designated as wetlands. Three-quarters are privately owned. Owing to a flurry of criticism by land owners, however, the government has narrowed its definition to exclude lands that are only periodically covered by water. Such a change will permit these lands to be converted to other uses. No one questions the value of wetlands as an environmental resource. The challenge is to balance human needs and desires with preservation of the environment.

In some cases where wetlands continue to exist, the discharge of fertilizers from adjacent agricultural areas has enabled undesirable plants to displace native plants, resulting in negative impacts on the normal beneficial functions of the ecosystem. In other cases, canals dug by engineers to channel the runoff from wetlands have led to flooding and widespread pollution. A notable example is the Florida Everglades, where engineers are attempting to return the wetlands to their original state (Culotta, 1995). Systems are also being developed to remove the nutrients prior to releasing wastes into these waters (Goforth, Jackson, and Fink, 1994). Another hopeful sign is the development of methods for discharging municipal sewage onto low-lying land areas to create artificial wetlands with many of the characteristics of natural areas.

Sustainable Development

One of the primary messages emerging from the above discussion is that present policies for meeting the needs of the peoples of the world and for protecting the environment are inadequate. The population of the world today totals almost 6 billion people. Unless future population control efforts are dramatically more successful than they have been in the past, this number could double before the year 2050. With such growth will come increasing demands for food, clothing, shelter, and manufactured goods, which in turn will place enormous pressures on the world's resources of energy, air, water, land, and biota. Unless the ways in which these needs are met change significantly, the resources of the earth will be depleted and the environment will become increasingly polluted.

What is needed are procedures and technologies such that the requisite goods and services can be provided in a sustainable manner. The concept of sustainable development has been defined by the World Commission on Environment and Development (1987) as "development that meets the needs of the present without compromising the ability of future generations to meet their own needs." The wide acceptance of this concept is exemplified by the number of industrial and political leaders throughout the world who increasingly are seeking policies that will achieve this goal. A significant portion of the economic summit held in Nova Scotia in July 1995 was devoted to a review of potential approaches.

As part of the need to develop sustainable technologies, industrial organizations recognize that it is not adequate simply to develop methods for treating waste after it has been produced; the current emphasis is on moving into industrial processes themselves to cope with problems of waste before they occur. Those responsible for the design of products and manufacturing processes must, early on, consider the disposal both of the wastes resulting from the manufacturing process and the products themselves at the end of their useful lives. In essence, for each newly proposed product manufacturers should conduct what is called a life-cycle assessment (Keoleian and Menerey, 1994). Savings are measured not only in the costs of production and materials, but also in the potential costs of waste disposal.

In some cases, planners are attempting to develop industrial systems that mimic natural ecological systems, in which any waste that contains usable materials or embodied energy is consumed by some organism. Materials move through a complex web of animals and plants, and almost nothing remains as unused waste. In essence, planners are developing what some call the science of industrial ecology (Frosch, 1993). To encourage such developments, the President's Council on Sustainable Development has recommended altering governmental policies to provide economic incentives that incorporate subsidy and tax policies to discourage pollution, encourage natural resource protection, and promote community redevelopment. The council has also endorsed voluntary measures for resource preservation and waste management, and has encouraged increased flexibility in regulatory compliance provided it is accompanied by requirements for increased environmental protection (Toman and Darmstadter, 1996).

Economic Growth and Environmental Quality

It is often stated that economic growth is a necessary condition for environmental improvement. Although this may be true under certain conditions and assumptions, the relationships are not simple. Sometimes one type of relationship is observed; at other times the relationship is entirely different. As income rises, environmental degradation tends to increase up to a certain point, after which environmental quality improves. The point at which the turnaround occurs varies widely, depending on a number of factors including the index of quality being evaluated. One explanation is that, during the initial stages of development, people in the poorer countries cannot afford to emphasize protection of the environment over personal well-being. Consequently, increased pollution is regarded as an acceptable side effect of economic growth. When a country has attained a sufficiently high standard of living, its people usually are willing to direct more attention to environmental protection, enacting appropriate regulations and controls.

As would be expected, a range of caveats must be applied in expressing these generalities. One is that these relationships between economic growth and environmental quality are valid primarily for pollutants having local short-term costs, for example, sulfur, particulates, and fecal coliforms. They do not apply to accumulated stocks of waste, and they do not apply to pollutants having long-term and more dispersed costs, such as those that lead to global warming. Discharges of such pollutants often continue to increase with income. Nor do these relationships apply to resource stocks—for example, those involving the soil and its cover, forests, and other ecosystems. These relationships also do not take into consideration system-wide consequences; for example, reductions in the releases of a pollutant in one country may lead to increases in the discharges of other pollutants in the same country, or to transfers of pollutants to other countries. Where emissions have declined with increasing income, the reductions often are due to local institutional reforms, which may ignore international and intergenerational consequences. Where the environmental costs of economic activity are borne by the poor, by future generations, or by other countries, incentives to correct the problem are likely to be weak.

The net effect is that the environmental consequences of increasing economic activity are complex, and those responsible for controlling the associated impacts must adopt a systems approach. Otherwise environmental problems may not be solved in time to avert severe global consequences. In short, economic growth is not a panacea for environmental quality. What matters is the *content* of the growth; that is, the composition of the inputs, including environmental resources, and the composition of the outputs, including the waste products generated (Arrow et al., 1995).

The General Outlook

In addition to global environmental health problems, each country in the world confronts a host of local challenges. To meet them successfully, policymakers need to seek out the root causes wherever possible and maintain perspective in setting priorities for action.

Prevention or control of environmental problems requires knowledge, technology, and incentive. Industrialists and industrial engineers must learn to incorporate sound environmental thinking in the initial selection and design of manufacturing processes and products (NAE, 1989). Pollution controls must be designed into industrial equipment, not added on later. As part of this effort, multinational corporations and financial institutions have an obligation to set a new moral tone for the world. They must commit themselves to a sustainable future and they must be prepared to sacrifice a portion of their profits to do so. Close examination will demonstrate that such an approach is in their best interest, as well as essential to their survival (Glaze, 1996).

In a similar manner, government agencies need to assume increased responsibility for protecting the environment and improving its impact on the world's population. As enumerated by the Science Advisory Board of the Environmental Protection Agency (Johnson, 1995; Loehr, 1995), the principal needs are as follows:

1. To develop programs that will provide continuing evaluation of key environmental areas, such as terrestrial ecosystem sustainability,

noncancer human health effects, total air pollutant loadings, nontraditional environmental stressors, and the health of the oceans;

2. To emphasize the avoidance of future environmental problems as much as the control of those that currently exist; an integral part of this effort should be the establishment of an early warning system to identify emerging environmental risks;
3. To stimulate coordinated efforts among federal, state, and local agencies and the private sector to develop the capability to anticipate and respond to environmental change;
4. To recognize that global environmental quality is a matter of strategic national interest, and to adopt policies that link security, foreign relations, environmental quality, and economic growth.

Society as a whole needs to develop a forward-looking attitude in dealing with environmental problems, instead of merely reacting when a crisis develops. Societal behavior will change when enough people become aware of environmental problems and act, both as individuals and through their elected representatives in government. The American Medical Association, in its policy statement on *Stewardship of the Environment,* suggests that the United States play a leading role in effecting change (1989, p. 2):

> The U.S. and the world at large appear to be facing environmental threats of unprecedented proportions, and scientists, environmental activists, health professionals, politicians and world leaders are beginning to realize the need for changes in societal behavior (i.e., human behavior as well as the conduct of business and industry) as a means of forestalling these potential threats. Societal changes must be initiated worldwide if they are to have any significant effect overall. However, their implementation will need a model, most suitably a national model . . . The U.S. could well become a model for environmental stewardship if a grassroots movement were to develop to encourage and endorse a protective and nurturing philosophy towards the environment at both the personal and societal levels.

Accomplishing these objectives will require action on several fronts, notably education. According to the EPA's National Advisory Council on Environmental Policy and Technology, the most productive such effort should be directed at young people; it should be designed to inculcate environmental values during primary and secondary school (EPA, 1989). Even then, such a program can be effective only if it has the support of the entire community—schools, churches, business and industry, trade associations and professional groups, advertising and news media, and government at all levels (AMA, 1989).

In spite of what may often seem an insurmountable task, progress is being made. About 80 percent of Americans are concerned about their environment, especially air and water quality; half to two-thirds are also worried about the greenhouse effect, ozone depletion, and hazardous substances at home and at work (Somerville, 1989). Worldwide, peoples in both the developing and industrialized nations have substantial concern about both the existing quality of the environment and its status in the future (Bloom, 1995). The will is there; the knowledge and technology continue to evolve. The primary need is for leadership to marshal existing support and to set priorities for action.

APPENDIX

Environmentally Related Journals

(An asterisk denotes a journal that provides broad coverage.)

American Industrial Hygiene Association Journal
American Institute of Chemical Engineering Journal
American Journal of Epidemiology
American Journal of Industrial Medicine
American Journal of Physiology: Lung Cellular and Molecular Physiology
*American Journal of Public Health
American Journal of Tropical Medicine and Hygiene
Annals of Biomedical Engineering
Annals of the ICRP
Annals of Occupational Hygiene
Applied and Environmental Microbiology
Applied Occupational and Environmental Hygiene
Archives of Environmental Contamination and Toxicology
Archives of Environmental Health
Archives of Toxicology
*Atlantic Monthly
Audubon
Aviation, Space, and Environmental Medicine

British Journal of Industrial Medicine
Bulletin of Environmental Contamination and Toxicology

Civil Engineering

Ecosystem Health
Ecotoxicology and Environmental Safety
*Environmental Engineer
*Environmental Health Perspectives
*Environmental Manager
Environmental and Molecular Mutagenesis
Environmental Research
*Environmental Science and Technology
Environmental Technology
Environmental Toxicology and Chemistry
*EPA Journal
Epidemiology
Epidemiology and Infection
*EPRI Journal
Estuaries
Estuarine Coastal and Shelf Science
European Respiratory Journal
European Respiratory Review
Experimental Lung Research

*Forum for Applied Research and Public Policy

Health Physics

Industrial and Engineering Chemistry Research
International Archives of Environmental and Occupational Health
International Journal of Epidemiology

*Journal of the Air and Waste Management Association
Journal of the American College of Nutrition
Journal of American College of Toxicology
Journal of the American Water Works Association
Journal of Analytical Toxicology
Journal of Applied Physiology
Journal of the Biomedical Engineering Society
Journal of Cell Biology
Journal of Environmental Engineering
*Journal of Environmental Health
Journal of Environmental Pathology, Toxicology and Oncology
Journal of Environmental Quality
Journal of Environmental Science and Health
Journal of Epidemiology and Community Health

Journal of Exposure Analysis and Environmental Epidemiology
Journal of Gerontology
Journal of Hydrology
Journal of Infectious Disease
Journal of Inherited Metabolic Disease
Journal of National Cancer Institute
Journal of Nuclear Biology and Medicine
Journal of Nutrition
Journal of Occupational and Environmental Medicine
Journal of Parasitology
Journal of Pharmacological and Toxicological Methods
Journal of Studies on Alcohol
Journal of Toxicology and Environmental Health
Journal of Trauma Injury, Infection, and Critical Care
Journal of Tropical Medicine and Hygiene
Journal of the Water Environment Federation
Journal of Water Resources Planning and Management

Marine Pollution Bulletin

*National Geographic
*Nature
New Solutions, A Journal of Environmental and Occupational Health Policy
Nuclear News
Nuclear Safety
Nutrition
Nutrition Reviews

Occupational and Environmental Medicine
Occupational Hazards

Particulate Science and Technology
Physiological Reviews
Physiological Zoology
*Priorities for Long Life and Good Health

Radiation Protection Management
Radiological Protection Bulletin
RadWaste Magazine
Resources
Respiratory Cell and Molecular Biology
Respiratory Physiology

Risk Analysis

Scandanavian Journal of Work, Environment and Health
*Science
Science of Total Environment
*Scientific American

Toxicology
Toxicology and Applied Pharmacology

Veterinary and Human Toxicology

*Water, Air, and Soil Pollution
Water Environment and Technology
Water Environmental Research
Water Research
Water Resources Bulletin
Water Resources Research
Water Science and Technology
*World-Watch

REFERENCES

1. The Scope

Anonymous. 1994. "CDC Shines Light on True Price of Smoking." *Nation's Health* 24, no. 7 (August), 13.
Canadian Public Health Association. 1992. "Human and Ecosystem Health—Canadian Perspectives, Canadian Action." Ottawa, Ontario, Canada.
Clinton, William J. 1994. Executive Order 12898, 11 February. "Federal Actions to Address Environmental Justice in Minority Populations and Low-Income Populations." The White House, Washington, D.C.
Doll, R., and R. Peto. 1981. "The Causes of Cancer: Quantitative Estimates of Avoidable Risks of Cancer in the United States Today." *Journal of the National Cancer Institute* 66, no. 6 (June), 1191–1309.
Easterling, J. Bennett. 1994. "Environmental Justice: Implications for Siting of Federal Radioactive Waste Management Facilities." Paper presented at the International High-Level Radioactive Waste Management Conference, Las Vegas, 22–26 May.
Graham, J. D., B.-H. Chang, and J. S. Evans. 1992. "Poorer Is Riskier." *Risk Analysis* 12, 333–337.
Hahn, R. W., and E. H. Males. 1990. "Can Regulatory Institutions Cope with Cross Media Pollution?" *Journal of the Air and Waste Management Association* 40, no. 1 (January), 24–31.
Jacobi, W. 1992. "Concepts of Radiation Protection: A Model for Environmental Protection?" *Progress in Radiation Protection,* pp. 72–83. Publication Series, International Radiation Protection Association. Hennopsmeer, South Africa.
Morris, R. D., and W. R. Hendee. 1992. "Environmental Stewardship: Exploring the Implications for Health Professionals." Draft report, Medical College of Wisconsin, Milwaukee.
New York Times. 1994. "Smokers Add to Los Angeles Smog." 9 August.
Potter, V. R. 1992. "Global Bioethics as a Secular Source of Moral Authority for Long-Term Human Survival." *Global Bioethics* 5, no. 1 (January–March), 5–11.
Surgeon General. 1989. "Executive Summary, The Surgeon General's 1989 Report on

Reducing the Health Consequences of Smoking: 25 Years of Progress." *Morbidity and Mortality Weekly Report* 38, no. S-2 (24 March), 8.
Swogger, Glenn, Jr. 1992. "Why Emotions Eclipse Rational Thinking about the Environment." *Priorities for Long Life and Good Health* (Fall), 7–10. American Council on Science and Health, New York.
Toman, Michael A., and Joel Darmstadter. 1996. "Grading 'Sustainable America,' The Report of the President's Council on Sustainable Development," *Resources* 123 (Spring), 18–19.
Train, Russell E. 1990. "Environmental Concerns for the Year 2000." *The Bridge* 20, no. 2 (Summer), 3–10.

2. Toxicology

Abelson, P. H. 1992. "Major Changes in the Chemical Industry." *Science* 255 (20 March), 1489.
Ames, B. N. 1971. "The Detection of Chemical Mutagens with Enteric Bacteria." In A. Hollander, ed., *Chemical Mutagens: Principles and Methods for Their Detection,* vol. 1, pp. 267–282. New York: Plenum Press.
Cullen, M. R. 1991. "Multiple Chemical Sensitivities." *New Solutions* 2, 16–24.
Doull, John. 1992. "Toxicology and Exposure Limits," *Applied Occupational and Environmental Hygiene* 7, 583–585.
Doull, J., and M. C. Bruce. 1986. "Origin and Scope of Toxicology." In C. D. Klaassen, Mary O. Amdur, and J. Doull, *Casarett and Doull's Toxicology: The Basic Science of Poisons,* 3rd ed., pp. 3–10. New York: Macmillan Publishing Company.
Garner, C. 1992. "Epidemiology: Molecular Potential." *Nature* 360 (19 November), 207–208.
Green, M. D. 1993. "Environmental Hazards: Real or Exaggerated?" Letter to the Editor. *Science* 262 (29 October), 637–638.
HHS, 1993. "Toxicological Profile for Aldrin/Dieldrin." Report TP-92/01. Agency for Toxic Substances and Disease Registry, Public Health Service, U.S. Department of Health and Human Services, Atlanta.
——— 1994. "Toxicological Profile for Zinc (Update)." Report TP-93/15. Agency for Toxic Substances and Disease Registry, Public Health Service, U.S. Department of Health and Human Services. Atlanta.
——— 1995. "Toxicological Profile for Asbestos (Update)." Agency for Toxic Substances and Disease Registry, Public Health Service, U.S. Department of Health and Human Services, Atlanta.
Holden, C. 1993a. "Toxicologists Watch as the Worm Turns." *Science* 261 (17 September), 1525.
——— 1993b. "Apple Growers vs. CBS: TV Wins." *Science* 262 (1 October), 35.
Kaiser, Jocelyn. 1996. "Environmental Estrogens—New Yeast Study Finds Strength in Numbers." *Science* 272 (7 June), 1418.
Klaassen, C. D. 1986. "Principles of Toxicology." In C. D. Klaassen, Mary O. Amdur, and J. Doull, *Casarett and Doull's Toxicology: The Basic Science of Poisons,* 3rd ed., pp. 11–32. New York: Macmillan Publishing Company.
Lippmann, Morton. 1992. "Introduction and Background." In Morton Lippmann, ed.,

Environmental Toxicants—Human Exposures and Their Health Effects, pp. 1–29. New York: Van Nostrand Reinhold.

Loomis, T. A. 1978. *Essentials of Toxicology*. 3rd ed. Philadelphia: Lea & Febiger.

Lu, Frank C. 1991. *Basic Toxicology: Fundamentals, Target Organs, and Risk Assessment*. 2nd ed. New York: Hemisphere Publishing Corporation.

Marshall, E. 1993. "Toxicology Goes Molecular." *Science* 259 (5 March), 1394–98.

Moriarty, F. 1988. *Ecotoxicology: The Study of Pollutants in Ecosystems*. 2nd ed. New York: Academic Press.

National Academy of Sciences. 1993. *Issues in Risk Assessment*. Committee on Risk Assessment Methodology. Washington, D.C.: National Academy Press.

NRC. 1977. *Drinking Water and Health*, Report of the Safe Drinking Water Committee, Advisory Center on Toxicology. Washington, D.C.: National Research Council.

——— 1983. *Drinking Water and Health*, vol. 5. Board on Toxicology and Environmental Health Hazards, National Research Council. Washington, D.C.: National Academy Press.

——— 1986. *Drinking Water and Health*, vol. 6. Board on Toxicology and Environmental Health Hazards, National Research Council. Washington, D.C.: National Academy Press.

Rhomberg, Lorenz. 1996. "Are Chemicals in the Environment Disrupting Hormonal Control of Growth and Development?" *Risk in Perspective* 4, no. 3 (April). Center for Risk Analysis, Harvard School of Public Health.

Rosenthal, Alon, George M. Gray, and John D. Graham. 1992. "Legislating Acceptable Cancer Risk from Exposure to Toxic Chemicals." *Ecology Law Quarterly* 19, 269–362.

SARA [Superfund Amendments and Reauthorization Act]. 1986. Public Law 99–499, 26 USC 4611 et seq., U.S. Congress, Washington, D.C.

Smith, K. 1990. "ALAR: One Year Later, A Media Analysis of a Hypothetical Health Risk." Special report, American Council on Science and Health, New York.

Smith, R. P. 1992. *A Primer of Environmental Toxicology*. Philadelphia: Lea & Febiger.

Stone, R. 1993a. "Toxicologists—and Snow—Descend on New Orleans." *Science* 260 (2 April), 30–31.

——— 1993b. "FCCSET Develops Neurotoxicology Primer." *Science* 261 (20 August), 975.

Walker, B. 1993. "PH Community Should Grab Podium in Superfund Debate" (Commentary). *Nation's Health* 23, no. 9, 6.

Weiss, Bernard. 1990. "Neurotoxic Risks in the Workplace." *Applied Occupational and Environmental Hygiene* 5, no. 9, 587–594.

Whelan, E. M. 1993. *Toxic Terror—The Truth behind the Cancer Scares*. Buffalo, N.Y.: Prometheus Books.

3. *Epidemiology*

BEIR. 1972. *The Effects on Populations of Exposure to Low Levels of Ionizing Radiation*. Report of the Advisory Committee on the Biological Effects of Ionizing Radiations. Washington, D.C.: National Academy of Sciences/National Research Council.

Bierbaum, Philip J., and John M. Peters, eds. 1991. *Proceedings of the Scientific Workshop on the Health Effects of Electric and Magnetic Fields on Workers,* Cincinnati: National Institute for Occupational Safety and Health, Centers for Disease Control, U.S. Department of Health and Human Services.

Cantor, K. P., A. Blair, G. Everett, R. Gibson, L. F. Burmeister, L. M. Brown, L. Schuman, and F. R. Dick. 1992. "Pesticides and Other Agricultural Risk Factors for Non-Hodgkin's Lymphoma among Men in Iowa and Minnesota." *Cancer Research* 52 (1 May), 2447–55.

Doll, Richard, and A. Bradford Hill. 1950. "Smoking and Carcinoma of the Lung—Preliminary Report." *British Medical Journal* 2 (30 September), 739–748.

Doll, Richard, and Richard Peto. 1976. "Mortality in Relation to Smoking: 20 Years' Observations on Male British Doctors." *British Medical Journal* 2 (25 December), 1525–36.

English, D. 1992. "Geographical Epidemiology and Ecological Studies." In P. Elliott, J. Guzick, D. English, and R. Stern, eds., *Geographical and Environmental Epidemiology: Methods for Small-Area Studies,* pp. 3–13. New York: World Health Organization Regional Office for Europe, Oxford University Press.

Goldsmith, John R. 1986. *Environmental Epidemiology: Epidemiological Investigation of Community Environmental Health Problems.* Boca Raton, Fla.: CRC Press.

Hill, A. B. 1965. "The Environment and Disease: Association or Causation?" *Proceedings of the Royal Society of Medicine* 58, 259–300.

Misch, Ann. 1994. "Assessing Environmental Health Risks." In Linda Starke, ed., *State of the World, 1994: A Worldwatch Institute Report on Progress toward a Sustainable Society,* pp. 117–136. New York: W. W. Norton.

Moeller, Dade W. 1990. "History and Perspective on the Development of Radiation Protection Standards." In *Radiation Protection Today: The NCRP at Sixty Years. Proceedings No. 11,* pp. 5–21. Bethesda, Md.: National Council on Radiation Protection and Measurements.

Monson, Richard. 1990. *Occupational Epidemiology.* 2nd ed. Boca Raton, Fla.: CRC Press.

Muller, Hermann J. 1927. "Artificial Transmutation of the Gene." *Science* 66 (11 July), 84–87.

NRC. 1991. *Environmental Epidemiology—Public Health and Hazardous Wastes* National Research Council. Washington, D.C.: National Academy Press.

Oak Ridge Associated Universities. 1992. "Health Effects of Low Frequency Electric and Magnetic Fields." Report ORAU 92/F9, prepared for the Committee on Interagency Radiation Research and Policy Coordination, Oak Ridge, Tenn.

Radiation Effects Research Foundation. 1994. "Cancer Incidence in Atomic Bomb Survivors." *Radiation Research* 137, no. 2, supplement (February), S1-S112.

Rowland, R. E. 1994. *Radium in Humans: A Review of U.S. Studies.* Report ANL/ER-3. Argonne, Ill.: Argonne National Laboratory.

Savitz, D. A., H. Wachtel, F. A. Barnes, E. M. John, and J. G. Tvrdik. 1988. "Case-Control Study of Childhood Cancer and Exposure to 60-Hz Magnetic Fields." *American Journal of Epidemiology* 128, 21–38.

Selvin, H. C. 1958. "Durkheim's 'Suicide' and Problems of Empirical Research." *American Journal of Sociology* 63, 607–619.

Straume, T., S. D. Egbert, W. A. Woolson, R. C. Finkel, P. W. Kubik, H. E. Gove, P.

Sharma, and M. Hoshi. 1992. "Neutron Discrepancies in the DS86 Hiroshima Dosimetry System." *Health Physics* 63, no. 4 (October), 421–426.

Surgeon General. 1989. "Executive Summary, The Surgeon General's 1989 Report on Reducing the Health Consequences of Smoking: 25 Years of Progress." *Morbidity and Mortality Weekly Report* 38, no. S-2 (24 March), 8.

Taubes, Gary. 1995. "Epidemiology Faces Its Limits." *Science* 269 (14 July), 164–169.

Terracini, B. 1992. "Environmental Epidemiology: A Historical Perspective." In P. Elliott, J. Cuzick, D. English, and R. Stern, eds., *Geographical and Environmental Epidemiology: Methods for Small-Area Studies,* pp. 253–263. New York: World Health Organization Regional Office for Europe, Oxford University Press.

Theriault, Gilles P. 1991. "Health Effects of Electromagnetic Radiation on Workers: Epidemiologic Studies." In Bierbaum and Peters (1991), pp. 91–124.

Trichopoulos, Dimitrios. 1994. "Risk of Lung Cancer from Passive Smoking." *Principles and Practice of Oncology: PPO Updates* 8, no. 8 (August), 1–8.

——— 1995. "The Discipline of Epidemiology." Letter to the Editor. *Science* 269 (8 September), 1326.

USPHS. 1964. *Smoking and Health: Report.* Surgeon General's Advisory Committee on Smoking and Health, Publication no. 1103, Washington, D.C.: U.S. Public Health Service.

Wertheimer, N., and E. Leeper. 1979. "Electrical Wiring Configurations and Childhood Cancer." *American Journal of Epidemiology* 109, 273–284.

WHO [World Health Organization]. 1983. "Guidelines on Studies in Environmental Epidemiology." *Environmental Criteria* 27.

Willett, W., S. Greenland, B. MacMahon, D. Trichopoulos, K. Rothman, D. Thomas, M. Thun, and N. Weiss. 1995. "The Discipline of Epidemiology." Letter to the Editor. *Science* 269 (8 September) 1325–26.

Williams, Ralph C. 1951. *The United States Public Health Service, 1798–1950.* Washington, D.C.: Commissioned Officers Association of the United States Public Health Service.

4. The Workplace

ACGIH. 1996. *1996 TLVs and BEIs—Threshold Limit Values for Chemical Substances and Physical Agents and Biological Exposure Indices.* Cincinnati: American Conference of Governmental Industrial Hygienists.

AIHA. 1994. "American Industrial Hygiene Association White Paper: Ergonomics." *American Industrial Hygiene Association Journal* 55, no. 7 (July), 601–602.

Associated Press. 1995. "OSHA Fails to Check Dangerous Workplaces." *Sun Journal,* New Bern, N.C. (5 September).

Burgess, William A. 1991. "Potential Exposures in the Manufacturing Industry—Their Recognition and Control." In G. D. Clayton and F. E. Clayton, eds., *Patty's Industrial Hygiene and Toxicology,* vol. 1, part A. New York: John Wiley and Sons.

——— 1994. "Philosophy and Management of Engineering Control." In R. L. Harris, L. J. Cralley, and L. V. Cralley, eds., *Patty's Industrial Hygiene and Toxicology,* vol. 3, part A. New York: John Wiley and Sons.

——— 1995. *Recognition of Health Hazards in Industry.* 2nd ed. New York: Wiley-Interscience.

Corn, M. 1989. "The Progression of Industrial Hygiene." *Applied Industrial Hygiene* 4, no. 6 (June), 153–157.

Cralley, Lewis, and Walter H. Konn. 1973. "The Significance and Uses of Guides, Codes, Regulations, and Standards for Chemical and Physical Agents." In *The Industrial Environment: Its Evaluation and Control.* Washington, D.C.: U.S. Department of Health, Education and Welfare.

Fine, Lawrence J. 1996. Editorial. "The Psychosocial Work Environment and Heart Disease." *American Journal of Public Health* 86, no. 3 (March), 301–303.

Gideon, J. A., E. R. Kennedy, D. M. O'Brien, and J. T. Talty. 1979. *Controlling Occupational Exposures: Principles and Practices.* Cincinnati: National Institute for Occupational Safety and Health, U.S. Department of Health, Education and Welfare.

Grubb, Gregg. 1990. "Occupational Health—More than Just Asbestos." *ECON: Environmental Contractor* 5, no. 7 (July), 50–53.

Hamilton, Alice. 1943. *Exploring the Dangerous Trades.* Boston: Little, Brown.

HHS. 1990. *National Health Goals for the Year 2000.* Washington, D.C.: U.S. Department of Health and Human Services.

Hickey, E. E., G. A. Stoetzel, D. J. Strom, G. R. Cicotte, C. M. Wiblin, and S. A. McGuire. 1993. "Air Sampling in the Workplace." Report NUREG-1400. U.S. Nuclear Regulatory Commission, Washington, D.C.

Jewett, D. L., P. Heinsohn, C. Bennett, A. Rosen, and C. Neuilly. 1992. "Blood-Containing Aerosols Generated by Surgical Techniques: A Possible Infectious Hazard." *American Industrial Hygiene Journal* 53, no. 4 (April), 228–231.

Kelsey, Timothy W. 1994. "The Agrarian Myth and Policy Responses to Farm Safety." *American Journal of Public Health* 84, no. 7 (July), 1171–77.

Macher, Janet. 1993. "Report of Workshop #8 on Bioaerosol Samplers." *Applied Occupational and Environmental Hygiene* 8, no. 4 (April), 410–411.

NIOSH. 1992. "Homicide in U.S. Workplaces: A Strategy for Prevention and Research." National Institute for Occupational Safety and Health, Centers for Disease Control. Morgantown, W. Va.

NSC. 1996. *Accident Facts, 1996 Edition.* Itasca, Ill.: National Safety Council.

OSHA [Occupational Safety and Health Administration]. 1992. *Enforcement Procedures for the Occupational Exposure to Bloodborne Pathogens Standard.* Code of Federal Regulations, Title 29, part 1910.1030. Washington, D.C.: U.S. Department of Labor.

Pasanen, A.-L., M. Nikulin, S. Berg, and E.-L. Hintikka. 1994. "Stachybotrys Atra Corda May Produce Mycotoxins in Respirator Filters in Humid Environments." *American Industrial Hygiene Association Journal* 55, no. 1 (January), 62–65.

Patty, F. A. 1978. "Industrial Hygiene: Retrospect and Prospect." In G. D. Clayton and F. E. Clayton, eds., *Patty's Industrial Hygiene and Toxicology,* 3rd rev. ed., vol. 1. New York: John Wiley and Sons.

Schneider, S. 1994. "Ergonomics." *Applied Occupational and Environmental Hygiene* 9, no. 3 (March), 180–181.

Snook, S. H. 1989. "The Control of Low Back Disability: The Role of Management."

In K. H. E. Kroemer, J. D. McGlothlin, and T. G. Bobick, eds., *Manual Material Handling: Understanding and Preventing Back Trauma,* pp. 97–101. Akron: American Industrial Hygiene Association.

Suter, A. H. 1994. "The Effects of Noise and the Benefits of Noise Cancelling Devices." *Industrial Hygiene News* 17, no. 6 (July), 26–27.

U.S. Congress. 1913. "An Act to Create a Department of Labor" (4 March). Public Law 426, 62nd Congress, Washington, D.C.

——— 1990. Pollution Prevention Act, Public Law 101–508, 42 USC 13101 et seq. Washington, D.C.

Virtanen, T. I., and R. A. Mantyjarvi. 1994. "Airborne Allergens and Their Quantification and Effect on the Development of Allergy in Occupational Environments." *Applied Occupational and Environmental Hygiene* 9, no. 1 (January), 65–70.

Walker, Bailus. 1988. "President's Column." *Nation's Health* (August), 2.

Weiss, Bernard. 1990. "Neurotoxic Risks in the Workplace." *Applied Occupational and Environmental Hygiene* 6, no. 9 (September), 587–594.

5. Air in the Home and Community

Abelson, Philip H. 1990. "New Technology for Cleaner Air." Editorial. *Science* 248, no. 4957 (18 May), 793.

AIHA. 1994. "Is Air Quality a Problem in My Home?" Operation Outreach, American Industrial Hygiene Association, Fairfax, Va.

AMA, Council on Scientific Affairs. 1987. "Radon in Homes." *Journal of the American Medical Association* 258, no. 5 (7 August), 668–672.

Brauer, M., and J. D. Spengler. 1994. "Nitrogen Dioxide Exposures inside Ice Skating Rinks." *American Journal of Public Health* 84, no. 3 (March), 429–433.

Brown, L. 1984. "National Radiation Survey in the U.K.: Indoor Occupancy Factors." *Radiation Protection Dosimetry* 5, no. 4, 203–208.

CEQ. 1995. *Twenty-Fourth Annual Report—The Council on Environmental Quality.* Washington, D.C.: Executive Office of the President.

EPA [U.S. Environmental Protection Agency]. 1994. "National Air Quality and Emissions Trends Report, 1993." Report EPA-454/R-94–026. Research Triangle Park, N.C.

EPA/NIOSH. 1991. *Building Air Quality—A Guide for Building Owners and Facility Managers.* U.S. Environmental Protection Agency and National Institute of Occupational Safety and Health. Washington, D.C.: U.S. Department of Health and Human Services.

Ferris, B. G., Jr., J. H. Ware, J. D. Spengler, D. W. Dockery, and F. E. Speizer. 1986. "The Harvard Six-Cities Study." In *Aerosols: Research, Risk Assessment and Control Strategies.* Proceedings of the Second U.S.-Dutch International Symposium, Williamsburg, Va., 19–25 May 1985. Chelsea, Mich.: Lewis Publishers.

Fotos, Christopher P. 1991. "Flight Attendants Question Health Risk of Recirculated Air in Newer Cabins." *Aviation Week and Space Technology,* (25 November), 79–81.

French, Hilary F. 1991. "Eastern Europe's Clean Break with the Past." *World-Watch* 4, no. 2 (March–April), 21–27.

Goldsmith, J. R. 1968. "Effects of Air Pollution on Human Health." In Arthur C. Stern, ed., *Air Pollution*, vol. 1: *Air Pollution and Its Effects*, 2nd ed. New York: Academic Press.

Graham, John D., and Douglas W. Dockery. 1994. "Rethinking Particle Pollution and the Clean Air Act," newsletter, *Risk in Perspective* 2, no. 2 (May),1–2.

Harrington, Winston, and Margaret A. Walls. 1994. "Shifting Gears: New Directions for Cars and Clean Air." *Resources* 115 (Spring), 2–6.

Highsmith, V. R., C. E. Rodes, and R. J. Hardy. 1988. "Indoor Particle Concentrations Associated with Use of Tap Water in Portable Humidifiers." *Environmental Science and Technology* 22, no. 9 (September), 1109.

Hofeldt, David L. 1996. "Mobile Source Emissions and Air Quality." In *Frontiers of Engineering*, Reports of Leading Edge Engineering from the 1995 NAE Symposium on Frontiers of Engineering, pp. 47–50. National Academy of Engineering. Washington, D.C.: National Academy Press.

Lave, Lester B., Chris T. Hendrickson, and Francis Clay McMichael. 1995. "Environmental Implications of Electric Cars." *Science* 268 (19 May), 993–995.

Lee, Bryan. 1991. "Highlights of the Clean Air Act Amendments of 1990." *Journal of the Air and Waste Management Association* 41, no. 1 (January), 16–19.

Mendell, Mark J., and Lawrence Fine. 1994. Editorial. "Building Ventilation and Symptoms—Where Do We Go from Here?" *American Journal of Public Health* 84, no. 3 (March), 346–348.

Mossman, B. T., J. Bignon, M. Corn, A. Seaton, and J. B. L. Gee. 1990. "Asbestos: Scientific Developments and Implications for Public Policy." *Science* 247 (19 January), 294–301.

Munson, Halsey. 1990. "Pollution in the Soviet Union." *ECON:Environmental Contractor* 5, no. 8 (August), 24–29.

NCRP. 1987. "Ionizing Radiation Exposure of the Population of the United States." Report no. 93. National Council on Radiation Protection and Measurements, Bethesda, Md.

——— In press. "Deposition, Retention and Dosimetry of Inhaled Radioactive Substances." National Council on Radiation Protection and Measurements, Bethesda, Md.

NRC. 1994. *Curbing Gridlock: Peak-Period Fees to Relieve Traffic Congestion*, vols. 1 and 2. Washington, D.C.: National Research Council.

Nriagu, Jerome O. 1996. "A History of Global Metal Pollution." *Science* 272, no. 5259 (12 April), 223–224.

Reichhardt, Tony. 1995. "Weighing the Health Risks of Airborne Particulates." *Environmental Science and Technology* 29, no. 8 (August), 360A-364A.

Ryan, P. Barry. 1991. "An Overview of Human Exposure Modeling." *Journal of Exposure Analysis and Environmental Epidemiology* 1, no. 4, 453–474.

Samet, Jonathan M. 1991. "The Environment and the Lung—Changing Perspectives." *Journal of the American Medical Association* 266, no. 5 (7 August), 670–675.

Samet, J., and J. D. Spengler, eds. 1991. *Indoor Air Pollution: A Health Perspective.* Baltimore: Johns Hopkins University Press.

Sexton, Ken. 1993. "An Inside Look at Air Pollution." *EPA Journal* 19, no. 4 (October–December), 9–12.

Shahan, James. 1994. "Energy-Efficient Health Buildings: The Role of Public Utilities." *IAQ Journal* 1, no. 3 (Spring), 7–14.

Spengler, John D. 1992. "Outdoor and Indoor Air Pollution." In Alyce Bezman Tarcher, ed., *Principles and Practices of Environmental Medicine,* pp. 21–41. New York: Plenum Medical Book Company.

——— 1994. "Faculty Research at the Harvard School of Public Health." School of Public Health, Harvard University, Boston.

Spengler, J. D., and K. Sexton. 1983. "Indoor Air Pollution: A Public Health Perspective." *Science* 221 (1 July), 9–17.

UNSCEAR [United Nations Scientific Committee on the Effects of Atomic Radiation]. 1993. *Sources, Effects and Risks of Ionizing Radiation, Report to the General Assembly.* Publication E.88.IX.7. United Nations, New York.

Valberg, Peter A. 1990. "The Respiratory Tract as a Portal of Entry for Toxic Particles." In T. R. Gerrity and C. J. Henry, eds., *Principles of Route-to-Route Extrapolation for Risk Assessment,* pp. 61–70. New York: Elsevier Science Publishing Company.

Walls, Margaret A., and Alan J. Krupnick. 1990. "Cost-Effectiveness of Methanol Vehicles." *Resources* 100 (Summer), 1–5.

6. Food

Ahmed, F. E. 1991. *Seafood Safety.* Committee on Evaluation of the Safety of Fishery Products, Food and Nutrition Board, National Research Council. Washington, D.C.: National Academy Press.

American Nuclear Society. 1994. "Food Industry Re-Examines Irradiation." *Re-Actions* 10 (September), 1–4.

Associated Press. 1993. "New Rules for Safe-Handling Labels on Raw Meat." *New York Times,* (31 October).

Bean, N. H., and P. M. Griffin. 1990. "Foodborne Disease Outbreaks in the United States, 1973–1987: Pathogens, Vehicles, and Trends." *Journal of Food Protection* 53, no. 9 (September), 804–817.

Benenson, Abram S., ed. 1995. *Control of Communicable Diseases Manual.* 16th ed. Washington, D.C.: American Public Health Association.

Bryan, F. L. 1980. *Foodborne Diseases and Their Control.* Atlanta: Centers for Disease Control, U.S. Department of Health and Human Services.

Cohen, Mitchell L., and Robert V. Tauxe. 1986. "Drug-Resistant *Salmonella* in the United States: An Epidemiologic Perspective." *Science* 234 (21 November), 964–969.

Etherton, Terry D. 1994. "The Efficacy, Safety and Benefits of Bovine Somatotropin and Porcine Somatotropin." American Council on Science and Health, New York.

FDA [Food and Drug Administration, U.S. Department of Health and Human Services]. 1990. "21 CFR Part 179, Irradiation in the Production, Processing, and Handling of Food: Final Rule." *Federal Register* (2 May) 18538–44.

——— 1995. *Food Code.* National Technical Information Service, Springfield, Va.

Fumento, Michael. 1994. "Irradiation: A Winning Recipe for Wholesome Beef," *Priorities* 6, no. 2, 37–39. American Council on Science and Health, New York.

Hoffman, John W. 1993. "Food Safety: A Unified, Risk-Based System Needed to Enhance Food Safety." Report GAO/T-RCED-94–71. U.S. General Accounting Office, Washington, D.C.

Huttner, Susanne L. 1996. "Biotechnology and Food." American Council on Science and Health, New York.

Institute of Food Technologists. 1992. "Government Regulation of Food Safety: Interaction of Scientific and Societal Forces." *Food Technology* 1 46(1) (January), 73–80.

Kaiser, Jocelyn. 1995. "Random Samples—Europe's Beef against Hormones." *Science* 270 (15 December), 1763.

Moffat, Anne Simon. 1994a. "Developing Nations Adapt Biotech for Own Needs." *Science* 265 (8 July), 186–187.

——— 1994b. "Plant Genetics: Mapping the Sequence of Disease Resistance." *Science* 265 (23 September), 1804–05.

Nelson, Norton, and James W. Whittenberger. 1977. *Human Health and the Environment: Some Research Needs.* DHEW Publication NIH 77–1277. Washington, D.C.: Government Printing Office.

Newberne, Paul M. 1994. "Naturally-Occurring Food-Borne Toxicants." Proceedings of the Second Asian Conference on Food Safety, Bangkok, 18–23 September.

NRC. 1993. "Pesticides in the Diets of Infants and Children," National Research Council, Washington, D.C.

——— 1996. *Carcinogens and Anticarcinogens in the Human Diet: A Comparison of Naturally Occurring and Synthetic Substances.* Washington, D.C.: National Academy Press.

O'Brien, Claire. 1996. "Mad Cow Disease—Scant Data Cause Widespread Concern." *Science* 271 (29 March), 1798.

Porter, J. D. H., C. Gaffney, D. Heymann, and W. Parkin. 1990. "Foodborne Outbreak of Giardia Lamblia." *American Journal of Public Health* 80, no. 10 (October), 1259–60.

Rosenblum, L. S., I. R. Mirkin, D. T. Allen, S. Safford, and S. C. Hadler. 1990. "A Multifocal Outbreak of Hepatitis A Traced to Commercially Distributed Lettuce." *American Journal of Public Health* 80, no. 9 (September), 1075–80.

Rubel, Judith, ed. 1987. *To Your Health!* Burlington, Mass.: Lahey Clinic.

Stone, Richard. 1994. "Science Scope: Analysis Questions BST's Safety in Cows." *Science* 266 (21 October), 355.

Walker, Bailus, Jr. 1993. "Statement to the Subcommittee on Human Resources and Intergovernmental Relations of the Committee on Governmental Operations, U.S. House of Representatives," 4 November. Washington, D.C.

Whelan, Elizabeth M. 1994. "Eat, Drink and Be Wary of Food Scares." *Priorities* 6, no. 3, 36–40.

Willett, Walter C. 1994. "Diet and Health: What Should We Eat?" *Science* 264 (22 April), 532–537.

Woods, Michael. 1993. "Genetically Engineered Livestock Offer Hope for 'Humanized' Milk." *Sun Journal*, New Bern, N.C., 25 August.

Wright, Karen. 1990. "The Policy Response: In Limbo" [on the use of antibiotics in animal feeds]. *Science* 249 (6 July), 24.

7. Drinking Water

AAVIM. 1973. *Planning for an Individual Water System.* Athens, Ga.: American Association of Vocational Instructional Materials.

Allen, L., and J. L. Darby. 1994. "Quality Control of Bottled and Vended Water in California: A Review and Comparison to Tap Water." *Journal of Environmental Health* 56, no. 8 (April), 17–22.

Allen, M. J., W. D. Bellamy, J. Bryck, D. W. Hendricks, R. M. Krill, and G. S. Logsdon. 1988. "Slow Sand Filtration." [Roundtable discussion.] *American Water Works Association Journal* 80, no. 12 (December), 12, 14, 18–19.

Anonymous. 1991. "Vital Signs." *World-Watch* 4, no. 2 (March–April), 6.

——— 1993. "NRDC Report Finds Dangerous Leaks in Drinking Water System." *Nation's Health* 23, no. 9 (November), 1, 20.

——— 1994. "News: At Maine Water-Treatment Plant, It's Ozone Only." *Civil Engineering* 64, no. 5 (May), 12–13.

AWWA. 1984. *Introduction to Water Treatment: Principles and Practices of Water Supply Operations,* vol. 2, p. 284. Denver: American Water Works Association.

Bradley, David J. 1977. "Health Aspects of Water Supplies in Tropical Countries." In Richard Feachem, Michael McGarry, and Duncan Mara, eds., *Water, Wastes and Health in Hot Climates.* New York: John Wiley and Sons.

Canby, Thomas Y. 1980. "Water—Our Most Precious Resource." *National Geographic* 158, no. 2 (August), 144–179.

CEQ [Council on Environmental Quality]. 1989. *Environmental Trends.* Washington, D.C.: Executive Office of the President.

——— 1993. *Environmental Quality, 23rd Annual Report.* Washington, D.C.: Executive Office of the President.

Chanlett, Emil T. 1979. *Environmental Protection.* 2nd ed. New York: McGraw-Hill.

Dimitriou, M. A. 1994. "Ozone Treatment Comes of Age in Water, Wastewater Applications." *Environmental Solutions* 7, no. 12 (December), 38–43.

Eaton, A. D., L. S. Clesceri, and A. E. Greenberg, eds. 1995. *Standard Methods for the Examination of Water and Wastewater.* 19th ed. Waldorf, Md.: American Public Health Association.

EPA. 1975. *National Primary Drinking Water Regulations.* Code of Federal Regulations, Title 40, part 141, 42 USC 300f-300j. 40 FR 59570 (24 December). Washington, D.C.: U.S. Environmental Protection Agency.

——— 1977. "Manual of Treatment Techniques for Meeting the Interim Primary Drinking Water Regulations." Report EPA-600/8–77–005, rev. ed. U.S. Environmental Protection Agency, Cincinnati.

——— 1979. *National Secondary Drinking Water Regulations.* Code of Federal Regulations, Title 40, part 143, 42 USC 300f et seq., 44 FR 42198 (19 July). Washington, D.C.: U.S. Environmental Protection Agency.

——— 1993. "Drinking Water Maximum Contaminant Level Goal: Fluoride." Code of Federal Regulations, Title 40, part 141, 58 FR 68826-27 (27 December). Washington, D.C.: U.S. Environmental Protection Agency.

FDA. 1996. "Bottled Water—Quality Standards." Federal Register, Vol. 61, part 3, no. 59, (26 March), 13258-70. Washington, D.C.: Food and Drug Administration.

Frederick, Kenneth D. 1996. "Water as a Source of International Conflict." *Resources* 123 (Spring), 9–12.

Frontinus, Sextus Julius. A.D. 97. *The Water Supply of the City of Rome.* Clemens Herschel, trans., 1973. Boston: New England Water Works Association.

Hicks, Craig. 1993. "Fluoridated Drinking Water No Cause for Concern." *NewsReport, National Research Council* 43, no. 4, 9.

Leland, D. E., and M. Damewood III. 1990. "Slow Sand Filtration in Small Systems in Oregon." *American Water Works Association Journal* 82, no. 6 (June), 50–59.

McKone, Thomas E. 1987. "Human Exposure to Volatile Organic Compounds in Household Tap Water: The Indoor Inhalation Pathway." *Environmental Science and Technology* 21, no. 12 (December), 1194–1201.

Mitchell, Ralph. 1996. "Emerging Frontiers: Molecular Biology, Membrane Biology." *Environmental Science and Technology* 30, no. 1 (January), 33A-34A.

Montgomery, James M. 1985. *Water Treatment Principles and Design.* New York: John Wiley and Sons.

Moore, Taylor. 1994. "Separable Feast—Membrane Applications in Food Processing." *EPRI Journal* 19, no. 6, 17–23.

Newman, Alan. 1995. "EPA Considering Quick Action on *Cryptosporidium*," *Environmental Science and Technology* 29, no. 1, 17A.

NRC. 1977–89. *Drinking Water and Health,* vols. 1–9. National Research Council, Safe Drinking Water Committee. Washington, D.C.: National Academy Press.

——— 1977. "Historical Note." In *Drinking Water and Health,* vol. 1, pp. 1–8. Washington, D.C.: National Academy Press.

Okun, Daniel A. 1993. "New York City's Drinking Water—What Is the Risk?" *Environmental Engineer* 29, no. 4 (October), 8–13, 34.

Parfit, Michael. 1993. "Water—The Power, Promise, and Turmoil of North America's Fresh Water—Map Supplement: Water." *National Geographic* special edition, 184, no. 5A (November).

Postel, Sandra. 1992. *Last Oasis: Facing Water Scarcity.* New York: W. W. Norton.

Postel, Sandra L., Gretchen C. Daily, and Paul R. Ehrlich. 1996. "Human Appropriation of Renewable Fresh Water." *Science* 271 (9 February), 785–788.

Schroeder, H. A. 1974. "Role of Trace Elements in Cardiovascular Diseases." *Medical Clinics of North America* 58, no. 2, 381–396.

Symons, James M. 1992. *"Plain Talk About Drinking Water."—Answers to 101 Important Questions about the Water You Drink.* Denver: American Water Works Association.

UNESCO [United Nations Educational, Scientific, and Cultural Organization]. 1971. *Scientific Framework of World Water Balance.* Technical Papers in Hydrology no. 7. Paris.

USGS. 1986. "Ground Water." Report 491–402/04. U.S. Geological Survey, Denver.

——— 1988. "Estimated Use of Water in the United States in 1985." Circular no. 1004, U.S. Geological Survey, Washington, D.C.

8. Liquid Waste

Anonymous. 1993. "Storm-Water Regulations." *Waterworld Review* 9, no. 5 (September/October), 7.

Burke, Maria. 1994. "Phosphorus Fingered as Coral Killer." *Science* 263, no. 5150 (25 February), 1086.

CEQ. 1993. *Twenty-Third annual report, Council on Environmental Quality,* together with the President's Message to Congress. Washington, D.C.: U.S. Government Printing Office.

Chanlett, Emil T. 1979. *Environmental Protection.* 2nd ed. New York: McGraw-Hill.

Culotta, Elizabeth. 1992. "Red Menace in the World's Oceans." *Science* 257, no. 5076 (11 September), 1476–77.

Edwards, Peter. 1992. "Reuse of Human Wastes in Aquaculture—A Technical Review." Water and Sanitation Report no. 2, UNDP-World Bank Water and Sanitation Program. World Bank, Washington, D.C.

EPA. 1991. *Protecting the Nation's Ground Water: EPA's Strategy for the 1990s.* Report 21Z-1020. Washington, D.C.: U.S. Environmental Protection Agency.

——— 1992a. *Small Community Water and Wastewater Systems.* Report EPA/625/R-92/010. Cincinnati: U.S. Environmental Protection Agency.

——— 1992b. *State Sludge Management Program Regulations.* Code of Federal Regulations, Title 40, part 501. Washington, D.C.: U.S. Environmental Protection Agency.

——— 1992c. "Hazardous Waste Injection Restrictions." Code of Federal Regulations, Title 40, part 148, Washington, D.C.: U.S. Environmental Protection Agency.

Federal Water Pollution Control Act. 1956. Public Law 92–500. U.S. Congress, Washington, D.C.

Findley, Roger W., and Daniel A. Farber. 1992. *Environmental Law in a Nutshell.* 3rd ed. St. Paul, Minn.: West Publishing Company.

Gloyna, Ernest. 1971. *Waste Stabilization Ponds.* Monograph Series no. 60. Geneva: World Health Organization.

Kleene, J. W., C. R. Mote, and J. S. Allison. 1993. "Environmental Impact of Irrigating Lawns with Treated Domestic Wastewater." *Journal of Environmental Health* 55, no. 5 (March), 17–23.

Mara, Duncan. 1982. *Appropriate Technology for Water Supply and Sanitation: Sanitation Alternative for Low-Income Communities—A Brief Introduction.* Washington, D.C.: World Bank.

Mitchell, John G. 1996. "Our Polluted Runoff." *National Geographic* 189, no. 2 (February), 106–125.

NRC. 1993. *Managing Wastewater in Coastal Urban Areas.* National Research Council, Committee on Wastewater Management for Coastal Urban Areas. Washington, D.C.: National Academy Press.

Olson, Betty H. 1992. "Environmental Water Pollution." In William N. Rom, ed., *Environmental and Occupational Medicine,* pp. 1255–73. 2nd ed. Boston: Little, Brown.

O'Mara, Lisa. 1991. "Deep Well Injection." *Environmental Protection* 2, no. 4 (June), 14–18.

Ouellette, Robert P. 1991. "A Perspective on Water Pollution." *National Environmental Journal* 1, no. 1 (September/October), 20–24.

Richards, William. 1991. "Finding Solutions—My Experience: The More the Residue, the Less the Pollution of Surface Water." *EPA Journal* 17, no. 5 (November/December), 42–46.

Satchell, Michael. 1996. "Hog Factories: Cheap Meat, Costly Problems." *U.S. News and World Report*, 22 January, pp. 58–59.

Skanavis, C., and W. A. Yanko. 1994. "Evaluation of Composted Sewage Sludge Based Soil Amendments for Potential Risks of Salmonellosis." *Journal of Environmental Health* 56, no. 7 (March), 19–23.

Sun, Marjorie. 1989. "Mud-Slinging over Sewage Technology." *Science* 246, no. 4929 (27 October), 440–443.

Vassos, Troy D., and Dean Reil. 1994. "Constructed Wetlands' Treatment of Municipal Wastewater." *Waterworld Review* 10, no. 5, 55–57.

WHO. 1989. *Health Guidelines for Use of Wastewater and Aquaculture.* Technical Report Series no. 778. Geneva: World Health Organization.

9. Solid Waste

Abelson, Philip H. 1990. "New Technology for Cleaner Air." Editorial. *Science* 248 (18 May), 793.

——— 1994. "Minimizing Wastes." Editorial. *Science* 265 (1 July), 11.

AIChE. 1993. "Garbage! The Story of Waste Management and Recycling." American Institute of Chemical Engineers, Washington, D.C.

Anderson, W. C. 1994. "Excellence in Environmental Engineering, Clean Fuels from Landfill Gas." *Environmental Engineer* 30, no. 3 (April), 22–23.

ATSDR [Agency for Toxic Substances and Disease Registry]. 1994. "Hazardous Waste Sites Seem to Present a Small to Moderate Increase in the Risks of Birth Defects and Cancer." *Environmental Science and Technology* 28, no. 5 (May), 212A.

Connor, D. M. 1988. "Breaking through the 'NIMBY' Syndrome." *Civil Engineering* 58, no. 12 (December), 69–72.

Consumer Reports. 1994. "Recycling: Is It Worth the Effort?" *Consumer Reports* 59, no. 2, (February), 92–98.

Engdahl, R. B., R. E. Barrett, and D. A. Trayser. 1986. "Process Emissions and Their Control: Part I." In A. C. Stern, ed., *Air Pollution, Supplement to Measurements, Monitoring, Surveillance, and Engineering Control,* 3rd ed. New York: Academic Press.

EPA [U.S. Environmental Protection Agency]. 1986a. *RCRA Orientation Manual.* Report EPA/530-SW-86–001. Washington, D.C.

——— 1986b. *Solving the Hazardous Waste Problem: EPA's RCRA Program.* Report EPA/530-SW-86–037. Washington, D.C.

——— 1989. *The Toxics-Release Inventory: A National Perspective.* Report EPA 560/4–89–005. Washington, D.C.

——— 1990. *Medical Waste Management in the United States.* First Interim Report to Congress. Washington, D.C.

——— 1993a. *Safer Disposal for Solid Waste: The Federal Regulations for Landfills.* Report EPA/530-SW-91-092. Washington, D.C.

——— 1993b. *Guidance to Hazardous Waste Generators on the Elements of a Waste Minimization Program.* Federal Register 58, no. 102 (28 May), 31114–20.

EPRI. 1992. "Municipal Waste-to-Energy Technology Assessment." Report EPRI TR-100058. Electric Power Research Institute, Palo Alto.

——— 1993. "Analysis of Markets for Coal Combustion By-Products Use in Agriculture and Land Reclamation." Report EPRI TR-102575. Electric Power Research Institute, Palo Alto.

Finstein, M. S. 1989. "Composting Solid Waste: Costly Mismanagement of a Microbial Ecosystem." *American Society for Microbiology News* 55, no. 11 (November), 599–602.

Fox, Robert D. 1996. "Physical/Chemical Treatment of Organically Contaminated Soils and Sediments." *EM* (Air and Waste Management Association's magazine for environmental managers (May), 28–34.

GAO. 1994. "Nuclear Waste: Foreign Countries' Approaches to High-Level Waste Storage and Disposal." Report GAO/RCED-94–172. U.S. General Accounting Office, Washington, D.C.

Golaine, Andrea. 1991. "Superfund: Money Squandered in the Name of Public Health." *Priorities* (Fall), 30–31.

Grove, Noel, 1994. "Recycling." *National Geographic* 186, no. 1 (July), 92–115.

Herman, Robert, Siamak A. Ardekani, and Jessie H. Ausubel. 1989. "Dematerialization." In Jesse H. Ausubel and Hedy E. Sladovich, eds., *Technology and Environment.* Washington, D.C.: National Academy Press.

Hughes, Joseph B. 1996. "Biological Treatment of Hazardous Waste." In *Frontiers of Engineering,* pp. 37–39. Reports of leading edge engineering from the 1995 NAE Symposium on Frontiers in Engineering, National Academy of Engineering. Washington, D.C.: National Academy Press.

Keoleian, Gregory A., and Dan Menerey. 1994. "Sustainable Development by Design: Review of Life Cycle Design and Related Approaches." *Air and Waste* 44, no. 5 (May), 645–668.

Lamarre, L. 1994. "Building from Ash." *EPRI Journal* 19, no. 3 (April/May), 22–28.

——— 1995. "Tapping the Tire Pile." *EPRI Journal* 20, no. 5 (September), 28–34.

Langone, John. 1989. "Waste: A Stinking Mess." *Time,* 2 January, pp. 44, 45, and 47.

Long, S. W. 1990. "The Incineration of Low-Level Radioactive Waste." Report NUREG-1393. U.S. Nuclear Regulatory Commission, Washington, D.C.

Lubenau, J. O., and J. G. Yusko. 1995. "Radioactive Materials in Recycled Metals." *Health Physics* 68, no. 4 (April), 440–451.

Melody, Mary. 1994. "Materials Exchanges Promote Waste, Recycling Markets." *Hazmat World* 7, no. 5 (May), 27–32.

Mitchell, Frank L. 1992. "Hazardous Waste." William N. Rom, ed., *Environmental and Occupational Medicine,* 2nd ed., pp. 1275–84. Boston: Little, Brown.

News Briefs. 1995. "Markets for Recyclables . . ." *Environmental Science and Technology* 29, no. 6 (June), 255A.

NRC. 1994. *Ranking Hazardous-Waste Sites for Remedial Action.* National Research

Council, Commission on Geo-sciences, Environment, and Resources. Washington, D.C.: National Academy Press.

——— 1995a. "Waste Incineration." *NewsReport* 45, no. 2 (Spring-Summer), 12.

——— 1995b. *Clean Ships, Clean Ports, Clean Oceans: Controlling Garbage and Plastic Wastes at Sea.* National Research Council. Washington, D.C.: National Academy Press.

Peelle, E., and R. Ellis. 1987. "Beyond the 'Not-in-My-Backyard' Impasse." *Forum for Applied Research and Public Policy* 2, no. 3, 68–77.

Renner, Rebecca. 1996. "Life-Cycle Analysis Stirs Continued Debate on Impact of Electric Cars." *Environmental Science and Technology* 30, no. 1 (January), 17A–18A.

Rose, Julian. 1994. "Incineration Regulations Set to Tighten." *Environmental Science and Technology* 28, no. 12 (December), 512A.

Shea, Cynthia. 1988. "Plastic Waste Proliferates." *World-Watch* 1, no. 2 (March–April), 7–8.

Travis, Curtis C. 1993. "Waste Remediation: Can We Afford It?" *Forum for Applied Research and Public Policy* (Spring), 57–59.

U.S. Congress. 1990. Pollution Prevention Act, Public Law 101–508, 42 USC 13101 et seq., Washington, D.C.

——— 1992. Federal Facility Compliance Act, Public Law 102–386, 42 USC 6901 et seq., Washington, D.C.

10. Rodents and Insects

Anonymous. 1994. "Lyme Disease Vaccine Trials." *Harvard Public Health Review* 5, no. 2 (Spring), 4.

——— 1995a. "Dengue Epidemic." *Environmental Health Letter* 34, no. 19 (11 September), 155.

——— 1995b. "WHO Reports on New, Re-emerging Diseases Threatening World Health." *Nation's Health* 25, no. 10 (November), 24.

APHA [American Public Health Association]. 1990. "Tropical Diseases Affect One-Tenth of the World's Population." *Nation's Health* 20, no. 7 (July), 8–9.

Associated Press. 1993. "Ag Scientists Trick Sap Beetle into Carrying Aflatoxin Enemy." *Sun Journal,* New Bern, N.C., 29 July.

Best, Cheryl. 1989. "Natural Pest Controls." *Garbage* 1, no. 1 (September/October), 40–46.

Canby, Thomas Y. 1977. "The Rat—Lapdog of the Devil." *National Geographic* 152, no. 1 (July), 60–87.

Carson, Rachel. 1962. *Silent Spring.* Boston: Houghton Mifflin.

Chege, Nancy. 1994. "Environmental Intelligence: Killing Malaria." *World-Watch,* 7, no. 3 (May/June), 7.

Collins, Frank H., and Nora J. Besansky. 1994. "Vector Biology and the Control of Malaria in Africa." *Science* 264, no. 5167, (24 June), 1874–75.

Conniff, Richard. 1977. "The Malevolent Mosquito." *Reader's Digest* 111, no. 664 (August), 153–157.

Consumer Reports. 1987. "Insect Repellents." *Consumer Reports* 52, no. 7 (July), 423–426.

Cromie, William J. 1994. "Genetic Weapons Being Used in War on Malaria." *Harvard University Gazette* 89, no. 43 (12 August), 1, 10.

Drollette, Dan. 1996. "Australia Fends Off Critic of Plan to Eradicate Rabbits." *Science* 272 (12 April), 191–192.

Duplaix, Nicole. 1988. "Fleas, the Lethal Leapers." *National Geographic* 173, no. 5 (May), 672–694.

Fischer, Karl C. 1969. *Environment and Health*, chap. 6. Loma Linda, Calif.: Loma Linda University Printing Service.

Georghiuo, G. P. 1985. "Pesticide Resistance Management." *NewsReport* (National Research Council) 35, no. 2 (February), 28–29.

Gianessi, L. P. 1987. "Lack of Data Stymies Informed Decisions on Agricultural Pesticides." *Resources* 89 (Fall), 1–4.

Goodman, Billy. 1993a. "Research Community Swats Grasshopper Control Trial" *Science* 260, no. 5110 (14 May), 887.

——— 1993b. "Debating the Use of Transgenic Predators." *Science* 262, no. 5139 (3 December), 1507.

Gould, Fred. 1991. "The Evolutionary Potential of Crop Pests." *American Scientist* 79, no. 6 (November–December), 496–507.

Guest, Charles. 1984. "The Pied Piper of Hamelin." *Lancet* (22/29 December), 1454–1455.

Holden, Constance. 1989. "Entomologists Wane as Insects Wax." *Science* 246, no. 4931 (10 November), 754–756.

Holden, Constance, ed. 1996. "Random Samples: Ominous Trends for Infectious Diseases." *Science* 272 (31 May), 1269.

Keiding, J. 1976. *The House-Fly—Biology and Control.* Report WHO/VBC/76.650. Geneva: World Health Organization.

Knipling, Edward F. 1960. "The Eradication of the Screw-Worm Fly." *Scientific American* 203, no. 4 (October), 54–61.

Lang, Leslie. 1993. "Are Pesticides a Problem?" *Environmental Health Perspectives* 101, no. 7 (December), 578–583.

Lore, Richard, and Kevin Flannelly. 1977. "Rat Societies." *Scientific American* 236, no. 5, (May), 106–116.

Matson, David O. 1996. "Release of RHD Virus in Australia." Letter to the editor. *Science* 273, no. 5271 (5 July), 16–18.

Matuschka, F-R., and A. Spielman. 1993. "Risk of Infection from and Treatment of Tick Bite." *Lancet* 342, no. 8870 (28 August), 529–530.

Maurice, John. 1995. "Malaria Vaccine Raises a Dilemma." *Science* 267, no. 5196 (20 January), 320–323.

OTA [Office of Technology Assessment]. 1995. *Biologically Based Technologies for Pest Control.* Report OTA-ENV-636. Washington, D.C.: Government Printing Office.

Palca, Joseph. 1990. "Libya Gets Unwelcome Visitor from the West." *Science* 249, no. 4965 (13 July), 117–118.

Pennisi, Elizabeth. 1996. "U.S. Beefs Up CDC's Capabilities." *Science* 272 (7 June), 1413.

Reed, George H. 1993. "Lyme Disease and Other Tick-Borne Diseases: A Review." *Journal of Environmental Health* 55, no. 8 (June), 6–9.

Richardson, R. H., J. R. Ellison, and W. W. Averhoff. 1982. "Autocidal Control of Screwworms in North America." *Science* 215, no. 4531 (22 January), 361–370.

Schmidt, Karen. 1994. "Genetic Engineering Yields First Pest-Resistant Seeds." *Science* 265, no. 5173 (5 August), 739.

Spielman, Andrew. 1994. "Why Entomological Antimalaria Research Should Not Focus on Transgenic Mosquitoes." *Parasitology Today* 10, no. 10, 374–376.

——— 1995. Private communication.

Spielman, Andrew, Sam R. Telford, III, and Richard J. Pollack. 1993. "The Origins and Course of the Present Outbreak of Lyme Disease." In Howard S. Ginsberg, ed., *Ecology and Environmental Management of Lyme Disease*, pp. 83–91. Rutgers, N.J.: Rutgers University Press.

Telford, Sam R. III, Richard J. Pollack, and Andrew Spielman. 1991. "Emerging Vector-Borne Infections." *Infectious Disease Clinics in North America* 5, no. 1, 7–17.

Watanabe, Myrna E. 1994. "Pollination Worries Rise as Honey Bees Decline." *Science* 265, no. 5176 (26 August), 1170.

Weinstein, Jack S. 1990. *Lyme Disease.* New York: American Council of Science and Health.

Yoder, Jay A., Richard J. Pollack, and Andrew Spielman. 1992. "An Ant-Diversionary Secretion of Ticks: First Demonstration of an Acarine Allomone." *Journal of Insect Physiology* 39, no. 5, 429–435.

11. Injury Control

Anonymous. 1991. "Profile: Remedy for Violence: Dr. Deborah Prothrow-Stith." *Injury Update, Harvard Injury Control Center* 1, no. 2 (Spring), 6–7.

Berl, Walter G., and Byron M. Halpin. 1979. "Human Fatalities from Unwanted Fires." *Fire Journal* 73, no. 5 (September), 105–115, 123.

CDC [Centers for Disease Control]. 1989. *National Traumatic Occupational Fatalities: 1980–1985.* Cincinnati: National Institute for Occupational Safety and Health, U.S. Department of Health and Human Services.

Christoffel, Katherine K. 1994. "Editorial: Reducing Violence—How Do We Proceed?" *American Journal of Public Health* 84, no. 4 (April), 539–541.

De Haven, H. 1942. "Mechanical Analysis of Survival in Falls from Heights of Fifty to One Hundred and Fifty Feet." *War Medicine* 2, 586–596.

FEMA [Federal Emergency Management Agency]. 1993. *Fire in the United States: 1983–1990.* 8th ed. Washington, D.C.

Ferguson, S. A., D. F. Preusser, A. K. Lund, P. L. Zador, and R. G. Ulmer. 1995. "Daylight Saving Time and Motor Vehicle Crashes: The Reduction in Pedestrian and Vehicle Occupant Fatalities." *American Journal of Public Health* 85, no. 1 (January), 92–95.

Gallagher, S. S., K. Finison, B. Guyer, and S. Goodenough. 1984. "The Incidence of Injuries among 87,000 Massachusetts Children and Adolescents: Results of the 1980–81 Statewide Childhood Injury Prevention Program Surveillance System." *American Journal of Public Health* 74, no. 12 (December), 1340–47.

GAO. 1996. "Motor Vehicle Safety: Comprehensive State Programs Offer Best Opportunity for Increasing Use of Safety Belts." In *Reports and Testimony: January 1996,* Report GAO/OPA-96-4, p. 15. U.S. General Accounting Office, Washington, D.C.

Gibson, J. J. 1961. "Contribution of Experimental Psychology to the Formulation of the Problem of Safety: A Brief for Basic Research." In *Behavioral Approaches to Accident Research.* New York: Association for the Aid of Crippled Children.

Haddon, William, Jr. 1970. "On the Escape of Tigers: An Ecologic Note." *American Journal of Public Health* 60, no. 12, 2229–34.

Hemenway, David. 1995. Private communication.

IIHS [Insurance Institute for Highway Safety]. 1991a. "Studies Quantify Effectiveness of Air Bags in Cars." *IIHS Status Report* 26, no. 10 (30 November), 1–2.

——— 1991b. "NHTSA Moves to Allow Daytime Running Lights." *IIHS Status Report* 26, no. 10 (30 November), 7.

——— 1992. "Driver Deaths Down Substantially in Cars Equipped with Air Bags." *IIHS Status Report* 27, no. 12 (3 October), 1–3.

——— 1993a. "Deer, Moose Collisions with Motor Vehicles Peak in Spring and Fall." *IIHS Status Report* 28, no. 4 (3 April), 6–7.

——— 1993b. "Helmet Use Is Up and Cycle Injuries Down under Texas Law." *IIHS Status Report* 28, no. 5 (24 April), 6.

——— 1993c. "North Carolina: First State to Launch Long-Term Program to Increase Safety Belt Use." *IIHS Status Report* 28, no. 6 (15 May), 1, 4.

——— 1993d. "New Facts about Highway Deaths Show the Lowest Toll in 31 Years." *IIHS Status Report* 28, no. 9 (24 July), 1–5.

——— 1993e. "First North Carolina Results Show Mix of Publicity, Enforcement Sends Belt Use to about 80 Percent." *IIHS Status Report* 28, no. 10 (21 August), 1, 4.

——— 1994a. "FHWA Finally Bans Radar Detector Use in Commercial Vehicles." *IIHS Status Report* 29, no. 1 (8 January), 2.

——— 1994b. "What Antilocks Can Do, What They Cannot Do." *IIHS Status Report* 29, no. 2 (29 January), 1–2.

——— 1994c. "Studies Find Drivers Might Change Ways with Antilock Brakes." *IIHS Status Report* 29, no. 2 (29 January), 3.

——— 1994d. "IVHS Business Booms with Infusion of Hundreds of Millions of Scarce Federal Dollars—But Safety Claims Aren't Backed by Science." *IIHS Status Report* 29, no. 8 (30 July), 1–4, 6.

——— 1994e. "Death Trend Is Down Overall since 1975 Except in Pickup Trucks and Utility Vehicles." *IIHS Status Report* 29, no. 9 (20 August), 1–3, 6.

——— 1994f. "Seven Straight Years: Deaths Higher after 65 MPH Speed Limits than Before." *IIHS Status Report* 29, no. 10 (10 September), 3.

——— 1994g. "Early Licensure Laws Increase Teenagers' Risk Behind the Wheel." *IIHS Status Report* 29, no. 10 (10 September), 4–5.

——— 1994h. "Five Crash Types Account for Most Motorcycle Fatalities." *IIHS Status Report* 29, no. 10 (10 September), 6.

——— 1994i. "Best and Worst—1988–92 Passenger Vehicles with Lowest and Highest Drive Death Rates during 1989–93." *IIHS Status Report* 29, no. 11 (8 October), 2.

——— 1995. "Highway Death Toll Tops 40,000 in 1994 for the Second Straight Year." *IIHS Status Report* 30, no. 7 (12 August), 3.

——— 1996a. "Kids and Air Bags" (pamphlet), Arlington, Va.

——— 1996b. "Side-Impact Air Bags New on Safety Marquee." *IIHS Status Report* 31, no. 4 (4 May), 1–3.

Koshland, Daniel E., Jr. 1994. Editorial. "The Spousal Abuse Problem." *Science* 265, no. 5171 (22 July), 455.

Menninga, Bert. 1994. "Violence in Schools: No Single Cause, No Simple Solutions." *Wingspread Journal* 12, no. 2 (Summer), 3.

NIOSH [National Institute for Occupational Safety and Health]. 1993. "Alert: Request for Assistance in Preventing Homicide in the Workplace." Publication no. 93–109. Centers for Disease Control and Prevention, Cincinnati.

——— 1996. "Violence in the Workplace: Risk Factors and Prevention Strategies." Current Intelligence Bulletin 57, DHHS Publication No. 96-100. National Institute for Occupational Safety and Health, Centers for Disease Control and Prevention, Cincinnati.

NRC [National Research Council]. 1996. *Shopping for Safety: Providing Customer Automotive Safety Information.* Special Report 248. Committee for Study of Consumer Automotive Safety Information, Transportation Research Board, Washington, D.C.

NSC. 1990. *Accident Facts, 1990 Edition.* Chicago: National Safety Council.

——— 1994. *Accident Facts, 1994 Edition,* Itasca, Ill.: National Safety Council.

——— 1995. *Accident Facts, 1995 Edition.* Itasca, Ill.: National Safety Council.

Ridley, Kimberly. 1994. "Crime: What Makes the Difference?" *Harvard Public Health Review* 5, no. 2 (Spring), 36–41.

Rowland, J., F. Rivara, P. Salzberg, R. Soderberg, R. Maier, and T. Koepsell. 1996. "Motorcycle Helmet Use and Injury Outcome and Hospitalization Costs from Crashes in Washington State." *American Journal of Public Health* 86, no. 1 (January) 41–45.

Scheidt, P. C., Y. Harel, A. C. Trumble, D. H. Jones, M. D. Overpeck, and P. E. Bijur. 1995. "The Epidemiology of Nonfatal Injuries among U.S. Children and Youth." *American Journal of Public Health* 85, no. 7 (July), 932–938.

Schenker, M. B., R. Lopez, and G. Wintemute. 1995. "Farm-Related Fatalities among Children in California, 1980 to 1989." *American Journal of Public Health* 85, no. 1 (January), 89–92.

Waller, Julian A. 1987. "Injury: Conceptual Shifts and Preventive Implications." *Annual Review of Public Health* 8, 21–49.

——— 1994. "Reflections on a Half Century of Injury Control." *American Journal of Public Health* 84, no. 4 (April), 664–670.

Whitfield, R. A., and Ian S. Jones. 1995. "The Effect of Passenger Load on Unstable Vehicles in Fatal, Untripped Rollover Crashes." *American Journal of Public Health* 85, no. 9 (September), 1268–71.

WHO. 1995. *In Point of Fact.* NRPB Bulletin no. 84. Geneva: World Health Organization.

Winsten, Jay A. 1995. "Public Health, 90210." *Harvard Public Health Review* 6, no. 2 (Spring), 5–7.

12. Electromagnetic Radiation

ACGIH. 1996. *1996 TLVs and BEIs—Threshold Limit Values for Chemical Substances and Physical Agents and Biological Exposure Indices.* Cincinnati: American Conference of Governmental Industrial Hygienists.

BEIR [Committee on the Biological Effects of Ionizing Radiation]. 1980. *The Effects on Populations of Exposure to Low Levels of Ionizing Radiation.* Report no. 3. Washington, D.C.: National Academy Press.

——— 1990. *Health Effects of Exposure to Low Levels of Ionizing Radiation.* Report no. 5. Washington, D.C.: National Academy Press.

Blix, H. 1989. "The Peaceful Applications of Nuclear Energy." Paper presented at the annual symposium of the Uranium Institute, London, 6–8 September.

BRWM. 1990. *Rethinking High-Level Radioactive Waste Management.* Board on Radioactive Waste Management, National Academy of Sciences, National Research Council. Washington, D.C.: National Academy Press.

CDRH [Center for Devices and Radiological Health]. 1992. "Center Distributes Update on Possible Hazards of Police Radar Devices." *Radiological Health Bulletin* 26, no. 9 (September), 1–2.

——— 1993. "Center Addresses Cellular Phone Issues." *Radiological Health Bulletin* 27, no. 1 (Spring), 2–4.

Duchene, A. S., and John Lakey, eds. 1990. *The IRPA Guidelines on Protection against Non-Ionizing Radiation.* New York: Pergamon Press.

EPRI. 1989. *Extremely Low Frequency Electric and Magnetic Fields and Cancer: A Literature Review.* Report EPRI EN-6674. Palo Alto, Calif.: Electric Power Research Institute.

Ferris, Benjamin G. 1966. "Environmental Hazards: Electromagnetic Radiation." *New England Journal of Medicine* 275 (17 November), 1100–5.

Gies, H. P., C. R. Roy, G. Elliott, and W. Zonli. 1994. "Ultraviolet Radiation Protection Factors for Clothing." *Health Physics* 67, no. 2 (August), 131–139.

Graham, John D., and Susan W. Putnam. 1994. "EMF and Human Health—An Open Question." Newsletter, *Risk in Perspective* 2, no. 3 (October), 1–2.

——— 1995. "Cellular Telephones and Brain Cancer." Newsletter, *Risk in Perspective* 3, no. 4 (June), 1–2.

HHS. 1970. "Microwave Ovens: Performance Standards." U.S. Department of Health and Human Services, Washington, D.C. Code of Federal Regulations, Title 42, pt. 78.

Hidy, George M. 1990. "EMG Research: A Commitment to Excellence." Editorial. *EPRI Journal* 15, no. 1 (January/February), 1.

ICNIRP [International Commission on Non-Ionizing Radiation Protection]. 1994. "Guidelines on Limits of Exposure to Static Magnetic Fields." *Health Physics* 66, no. 1 (January), 100–106.

——— 1996. "Health Issues Related to the Use of Hand-Held Radiotelephones and Base Transmitters." *Health Physics* 70, no. 4 (April), 587–593.

ICRP. 1991a. *1990 Recommendations of the International Commission on Radiological Protection.* Publication 60. Annals of the ICRP 21, no. 1–3. New York: Pergamon Press.

——— 1991b. *Annual Limits on Intake of Radionuclides by Workers Based on the 1990 Recommendations.* Publication 61. Annals of the ICRP 21, no. 4. New York: Pergamon Press.

——— 1989–95. *Age-Dependent Doses to Members of the Public from Intake of Radionuclides: Parts 1–4.* Publications 56, 67, 69, 71. Annals of the ICRP 20, no. 2 (1989); 23, no. 3–4 (1993); 25, no. 1 (1995); 25, no. 3–4 (1995). New York: Pergamon Press.

INPO [Institute of Nuclear Power Operations]. 1996. "1995 WANO Performance Indicators for the U.S. Nuclear Utility Industry." *Nuclear Professional* 11, no. 2 (Spring), 23–25.

IRPA [International Non-Ionizing Radiation Committee of the International Radiation Protection Association]. 1985a. "Guidelines on Limits of Exposure to Ultraviolet Radiation of Wavelengths between 180 nm and 400 nm (Incoherent Optical Radiation)." *Health Physics* 49, no. 2 (August), 331–340.

——— 1985b. "Guidelines on Limits of Exposure to Laser Radiation of Wavelengths between 180 nm and 1 mm." *Health Physics* 49, no. 2 (August), 341–359.

——— 1988. "Guidelines on Limits of Exposure to Radiofrequency Electromagnetic Fields in the Frequency Range from 100 kHz to 300 GHz." *Health Physics* 54, no. 1 (January), 115–123.

——— 1991. "Health Issues of Ultraviolet 'A' Sunbeds Used for Cosmetic Purposes." *Health Physics* 61, no. 2 (August), 285–288.

Johnson, D. W., and W. A. Goetz. 1985. *Patient Exposure Trends in Medical and Dental Radiography.* Rockville, Md.: Center for Devices and Radiological Health, U.S. Department of Health and Human Services.

Johnson, G. Timothy, ed. 1982. "Microwaves." *Harvard Medical School Health Letter* 7, no. 10 (August), 1–2, 5.

Little, John B. 1993. "Biologic Effects of Low-Level Radiation Exposure." In J. M. Taveras, M. Juan, and J. T. Ferrucci, eds., *Radiologic Physics and Pulmonary Radiology,* vol. 1, ch. 13. Philadelphia: J. B. Lippincott.

Maher, Edward F., Stephen E. Rudnick, and Dade W. Moeller. 1987. "Effective Removal of Airborne ^{222}Rn Decay Products inside Buildings." *Health Physics* 53, no. 4 (October), 351–356.

Mettler, Fred A., Jr. 1993. "Exposure of the United States Population from Medical Radiation." In *Radiation Protection in Medicine,* pp. 11–14. Bethesda, Md.: Proceedings of the twenty-eighth annual meeting of the National Council on Radiation Protection and Measurements.

Mettler, F. A., Jr., M. Davis, C. A. Kelsey, R. Rosenberg, and A. Williams. 1987. "Analytical Modeling of Worldwide Medical Radiation Use." *Health Physics* 52, no. 2 (February), 133–141.

Mettler, F. A., Jr., W. K. Sinclair, L. Anspaugh, C. Edington, J. H. Harley, R. C. Ricks, P. B. Selby, E. W. Webster, and H. O. Wyckoff. 1990. "The 1986 and 1988 UNSCEAR Reports: Findings and Recommendations." *Health Physics* 58, no. 3 (March), 241–250.

Moeller, Dade W. 1996a. "Radiation Sources: Natural Background." In William R. Hendee and F. Marc Edwards, eds., *Health Effects of Exposure to Low-Level Ionizing Radiation,* pp. 269–286. Philadelphia: Institute of Physics Publishing.

——— 1996b. "Radiation Sources: Consumer Products." In William R. Hendee and

F. Marc Edwards, eds., *Health Effects of Exposure to Low-Level Ionizing Radiation,* pp. 287–313. Philadelphia: Institute of Physics Publishing.

Nair, I., M. G. Morgan, and H. K. Florig. 1989. *Biological Effects of Power Frequency Electric and Magnetic Fields.* Washington, D.C.: Office of Technology Assessment.

NCRP [National Council on Radiation Protection and Measurements]. 1987a. *Exposure of the Population in the United States and Canada from Natural Background Radiation.* Report no. 94. Bethesda, Md.

——— 1987b. *Ionizing Radiation Exposure of the Population of the United States.* Report no. 93. Bethesda, Md.

——— 1987c. *Radiation Exposure of the U.S. Population from Consumer Products and Miscellaneous Sources.* Report no. 95. Bethesda, Md.

——— 1989. *Exposure of the U.S. Population from Diagnostic Medical Radiation.* Report no. 100. Bethesda, Md.

——— 1994. *Dose Control at Nuclear Power Plants.* Report no. 120. Bethesda, Md.

——— 1995. *Radiation Exposure and High-Altitude Flight.* Commentary no. 12. Bethesda, Md.

NRPB [National Radiological Protection Board]. 1991. "Non-Ionizing Radiation." At-a-Glance Series, U.K.

——— 1993a. "Natural Radiation Maps of Western Europe." At-a-Glance Series, U.K.

——— 1993b. "Ultraviolet Radiation." At-a-Glance Series, U.K.

——— 1996. "Effects of UVR on Human Health." *Radiological Protection Bulletin* 173 (January), 8–13, U.K.

Stone, Richard. 1994. "Science Scope: Chernobyl Cancers to Come Under Scrutiny." *Science* 263, no. 5144 (14 January), 163.

Tanner, Melissa. 1990. "Increasing Use, Power of Lasers Make Eye Protection Essential." *Occupational Health and Safety,* (July), 44–46.

UNSCEAR. 1993. "Sources and Effects of Ionizing Radiation." United Nations Scientific Committee on the Effects of Atomic Radiation, United Nations, New York.

USNRC. 1990. *Population Dose Commitments due to Radioactive Releases from Nuclear Power Plant Sites in 1987.* Report NUREG/CR-2850, vol. 9. Washington, D.C.: U.S. Nuclear Regulatory Commission.

——— 1991. *Information Digest, 1991 Edition.* Report NUREG-1350, vol. 3. Washington, D.C.: Office of the Controller, U.S. Nuclear Regulatory Commission.

——— 1993. *Chief Financial Officer's Annual Report—1993.* Report NUREG-1470, vol. 2. Washington, D.C.: U.S. Nuclear Regulatory Commission.

——— 1995. *Nuclear Regulatory Commission Information Digest, 1995 Edition.* Report NUREG-1350, vol. 7. Washington, D.C.: U.S. Nuclear Regulatory Commission.

13. Environmental Law

Becker, Julie C. 1991. "The 'Least Favored Method'—A Primer on the RCRA Land Disposal Regulations." *Journal of the Air and Waste Management Association* 41, no. 4 (April), 414–417.

Bingham, Eula. 1992. "The Occupational Safety and Health Act." In William N. Rom, ed., *Environmental and Occupational Medicine,* 2nd ed., pp. 1325–31. Boston: Little, Brown and Company.

BNA. 1993. "Guide to Federal Environmental Laws." Bureau of National Affairs, Washington, D.C.

Boyd, J., and M. K. Macauley. 1994. "The Impact of Environmental Liability on Industrial Real Estate Development." *Resources,* no. 114 (Winter) 19–23.

EPA. 1986. "RCRA Orientation Manual." Report EPA/530-SW-86–001. U.S. Environmental Protection Agency, Washington, D.C.

Findley, R. W., and D. A. Farber. 1992. *Environmental Law in a Nutshell,* 3rd ed. St. Paul, Minn.: West Publishing Company.

Graham, John D., and George M. Gray. 1993. "Optimal Use of 'Toxic Chemicals.'" Newsletter, *Risk in Perspective* 1, no. 2 (May).

Greaves, William W. 1992. "Toxic Substances Control Act." In William N. Rom, ed., *Environmental and Occupational Medicine,* 2nd ed., pp. 1333–38. Boston: Little, Brown and Company.

Hutt, Peter, and Richard Merrill. 1991. *Food and Drug Law,* 2nd ed. Westbury, N.Y.: Foundation Press.

Merrill, Richard A. 1986. "Regulatory Toxicology." In C. D. Klaassen, Mary O. Amdur, and J. Doull, eds., *Casarett and Doull's Toxicology, The Basic Science of Poisons,* 3rd ed., pp. 917–932. New York: Macmillan Publishing Company.

NAE [National Academy of Engineering]. 1993. *Keeping Pace with Science and Engineering: Case Studies in Environmental Regulation.* Washington, D.C.: National Academy Press.

OSHA [Occupational Safety and Health Administration]. 1980. "All about OSHA," Publication no. 2056, U.S. Department of Labor, Washington, D.C.

Sun, Marjorie. 1990. "Emissions Trading Goes Global." *Science* 247, (2 February), 520–521.

USNRC [U.S. Nuclear Regulatory Commission]. 1991. "Nuclear Regulatory Legislation, 101st Congress." Report NUREG-0980, vol. 2, no. 1, Washington, D.C.

14. Standards

AAEE. 1983. *Proceedings of the Seminar on Development and Assessment of Environmental Standards.* Annapolis, Md.: American Academy of Environmental Engineers.

ACGIH. 1996. *1996 TLVs and BEIs—Threshold Limit Values for Chemical Substances and Physical Agents and Biological Exposure Indices (BEIs).* Cincinnati: American Conference of Governmental Industrial Hygienists.

BEIR [Committee on Biological Effects of Ionizing Radiation]. 1972. *The Effects on Populations of Exposure to Low Levels of Ionizing Radiation.* Report no. 1. Washington, D.C.: National Academy Press.

——— 1980. *The Effects on Populations of Exposure to Low Levels of Ionizing Radiation.* Report no. 3. Washington, D.C.: National Academy Press.

——— 1990. *The Effects on Populations of Exposure to Low Levels of Ionizing Radiation.* Report no. 5. Washington, D.C.: National Academy Press.

Buonicore, Anthony J. 1991. "How Clean Is Clean? Cleanup Criteria for Contaminated Soil and Groundwater." *Journal of the Air and Waste Management Association* 41, no. 11 (November), 1446–1449.

Eisenbud, M. 1978. *Environment, Technology, and Health: Human Ecology in Historical Perspective.* New York: New York University Press.

FRC [Federal Radiation Council]. 1960. *Background Material for the Development of Radiation Protection Standards.* Report no. 1. Washington, D.C.: U.S. Department of Health, Education and Welfare.

GAO. 1994. "Nuclear Health and Safety: Consensus on Acceptable Radiation Risk to the Public Is Lacking." Report GAO/RCED-94–190, U.S. General Accounting Office, Washington, D.C.

ICRP. 1977. *Recommendations of the International Commission on Radiological Protection.* Publication 26, Annals of the ICRP, vol. 1, no. 3. New York: Pergamon Press.

——— 1985. *Quantitative Basis for Developing a Unified Index of Harm.* Publication 45, Annals of the ICRP, vol. 15, no. 3. New York: Pergamon Press.

——— 1991. *1990 Recommendations of the International Commission on Radiological Protection.* Publication 60, Annals of the ICRP, vol. 21, no. 1–3. New York: Pergamon Press.

——— 1994a. *Age-Dependent Doses to Members of the Public from Intake of Radionuclides: Part 2—Ingestion Dose Coefficients.* Publication 67, Annals of the ICRP, vol. 23, no. 3/4. New York: Pergamon Press.

——— 1994b. *Dose Coefficients for Intakes of Radionuclides by Workers.* Publication 68, Annals of the ICRP, vol. 24, no. 4. New York: Pergamon Press.

Kocher, D. C. 1988. "Review of Radiation Protection and Environmental Radiation Standards for the Public." *Nuclear Safety* 29, no. 4 (October-December), 463–475.

Lefohn, Allen S., and Janell K. Foley. 1993. "Establishing Relevant Ozone Standards to Protect Vegetation and Human Health: Exposure/Dose-Response Considerations." *Air and Waste* 43, no. 1 (January), 106–112.

Moeller, Dade W. 1990. "History and Perspective on the Development of Radiation Protection Standards." In *Radiation Protection Today: The NCRP at Sixty Years.* Proceedings no. 11. Bethesda, Md.: National Council on Radiation Protection and Measurements.

Muller, Hermann J. 1927. "Artificial Transmutation of the Gene." *Science* 66, no. 1699 (11 July), 84–87.

NCRP [National Council on Radiation Protection and Measurements]. 1971. *Basic Radiation Protection Criteria.* Report no. 39. Bethesda, Md.

——— 1987. *Recommendations on Limits for Exposure to Ionizing Radiation.* Report no. 91. Bethesda, Md.

——— 1993. *Limitation of Exposure to Ionizing Radiation.* Report no. 116. Bethesda, Md.

——— 1995. *Principles and Application of Collective Dose in Radiation Protection.* Report no. 121. Bethesda, Md.

Pompili, Michael J. 1995. "Environmental Mandates: The Impact on Local Government." *Journal of Environmental Health* 57, no. 6 (January/February), 6–12.

Robinson, James C., and William S. Pease. 1991. "From Health-Based to Technology-Based Standards for Hazardous Air Pollutants." *American Journal of Public Health* 81, no. 11 (November), 1518–1523.

USNRC. 1991. "Standards for Protection against Radiation." Code of Federal Regulations, Title 10, pt. 20. U.S. Nuclear Regulatory Commission, Washington, D.C.

15. Monitoring

APHA. 1995. *Standard Methods for the Evaluation of Water and Wastewater, 19th ed.* Published jointly by the American Public Health Association, American Water Works Association, and the Water Environmental Federation. Washington, D.C.

Baker, Chuck. 1992. "Sampling for Water Quality." *Waterworld News* (May/June), 31–32.

Baker, D. A. 1994. "Dose Commitments due to Radioactive Releases from Nuclear Power Plant Sites in 1990." Report NUREG/CR-2850. U.S. Nuclear Regulatory Commission, Washington, D.C.

Bromberg, Steven M. 1990. "Identifying Ecological Indicators: An Environmental Monitoring and Assessment Program." *Journal of the Air and Waste Management Association* 40, no. 7 (July), 976–978.

EPA. 1989. *The Toxics-Release Inventory: A National Perspective.* Report EPA 560/4–89–005. Washington, D.C.: U.S. Environmental Protection Agency.

———. 1993. *R-EMAP: Regional Environmental Monitoring and Assessment Program.* Report EPA/625/R-93/012. Washington, D.C.: U.S. Environmental Protection Agency.

———. 1994. *National Air Quality and Emissions Trends Report, 1993.* Report EPA 454R-94–026. Washington, D.C.: U.S. Environmental Protection Agency.

Griffith, Jerry A., and Carolyn T. Hunsaker. 1994. *Ecosystem Monitoring and Ecological Indicators: An Annotated Bibliography.* Report EPA/620/R-94/021. Washington, D.C.: U.S. Environmental Protection Agency.

ICRP. 1985. *Principles of Monitoring for the Radiation Protection of the Population,* International Commission on Radiological Protection, Annals of the ICRP, vol. 15, no. 1. New York: Pergamon Press.

Lawler, Andrew. 1995. "From Russia with Love: U.S. Cloud Data." *Science* 269, no. 5223 (28 July), 473–474.

NCRP [National Council on Radiation Protection and Measurements]. 1989. *Screening Techniques for Determining Compliance with Environmental Standards.* Commentary no. 3, revision 1. Bethesda, Md.

———. 1991. *Effects of Ionizing Radiation on Aquatic Organisms.* Report no. 109. Bethesda, Md.

———. 1993. *Limitation of Exposure to Ionizing Radiation.* Report no. 116. Bethesda, Md.

———. 1994. *A Practical Guide to the Determination of Human Exposure to Radiofrequency Fields.* Report no. 119. Bethesda, Md.

——— 1996. *Screening Models for Releases of Radionuclides to Atmosphere, Surface Water, and Ground.* Report no. 123. Bethesda, Md.

Newman, Alan. 1995a. "EMAP Shifts Focus to Research." *Environmental Science and Technology* 29, no. 3 (March), 113A.

——— 1995b. "Major U.S. Human Exposure Assessment Survey Gets under Way." *Environmental Science and Technology* 29, no. 9 (September), 398A–399A.

Sioutas, Constantinos. 1993. Personal communication.

Stone, Richard. 1995. "Environmental Monitoring—EPA Streamlines Troubled National Ecological Survey." *Science* 269, no. 5216 (9 June), 1427–28.

Thornton, Kent W., D. Eric Hyatt, and Cynthia B. Chapman. 1993. *Environmental*

Monitoring and Assessment Program Guide. Report EPA/620/R-93/012. Washington, D.C.: U.S. Environmental Protection Agency.

Till, J. E. 1988. "Modeling the Outdoor Environment—New Perspectives and Challenges." *Health Physics* 55, no. 2 (August), 331–338.

U.S. Navy. 1995. *Environmental Monitoring and Disposal of Radioactive Wastes from U.S. Naval Nuclear Powered Ships and Their Support Facilities.* Report NT-95–1. Washington, D.C.

USNRC. 1977. *Calculation of Annual Doses to Man from Routine Releases of Reactor Effluents for the Purpose of Evaluating Compliance with 10 CFR Part 50, Appendix I.* Regulatory Guide 1.109. Washington, D.C.: U.S. Nuclear Regulatory Commission.

——— 1991. "Standards for Protection against Radiation." Code of Federal Regulations, Title 10, pt. 20. Washington, D.C.

Wilhelmsen, R. N., K. C. Wright, D. W. McBride, and B. W. Borsella. 1994. *Annual Report—1993 Environmental Surveillance for EG&G Idaho Waste Management Facilities at the Idaho National Engineering Laboratory.* Report EGG-2679(93). Idaho Falls, Idaho.

Wolman, M. Gordon. 1995. "Human and Ecosystem Health: Management Despite Some Incompatibility." *Ecosystem Health* 1, no. 1 (March), 35–40.

Young, John E. 1994. "Using Computers for the Environment." In Linda Starke, ed., *State of the World—1994,* pp. 99–116. New York: W. W. Norton and Company.

16. Risk Assessment

Albert, Roy E. 1980. "Federal Regulatory Agency Approaches to the Assessment and Control of Risk from Carcinogens." In *Perceptions of Risk.* Proceedings of the fifteenth annual meeting of the National Council on Radiation Protection and Measurements, pp. 6–14. Bethesda, Md.: National Council on Radiation Protection and Measurements.

Breyer, Stephen. 1993. *Breaking the Vicious Circle—Toward Effective Risk Regulation* Cambridge, Mass.: Harvard University Press.

Caplan, Knowlton J., and Jeremiah Lynch. 1996. "A Need and an Opportunity: AIHA Should Assume a Leadership Role in Reforming Risk Assessment." *American Industrial Hygiene Association Journal* 57, no. 3 (March), 231, 233–237.

Davies, Terry. 1995. "Message from the Director." Newsletter, *Resources for the Future,* no. 8 (Summer), 1. Washington, D.C.: Center for Risk Management.

Douglas, J. 1985. "Measuring and Managing Environmental Risk." *EPRI Journal,* 10, 6 (July/August), 6–13.

Dwyer, J. P. 1990. "The Limits of Environmental Risk Assessment." Boalt Hall transcript, pp. 20–25. In press, *Journal of Energy Engineering.*

EPA. 1987. "The Risk Assessment Guidelines of 1986." Report EPA/600/8–87/045, U.S. Environmental Protection Agency, Washington, D.C.

Falk, H. 1992. "The Risk of Risk Assessment." In *Proceedings of a Conference on Chemical Risk Assessment in the DoD: Science Policy and Practices,* pp. 207–210. Cincinnati: American Conference of Governmental Industrial Hygienists.

Federal Register. 1993. "The President: Executive Order 12866—Regulatory Planning and Review." Federal Register, 58:190, 51735–44 (4 October), Washington, D.C.

Garrick, B. John. 1989. "Risk Assessment Practices in the Space Industry: The Move Toward Quantification." Guest editorial. *Risk Analysis* 9, no. 1, 1–7.

Graham, John D. 1991. "OMB vs. the Agencies: The Future of Cancer Risk Assessment." Report of a workshop to peer review the OMB report on risk assessment and risk management, Center for Risk Analysis, Harvard School of Public Health, Boston, June.

——— 1993a. "The Legacy of One in a Million." Newsletter, *Risk in Perspective* 1, no. 1 (March), 1–2.

——— 1993b. "Annual Report, 1992." Center for Risk Analysis, School of Public Health, Harvard University, Boston.

Gray, George M. 1994. "Complete Risk Characterization." Newsletter, *Risk in Perspective* 2, no. 4 (November), 1–2.

Hamilton, D. P. 1992. "Environmental Agency Launches a Study in 'Ecological Risk Assessment,'" *Science* 255, no. 5051 (20 March), 1499.

Herrick, R. F. 1992. "Exposure Assessment in Risk Assessment." In *Proceedings of a Conference on Chemical Risk Assessment in the DoD: Science Policy, and Practices*, pp. 53–57. Cincinnati: American Conference of Governmental Industrial Hygienists.

Hunter, R. L., R. M. Cranwell, and M. S. Y. Chu. 1986. "Assessing Compliance with the EPA High-Level Waste Standard: An Overview." Report NUREG/CR-4510, U.S. Nuclear Regulatory Commission, Washington, D.C.

ICF. 1986. "Risk Assessment Information Directory." Washington, D.C.

Jasanoff, Sheila. 1993. "Bridging the Two Cultures of Risk Analysis." Guest editorial. *Risk Analysis* 13, no. 2, 123–129.

Johnson, B. L. 1992. "Principles of Chemical Risk Assessment: The ATSDR Perspective." In *Proceedings of a Conference on Chemical Risk Assessment in the DoD: Science Policy, and Practices*, pp. 29–35. Cincinnati: American Conference of Governmental Industrial Hygienists.

Kaplan, Stanley, and B. John Garrick. 1981. "On the Quantitative Definition of Risk." *Risk Analysis* 1, no. 1, 11–27.

Naugle, D. F., and T. K. Pierson. 1991. "A Framework for Risk Characterization of Environmental Pollutants." *Journal of the Air and Waste Management Association* 41, no. 10, 1298–1307.

NCRP [National Council on Radiation Protection and Measurements]. 1993. "Research Needs for Radiation Protection." Report no. 117. Bethesda, Md.

——— 1996. "A Guide for Uncertainty Analysis in Dose and Risk Assessments Related to Environmental Contamination." Commentary no. 14. Bethesda, Md.

NRC. 1994. "Building Consensus through Risk Assessment and Management of the Department of Energy's Environmental Remediation Program." Committee to Review Risk Management in the DOE's Environmental Remediation Program, National Research Council, Washington, D.C.

——— 1996. "Understanding Risk: Informing Decisions in a Democratic Society." Committee on Risk Characterization, National Research Council, Washington, D.C.

Renner, Rebecca. 1996. "Ecological Risk Assessment Struggles to Define Itself." *Environmental Science and Technology* 30, no. 4 (April), 172A–174A.
Rosenthal, A., G. M. Gray, and J. D. Graham. 1992. "Legislating Acceptable Cancer Risk from Exposure to Toxic Chemicals." *Ecology Law Quarterly* 19, no. 2, 269–362.
Smith, C. M., K. T. Kelsey, and D. C. Christiani. 1993. "Risk Assessment and Occupational Health." *New Solutions* 3, no. 2, 26–38.
Starr, Chauncey. 1993. "Unintended Consequences of Single-Issue Risk Assessment—A Proposal for Regulatory Restructure." *Proceedings of a Topical Meeting on Risk Management—Expanding Horizons*, pp. 32–34. La Grange, Ill.: American Nuclear Society.
USNRC. 1975. "Reactor Safety Study: An Assessment of Accident Risks in U.S. Commercial Nuclear Power Plants." Report WASH-1400, NUREG-75/014. U.S. Nuclear Regulatory Commission, Washington, D.C.
Wilson, Richard, and William Clark. 1991. "Risk Assessment and Risk Management: Their Separation Should Not Mean Divorce." In C. Zervos, ed., *Risk Analysis*, pp. 187–196. New York: Plenum Press.

17. Energy

Abelson, Philip H. 1989. "Effects of Electric and Magnetic Fields." Editorial. *Science* 245, no. 4915 (21 July), 241.
——— 1993. "Power from Wind Turbines." Editorial. *Science* 261, no. 5126 (3 September), 1255.
Abrahamson, Dean E. 1970. *Environmental Cost of Electric Power.* New York: Scientists' Institute for Public Information.
Anonymous. 1981. "Geothermal—Tapping the Earth's Furnace." *National Geographic*, Special Report, February, pp. 66–67.
——— 1995. "Landfill Methane Recovery." *Environmental Science and Technology* 29, no. 2 (February), 67A.
ASME. 1989. *Energy and the Environment: A General Position Paper.* New York: American Society of Mechanical Engineers.
AT&T [American Telephone and Telegraph]. 1994. "Environmental Tips from AT&T—Simple Things You Can Do to Help Save the Earth." Insert in *Reader's Digest*, 144, no. 864 (April).
Ayres, Ed. 1995. "The Search for an Alternative Firewood." *World-Watch* 8, no. 2 (March/April), 6.
Blatt, Morton. 1994. "The Energy-Efficient Office." *EPRI Journal* 19, no. 5 (July/August), 16–23.
Carrier Corporation, 1994. "Home Comfort and Energy Savings Guide." Syracuse, N.Y.
CEQ. 1995. *Environmental Quality, Twenty-Fourth Annual Report—The Council on Environmental Quality*, Washington, D.C.: Executive Office of the President.
DOE. 1993. "World Record Set in Fusion Energy." News release. U.S. Department of Energy (10 December).
EPA. 1994. *National Air Quality and Emissions Trends Report, 1993.* Report EPA 454/R-94–026. Washington, D.C.: U.S. Environmental Protection Agency.

EPRI [Electric Power Research Institute]. 1989. *Photovoltaic Field Test Performance Assessment: 1987.* Report EPRI GS-6251. Palo Alto, Calif.: Research Reports Center.

Evans, Michael. 1994. "Efficiency—A Hard Sell." *EPRI Journal* 19, no. 7 (October/November), 16–23.

Galvin, Joseph G., Jr. 1994. "The United States Needs a Long-Term Energy Policy." *The Bridge* 24, no. 2 (Summer), 35–36.

GAO. 1994. "Geothermal Energy: Outlook Limited for Some Uses but Promising for Geothermal Heat Pumps." Reports and Testimony, July 1994, Report GAO/OPA-94-10. U.S. General Accounting Office, Washington, D.C.

Gotchy, Reginald L. 1987. *Potential Health and Environmental Impacts Attributable to the Nuclear and Coal Fuel Cycles.* Report NUREG-0332 (final). Washington, D.C.: U.S. Nuclear Regulatory Commission.

Hellums, W. E. 1990. "Let's All Play by the Same Rules!" Letter to the editor. *Health Physics* 58, no. 3 (March), 377.

Hubbert, M. King. 1973. *Survey of World Energy Resources.* Washington, D.C.: U.S. Geographical Survey.

Jaret, Peter. 1992. "Electricity for Increasing Energy Efficiency." *EPRI Journal* 17, no. 3 (April/May), 4–5.

Lamarre, Leslie. 1989. "Lighting the Commercial World." *EPRI Journal* 14, no. 8 (December), 4–15.

Lawler, Andrew. 1995. "Future Grows Dim for Fusion." *Science* 267, no. 5194 (13 January), 164–165.

Martino, P. 1980. "LNG Risk Management." *Environmental Science and Technology* 14, no. 12 (December), 1446–54.

Mettler, F. A., W. K. Sinclair, L. Anspaugh, C. Edington, J. H. Harley, R. C. Ricks, P. B. Selby, E. W. Webster, and H. O. Wyckoff. 1990. "The 1986 and 1988 UNSCEAR Reports: Findings and Implications." *Health Physics* 58, no. 3 (March), 241–250.

Moore, Taylor. 1996. "Harvesting the Benefits of Biomass." *EPRI Journal* 21, no. 3 (May/June), 16–25.

NCI. 1990. *Cancer in Populations Living near Nuclear Facilities.* Washington, D.C.: National Cancer Institute, U.S. Department of Health and Human Services.

NRC [National Research Council]. 1979. "Geothermal Energy." In *Energy in Transition, 1985–2010.* San Francisco: W. H. Freeman.

Service, Robert F. 1996. "Photovoltaics—New Solar Cells Seem to Have Power at the Right Price." *Science* 272, no. 5269 (21 June), 1744–45.

Somerville, Diana. 1989. "Whatever Happened to Solar Energy?" *American Scientist* 77, no. 4 (July-August), 328–329.

Stover, Dawn. 1995. "The Forecast for Wind Power." *Popular Science* 247, no. 1 (July), 66–72, 85.

USNRC. 1991. "Standards for Protection against Radiation." Code of Federal Regulations, Title 10, pt. 20, app. B, table 2. U.S. Nuclear Regulatory Commission, Washington, D.C.

White, Gilbert F. 1988. "Environmental Effects of the High Dam at Aswan (Egypt)." *Environment* 30, no. 7 (September), 4–11, 34–40.

Wilson, R., S. D. Colome, J. D. Spengler, and D. G. Wilson. 1981. *Health Effects of Fossil Fuel Burning.* Cambridge, Mass.: Ballinger.

World Resources Institute. 1992. *World Resources, 1992–1993.* Washington, D.C.

Wrenn, McDonald E. 1979. "A Comparison of Occupational Human Health Costs of Energy Production: Coal and Nuclear Electric Generation." In Norman E. Breslow and Alice S. Whittemore, eds., *Energy and Health.* Philadelphia: SIAM Institute for Mathematics and Society.

Wuethrich, Bernice. 1996. "Deliberate Flood Renews Habitats." *Science* 272, no. 5260 (19 April), 344–345.

18. Disaster Response

Abelson, Philip H. 1992. "Major Changes in the Chemical Industry." Editorial. *Science* 255, no. 5051 (20 March), 1489.

APHA [American Public Health Association]. 1994. "Public Health Responds to Disaster: The Los Angeles Earthquake." *Nation's Health* 24, no. 3 (March), 1, 6–7.

ATSDR. Agency for Toxic Substances and Disease Registry. 1992. "Preparing for the Unthinkable." *Hazardous Substances and Public Health* 2, no. 3 (July/August), 1.

———. 1994. "Emergency Incident Risk Communication: The Cantara Loop Spill." *Hazardous Substances and Public Health* 3, no. 4 (January), 1–4.

Buist, A. S., and R. S. Bernstein, eds. 1986. "Health Effects of Volcanoes: An Approach to Evaluating the Health Effects of an Environmental Hazard." *American Journal of Public Health* suppl. to vol. 76 (March).

Canby, Thomas Y. 1990. "Earthquake—Prelude to the Big One." *National Geographic* 177, no. 5 (May), 76–105.

CDC [Centers for Disease Control and Prevention]. 1993. "Public Health Consequences of a Flood Disaster—Iowa, 1993." *Journal of the American Medical Association* 270, no. 12 (22/29 September), 1406–7.

Davidson, Leslie L. 1996. Editorial. "Preventing Injuries from Violence towards Women." *American Journal of Public Health* 86, no. 1 (January), 12–14.

de Goyet, Claude de Ville. 1993. "Post Disaster Relief: The Supply—Management Challenge." *Disasters* 17, no. 2, 169–176.

DOE. 1994. *DOE Standard Preparation Guide for U.S. Department of Energy Nonreactor Nuclear Facility Safety Analysis Reports,* Report DOE STD-3009–94. Washington, D.C.: U.S. Department of Energy.

Douglas, Janet. 1994. "A Safer World for the Twenty-First Century." *Civil Protection* 32 (Autumn), 8.

Douglas, John. 1988. "Cleaning Up with Biotechnology." *EPRI Journal* 13, no. 6 (September), 14–21.

Filyushkin, I. V. 1996. "The Chernobyl Accident and the Resultant Long-Term Relocation of People." *Health Physics* 71, no. 1 (July), 4–8.

Holder, Brenda W., and Leo H. Munson. 1996. "U.S. Department of Energy Participation in the Development of Emergency Response Planning Guidelines: The Program and Process." *Applied Occupational and Environmental Hygiene* 11, no. 4 (April), 380–382.

Housner, George W. 1991. "Natural Disasters and Engineering." *The Bridge* 21, no. 4 (Winter), 12–15.

Lakey, John R. A. 1993. *Off-Site Emergency Response to Nuclear Accidents.* A joint project of SCK/CEN and the Commission of European Communities. Mol, Belgium.

Merchant, J. A. 1986. "Preparing for Disaster." Editorial. *American Journal of Public Health* 76, no. 3 (March), 233–235.

Mukerjee, Madhusree. 1995a. "Persistently Toxic—The Union Carbide Accident in Bhopal Continues to Harm." *Scientific American* 272, no. 6 (June), 16, 18.

——— 1995b. "Toxins Abounding—Despite the Lessons of Bhopal, Chemical Accidents Are on the Rise." *Scientific American* 273, no. 1 (July), 22–23.

NAE [National Academy of Engineering]. 1988. *Confronting Natural Disasters: An International Decade for Natural Hazard Reduction.* Washington, D.C.: National Academy Press.

NCRP [National Council on Radiation Protection and Measurements]. 1977. *Protection of the Thyroid Gland in the Event of Releases of Radioiodine.* Report no. 55. Bethesda, Md.

NIOSH. 1994. "NIOSH Warns of Hazards of Flood Cleanup Work." Update, National Institute for Occupational Safety and Health. Publication no. 94–123.

Orme, Thomas W. 1994. "The Great Midwest Flood—A Public Health Triumph." *Priorities* 6, no. 1, 29–31.

PAHO. 1982. *Emergency Vector Control after Natural Disaster.* Washington, D.C.: Pan American Health Organization.

———. 1985. *Emergency Health Management after Natural Disaster.* Washington, D.C.: Pan American Health Organization.

———. 1994. "Toward a New Spirit of Partnership: Regional Cooperation Highlighted at World Conference." *Disasters Preparedness and Mitigation in the Americas,* no. 59 (July), 1, 7.

Platt, Anne E. 1994. "Natural Disasters: A Call for Action." *World Watch* 7, no. 6 (November/December), 6.

Sheets, Robert C. 1995. "Stormy Weather." *Forum for Applied Research and Public Policy* 10, no. 1 (Spring), 5–15.

Sidel, Victor W., Erol Onel, H. Jack Geiger, Jennifer Leaning, and William H. Foege. 1992. "Public Health Responses to Natural and Human-made Disasters." In John Last and Robert Wallace, eds., *Maxcy-Rosenau-Last Public Health and Preventive Medicine,* 13th ed., pp. 1173–86. Norwalk, Conn.: Appleton and Lange.

UN. 1989. "International Decade for Natural Disaster Reduction." Resolution 1989/99, 26 July, General Assembly and Economic and Social Council, United Nations, New York.

USPHS. 1962. *Radioactive Contamination of the Environment: Public Health Action.* Washington, D.C.: National Advisory Committee on Radiation, U.S. Public Health Service, U.S. Department of Health, Education and Welfare.

Waeckerle, Joseph F. 1991. "Disaster Planning and Response." *New England Journal of Medicine* 324, no. 12 (21 March), 815–821.

Ward, Kaari, ed. 1989. *Great Disasters: Dramatic True Stories of Nature's Awesome Powers.* Pleasantville, N.Y.: Reader's Digest Association.

Wasserman, Ellen. 1985. "Technological Disasters in the Americas: A Public Health Challenge." *WHO Chronicle* 30, no. 3, 95–97.

19. A Macroscopic View

AMA. 1989. *Stewardship of the Environment.* Council on Scientific Affairs. Chicago: American Medical Association.

Arrow, Kenneth, Bert Bolin, Robert Costanza, Partha Dasgupta, Carl Folke, C. S. Holling, Bengt-Owe Jansson, Simon Levin, Karl-Goran Maler, Charles Perrings, and David Pimentel. 1995. "Economic Growth, Carrying Capacity, and the Environment." *Science* 268, no. 5210 (28 April), 520–521.

Ayres, Ed. 1994. "Killing the Forest for the Trees." *World-Watch* 7, no. 3 (May–June), 36–38.

Babbitt, Bruce. 1995. "Science: Opening the Next Chapter of Conservation History." *Science* 267, no. 5206 (31 March), 1954–55.

Bertrand, Gerard. 1990. Private communication.

Bloom, David E. 1995. "International Public Opinion on the Environment." *Science* 269, no. 5222 (21 July), 354–358.

Canby, Thomas Y. 1980. "Water—Our Most Precious Resource." *National Geographic* 158, no. 2 (August), 144–179.

CEQ [Council on Environmental Quality]. 1989. *Environmental Trends* Washington, D.C.: Executive Office of the President.

———. 1993. *Environmental Quality, 23rd Annual Report—The Council on Environmental Quality.* Washington, D.C.: Executive Office of the President.

———. 1995. *Environmental Quality, Twenty-Fourth Annual Report—The Council on Environmental Quality,* Washington, D.C.: Executive Office of the President.

Chadwick, Douglas H. 1995. "Dead or Alive: The Endangered Species Act." *National Geographic* 187, no. 3 (March), 2–41.

Cromie, William J. 1996. "Sand Fly Saliva May Prevent Blood Clots." *Harvard University Gazette* 91, no. 28 (11 April), 1–6.

Culotta, Elizabeth. 1995. "Bringing Back the Everglades." *Science* 268, no. 5218 (23 June), 1688–90.

Daily, Gretchen C. 1995. "Restoring Value to the World's Degraded Lands." *Science* 269, no. 5222 (21 July), 350–354.

Douglas, John. 1994. "Understanding the Global Carbon Cycle." *EPRI Journal* 19, no. 5 (July/August), 34–41.

EPA. 1989. "Report of the Committee on Training and Education." National Advisory Council on Environmental Policy and Technology, Washington, D.C.

French, Hilary F. 1995. "Forging a New Global Partnership." In *State of the World: 1995,* pp. 170–189. A Worldwatch Institute report. New York: W. W. Norton and Company.

Frosch, Robert A. 1993. "Sustainable Development and Industrial Ecology." Editorial. *Bridge* 23, no. 2 (Summer), 2.

Glanz, James. 1995. "Erosion Study Finds High Price for Forgotten Menace." *Science* 267, no. 5201 (24 February), 1088.

Glaze, William H. 1996. "The Leadership Vacuum." Editorial. *Environmental Science and Technology* 30, no. 1 (January), 7A.

Goforth, Gary, James Best Jackson, and Larry Fink. 1994. "Restoring the Everglades." *Civil Engineering* 64, no. 3 (March), 52–55.

Haberern, John F. 1995. "The Earth's Soil: Neglected Resource." *Forum for Applied Research and Public Policy* 10, no. 5 (Winter), 80–82.

Hammitt, James K. 1995. "Global Climate Change: Are We Over-Driving Our Headlights?" Newsletter, *Risk in Perspective* 3, no. 5 (July).

Hanson, Hans, and Gunnar Lindh. 1996. "The Rising Risks of Rising Tides." *Forum for Applied Research and Public Policy* 11, no. 2 (Summer), 86–88.

Harrington, W. 1988. "Breaking the Deadlock on Acid Rain Control." *Resources,* no. 93 (Fall), 1–4.

Hill, David K. 1995. "Pacific Warming Unsettles Ecosystems." *Science* 267, no. 5206 (31 March), 1911–12.

Johnson, Jeff. 1995. "EPA Must Look to the Future, Says Science Advisory Board." *Environmental Science and Technology* 29, no. 3 (March), 112A–113A.

Keoleian, Gregory A., and Dan Menerey. 1994. "Sustainable Development by Design: Review of Life Cycle Design and Related Approaches." *Journal of the Air and Waste Management Association* 44, no. 5 (May), 645–668.

Kerr, Richard A. 1995. "It's Official: First Glimmer of Greenhouse Warming Seen." *Science* 270, no. 5242 (8 December), 1565–67.

——— 1996. "Ozone-Destroying Chlorine Tops Out." *Science* 271, no. 5245 (5 January), 32.

Koenig, Robert. 1995. "Rio Signatories to Negotiate New Goals." *Science* 268, no. 197 (14 April), 197.

Linden, Eugene. 1989. "The Death of Birth." *Time,* 2 January, pp. 32–35.

Loehr, Raymond C. 1995. "Protecting the Environmental Future: Looking beyond the Horizon." *Environmental Engineer* 31, no. 2 (April), 7–9.

Manzer, L. E. 1990. "The CFC-Ozone Issue: Progress on the Development of Alternatives to CFCs." *Science* 249, no. 4964 (6 July), 31–35.

Myers, Norman. 1995. "The World's Forests: Need for a Policy Appraisal." *Science* 268, no. 5212 (12 May), 823–824.

NAE. 1989. *Technology and Environment,* ed. J. H. Ausubel and H. E. Sladovich. National Academy of Engineering. Washington, D.C.: National Academy Press.

Nierenberg, William A. 1995. "Progress and Problems: A Decade of Research on Global Warming." *Bridge* 25, no. 2 (Summer), 4–9.

NRC. 1995. *Understanding Marine Biodiversity: A Research Agenda for the Nation.* Committee on Biological Diversity in Marine Systems, National Research Council. Washington, D.C.: National Academy Press.

O'Connor, P. A., ed. 1977. *Congress and the Nation, 1973–1976,* vol. 4. Washington, D.C.: Congressional Quarterly.

Pimm, Stuart L., Gareth J. Russell, John L. Gittleman, and Thomas M. Brooks. 1995. "The Future of Biodiversity." *Science* 269, no. 5222 (21 July), 347–350.

Somerville, J. 1989. "Concerns about Environmental Hazards Growing Nationwide." *American Medical News,* 4 August, p. 21.

Speth, J. G. 1989. "The Greening of Technology." *Bridge* 19, no. 2 (Summer), 3–5.

Stone, Richard. 1995a. "If the Mercury Soars, So May Health Hazards." *Science* 267, no. 5200 (17 February), 957–958.

——— 1995b. "Endangered Species Act—Incentives Offer Hope for Habitat." *Science* 269, no. 5228 (1 September), 1212–1213.

Taubes, Gary. 1995. "Is a Warmer Climate Wilting the Forests of the North?" *Science* 267, no. 5204 (17 March), 1595.

Toman, Michael A., and Joel Darmstadter. 1996. "Grading 'Sustainable America,' the Report of the President's Council on Sustainable Development," *Resources* 123 (Spring), 18–19. Resources for the Future, Washington, D.C.

Warrick, Richard A., and Philip D. Jones. 1988. "The Greenhouse Effect: Impacts and Policies." *Forum for Applied Research and Public Policy* 3, no. 3 (Fall), 48–62.

Wilson, Edward O. 1992. *The Diversity of Life.* Cambridge, Mass.: Belknap Press, Harvard University Press.

World Commission on Environment and Development. 1987. *Our Common Future,* p. 393. Oxford: Oxford University Press.

World Resources Institute. 1990. *World Resources—1990–91: A Guide to the Global Environment.* New York: Oxford University Press.

——— 1995. *The United States Needs a National Biodiversity Policy.* New York: Oxford University Press.

CREDITS

Tables

Table 1.2 Adapted from Richard Doll, "Special Reference to the Effects of Ionising Radiation—The Hazards Forum 1995 Lecture and Dinner," *Hazards Forum Newsletter*, no. 18 (January 1996), Oxford, table 2, p. 3.

Table 3.1 Based on National Research Council, *Environmental Epidemiology—Public Health and Hazardous Wastes* (Washington, D.C.: National Academy Press, 1991), table 3-4, p. 120.

Table 4.2 Centers for Disease Control, "Prevention of Leading Work-Related Diseases and Injuries," *Morbidity and Mortality Weekly Report* (21 January 1983), table 1, p. 25.

Tables 4.3–4.5 Based on data in American Conference of Governmental Industrial Hygienists, 1996. *1996 TLVs and BEIs—Threshold Limit Values for Chemical Substances and Physical Agents and Biological Exposure Indices* (Cincinnati, 1995).

Table 5.1 Based on data in U.S. Environmental Protection Agency, *National Air Quality and Emissions Trends Report, 1993*, Report EPA 454/R-94-026 (Research Triangle Park, N.C., 1994).

Table 5.2 Based on Paul Urone, "The Pollutants," in Arthur C. Stern, ed., *Air Pollution*, 3rd ed. (New York: Harcourt Brace Jovanovich, 1986), vol. 6, table 8, pp. 24–25; and U.S. Environmental Protection Agency, *National Air Quality and Emissions Trends Report, 1993*, Report EPA 454/R-94-026 (Research Triangle Park, N.C., 1994).

Table 5.3 Based on Susan M. Zummo and Meryl H. Karol, "Indoor Air Pollution: Acute Adverse Health Effects and Host Susceptibility," *Journal of Environmental Health* 58, no. 6 (January/February 1996), table 3, pp. 27–28.

Table 6.1 Based on data in Abram S. Benenson, ed., *Control of Communicable Diseases Manual*, 16th ed. (Washington, D.C.: American Public Health Association, 1995); F. L. Byran, *Foodborne Diseases and Their Control* (Atlanta: Centers for Disease Control and Prevention, U.S. Department of Health and Human Services, 1980); and Norton Nelson and James L. Whittenberger, *Human Health and the Environment: Some Research*

Needs, DHEW Publication NIH 77-1277 (Washington, D.C.: Government Printing Office, 1977).

Table 6.2 Based on Abram S. Benenson, ed., *Control of Communicable Diseases Manual,* 16th ed. (Washington, D.C.: American Public Health Association, 1995), p. 184.

Table 6.3, 6.4 Based on information from Institute of Food Technologists, "Government Regulation of Food Safety: Interaction of Scientific and Societal Forces," *Food Technology,* 146(1) (January 1992).

Table 7.2 Sandra Postel, "Increasing Water Efficiency," in Lester R. Brown et al., eds., *State of the World,* Worldwatch Institute Report (New York: W. W. Norton, 1986), table 3-6, p. 55. Copyright 1986 by Worldwatch Institute. Reprinted by permission of W. W. Norton and Company, Inc.

Table 8.1 Based on National Research Council, *Managing Wastewater in Coastal Urban Areas.* Committee on Wastewater Management for Coastal Urban Areas (Washington, D.C.: National Academy Press, 1993), table ES.1, p. 5.

Table 9.2 U.S. Environmental Protection Agency, *RCRA Orientation Manual,* Report EPA/530-SW-86-001 (Washington, D.C., 1986), pp. II-9, III-4, IV-3.

Table 9.3 Based on American Institute of Chemical Engineers, "Garbage! The Story of Waste Management and Recycling" (Washington, D.C., 1993), unnumbered table, p. 3.

Table 9.4 U.S. Environmental Protection Agency, *Solving the Hazardous Waste Problem: EPA's RCRA Program,* Report EPA/530-SW-86-037 (Washington, D.C., 1986), p. 8.

Table 9.5 Based on U.S. Nuclear Regulatory Commission, *Regulation of the Disposal of Low-Level Radioactive Waste: A Guide to the Nuclear Regulatory Commission's 10 CFR Part 61* (Washington, D.C.: Office of Nuclear Material Safety and Safeguards, 1989), fig. 1, p. 2a; and idem, *Information Digest—1995 Edition,* Report NUREG-1350, vol. 7 (Washington, D.C., 1995).

Table 9.6 U.S. Environmental Protection Agency, *Low-Level Mixed Waste: A RCRA Perspective for NRC Licensees,* Report EPA/530-SW-90-057 (Washington, D.C., 1990), p. 23.

Table 9.7 Based on Noel Grove, "Recycling," *National Geographic,* vol. 186, no. 1 (July 1994), 92–115.

Table 9.8 Based on American Institute of Chemical Engineers, "Garbage! The Story of Waste Management and Recycling" (Washington, D.C., 1993), unnumbered table, p. 6.

Table 9.10 E. Peelle and R. Ellis, "Beyond the 'Not-in-My-Backyard' Impasse," *Forum for Applied Research and Public Policy* 2, no. 3 (1987), table 1, p. 72.

Table 10.1 Based on data in Pan American Health Organization, *Emergency Vector Control after Natural Disaster,* Scientific Publication no. 419 (Washington, D.C., 1982), p. 18.

Table 10.2 Based on data in American Public Health Association, "Tropical Diseases Affect One-Tenth of the World's Population," *Nation's Health* 20, no. 7 (July 1990), 8–9.

Table 11.1 Based on National Safety Council, *Accident Facts—1994 Edition* (Itasca, Ill., 1994), p. 34.

Table 12.2 Adapted from Federal Radiation Council, *Background Material for the*

Development of Radiation Protection Standards, Report no. 1 (Washington, D.C., 1960), table 2.1, p. 5.

Table 12.3 Adapted from National Council on Radiation Protection, *Ionizing Radiation Exposure of the Population of the United States,* Report no. 93 (Bethesda, Md., 1987), table 8.1, p. 53.

Table 14.2 Based on data in ICRP, *Recommendations of the International Commission on Radiological Protection,* Publications 1, 9, 26 (New York: Pergamon Press, 1959, 1966, 1977); and idem, *1990 Recommendations of the International Commission on Radiological Protection,* Publication 60, Annals of the ICRP, vol. 21, no. 1-3 (New York: Pergamon Press, 1991).

Table 14.3 Based on data in idem, *1990 Recommendations of the International Commission on Radiological Protection,* Publication 60, Annals of the ICRP, vol. 21, no. 1-3 (New York: Pergamon Press, 1991); and National Council on Radiation Protection and Measurement, *Recommendations on Limits for Exposure to Ionizing Radiation,* Report no. 91 (Bethesda, Md., 1987).

Tables 14.4 and 14.5 Based on data in ICRP, *1990 Recommendations of the International Commission on Radiological Protection,* ICRP Publication 60, Annals of the ICRP, vol. 21, no. 1-3 (New York: Pergamon Press, 1991).

Table 14.6 Based on data in International Commission on Radiation Protection, *Dose Coefficients for Intakes of Radionuclides by Workers,* Publication 68, Annals of the ICRP, vol. 24, no. 4 (New York: Pergamon Press, 1994).

Table 15.2 Based on data in J. P. Corley, D. H. Denham, R. E. Jaquish, D. E. Michels, A. R. Olsen, and D. A. Waite, *A Guide for Environmental Radiological Surveillance at ERDA Installations,* Report ERDA-77-24 (Springfield, Va.: Department of Energy, National Technical Information Service, 1977); and D. W. Moeller, J. M. Selby, D. A. Waite, and J. P. Corley, "Environmental Surveillance for Nuclear Facilities," *Nuclear Safety* 19, no. 2 (January–February 1978), table 3, p. 73.

Tables 15.3 and 15.4 Based on Jerry A. Griffith and Carolyn T. Hunsaker, *Ecosystem Monitoring and Ecological Indicators: An Annotated Bibliography,* Report EPA/620/R-94/021 (July 1994), Washington, D.C., p. 5, and table 1, p. 8.

Table 16.1 Based on John D. Graham, "Annual Report, 1990," Center for Risk Analysis, Harvard School of Public Health (Boston, 1991), unnumbered table, p. 3.

Table 16.2 Based on idem, "Annual Report, 1991," Center for Risk Analysis, Harvard School of Public Health (Boston, 1992), unnumbered table, p. 3.

Table 16.3 Based on E. A. C. Crouch, and R. Wilson, *Risk/Benefit Analysis* (Cambridge, Mass.: Ballinger Publishing Company, 1982), table 7-1, pp. 174–175.

Table 16.4 Based on Bernard L. Cohen, "Catalog of Risks Extended and Updated," *Health Physics* 61, no. 3 (September 1991), tables 3, 7, and 8 (pp. 319, 325, 327).

Table 16.5 Based on John D. Graham, "Annual Report, 1991," Center for Risk Analysis, Harvard School of Public Health (Boston, 1992), unnumbered table, p. 8; and personal communication, January 1994.

Table 16.6 U.S. Environmental Protection Agency, "Seven Cardinal Rules of Risk Communication," Report EPA 230-K-92-001 (Washington, D.C. 1992).

Table 16.7 Based on John D. Graham, "Annual Report, 1993," Center for Risk Analysis, Harvard School of Public Health (Boston, 1994), inside front cover.

Table 17.1 Based on Morton Blatt, "The Energy-Efficient Office," *EPRI Journal* 19, no. 5 (July/August), 16–23.

Table 17.2 Based on Science Concepts, Inc., "The Impact of Nuclear Energy on Utility Fuel Use and Utility Atmospheric Emissions, 1973–1990," (Chevy Chase, Md., December 1991), table 3, p. 14.

Table 17.3 Based on data in National Council on Radiation Protection and Measurement, *Carbon-14 in the Environment,* Report no. 81 (Bethesda, Md., 1985); and J. Tichler, K. Norden, and J. Congemi, *Radioactive Materials Released from Nuclear Power Plants: Annual Report, 1987,* Report NUREG/CR-2907, vol. 8 (Washington, D.C.: U.S. Nuclear Regulatory Commission, 1989).

Table 17.4 Limits on pollutants in the ambient air based on U.S. Environmental Protection Agency, *National Air Quality and Emissions Trends Report, 1993,* Report 454/R-94-026, Office of Air and Radiation, Office of Air Quality Planning and Standards (Research Triangle Park, N.C., 1994), table 2-1, p. 19, for particulates, SO_2, and NO_2; and U.S. Nuclear Regulatory Commission, "Standards for Protection against Radiation," Code of Federal Regulations, Title 10, pt. 20, app. B, table II, for Xe-133.

Table 18.2 Adapted from chart provided by Pan American Health Organization, Washington, D.C.

Table 18.3 Adapted from Federal Radiation Council, *Background Material for the Development of Radiation Protection Standards,* Report no. 5 (Washington, D.C., 1964), pp. 14, 16; and idem, *Background Material for the Development of Radiation Protection Standards: Protective Action Guides for Strontium-89, Strontium-90, and Cesium-137,* Report no. 7 (Washington, D.C., 1965), table 3, p. 16.

Table 19.1 Edward C. Wolf, "Survival of the Rarest," *World-Watch* 4, no. 2 (March–April 1991), 16.

Figures

Figure 1.2 Wayne R. Ott, "Total Human Exposure: Basic Concepts, EPA Field Studies, and Future Research," *Journal of the Air and Waste Management Association* 40, no. 7 (July 1990), fig. 1, p. 968.

Figure 1.3 Based on R. D. Morris and W. R. Hendee, "Environmental Stewardship: Exploring the Implications for Health Professionals," draft report, Medical College of Wisconsin, Milwaukee, 1992, figs. 1a–1c, pp. 3–5.

Figure 2.1 T. A. Loomis, *Essentials of Toxicology* (Philadelphia: Lea & Febiger, 1968), fig. 2-1, p. 15.

Figure 2.2 Based on ibid., fig. 2-2, p. 16.

Figure 3.2 Reprinted with permission from National Research Council, *Environmental Epidemiology—Public Health and Hazardous Wastes* (Washington, D.C.: National Academy Press, 1991), fig. 3-5, p. 122. Copyright 1990 by the American Chemical Society.

Figure 3.2 National Research Council, *Environmental Epidemiology—Public Health and Hazardous Wastes* (Washington, D.C.: National Academy Press, 1991), fig. 7-1, p. 221.

Figure 4.1 Adapted from Barry S. Levy and David H. Wegman, eds., *Occupational Health: Recognizing and Preventing Work-Related Disease,* 2nd ed. (Boston: Little, Brown, 1988), fig. 22-6, p. 21.

Figure 4.3 V. Crisp, "Use of Passive Dosimeters in Monitoring Exposure to Toxic Chemicals," *Respiratory Protection Newsletter* 8, no. 1 (Hebron, Conn.: RSA Publications, 1992), fig. 1, p. 4.

Figure 4.4 Adapted from American Conference of Governmental Industrial Hygienists, *Industrial Ventilation: A Manual of Recommended Practice,* 20th ed. (Lansing, Mich.: Committee on Industrial Ventilation, 1988), fig. VS-202, p. 5–21.

Figure 4.5 G. W. Morgan, "Laboratory Design," in H. Blatz, ed., *Radiation Hygiene Handbook* (New York: McGraw-Hill, 1959), fig. 9-1, p. 9–5.

Figure 4.6 National Institute for Occupational Safety and Health, *Occupational Safety and Health Guidance Manual for Hazardous Waste Site Activities* (Washington, D.C.: U.S. Department of Health and Human Services, 1985), fig. 8-2, p. 8–3.

Figure 5.1 Adapted from National Council on Radiation Protection and Measurements, "Deposition, Retention and Dosimetry of Inhaled Radioactive Substances," draft report (Bethesda, Md., in press), fig. 9-1, p. 250.

Figure 5.2 Based on personal communication August 1996 from Ray Clark, Associate Director for NEPA Oversight, Council on Environmental Quality, Executive Office of the President, Washington, D.C.

Figure 6.1 U.S. Department of Agriculture, Washington, D.C.

Figure 6.2 Adapted from F. L. Bryan, *Foodborne Diseases and Their Control* (Atlanta: Centers for Disease Control and Prevention, U.S. Department of Health and Human Services, 1980), fig. 4-1, p. 2.

Figure 7.1 Council on Environmental Quality, *Environmental Quality 23rd Annual Report* (Washington, D.C.: Executive Office of the President, 1993), unnumbered figure, p. 226.

Figure 7.2 Roger M. Waller, *Ground Water and the Rural Home Owner* (Washington, D.C.: U.S. Geological Survey, 1989), unnumbered figure, p. 13.

Figure 7.4 American Association of Vocational Instructional Materials. *Planning for an Individual Water System* (Athens, Ga., 1973), fig. 48, p. 58.

Figure 8.3 CEQ, *Twenty-Third Annual Report, Council on Environmental Quality* (Washington, D.C.: Government Printing Office, 1993), unnumbered figure, p. 5.

Figure 9.1 Washington State Department of Ecology, *Solid Waste Landfill Design Manual* (Olympia, 1987), fig. 4.26.

Figure 9.2 EPA staff.

Figure 9.3 Adapted from U.S. Nuclear Regulatory Commission, *Recommendations to the NRC for Review Criteria for Alternative Methods for Low-Level Radioactive Waste Disposal,* NUREG/CR-5041, vol. 1 (Washington, D.C., 1987), fig. 2.8.1, p. 2.8.2.

Figure 10.1 "Publications for Rat Control and Prevention Programs," *Public Health Reports* 80, no. 1 (January 1970), p. 40.

Figure 11.1 Adapted from Insurance Institute for Highway Safety, *Twenty Years of Accomplishment by the Insurance Institute for Highway Safety* (Arlington, Va., 1989), p. 2.

Figure 11.2 Idem, *IIHS Status Report* 28, no. 9 (24 July 1993), p. 3.

Figure 11.3 Adapted from idem, "Death Trend Is Down Overall since 1975 Except in Pickup Trucks and Utility Vehicles," *IIHS Status Report* 29, no. 9 (20 August, 1994), p. 3.

Figure 11.4 Logo by Sig Purwin, Center for Health Communications, Harvard School of Public Health.

Figure 12.3 Based on John B. Little, "Biologic Effects of Low-Level Radiation Exposure," in J. M. Taveras, M. Juan, and J. T. Ferrucci, eds., *Radiologic Physics and Pulmonary Radiology,* vol. 1, (Philadelphia: J. B. Lippincott, 1993), fig. 2, p. 7.

Figure 12.5 National Council on Radiation Protection, *Ionizing Radiation Exposure of the Population of the United States,* Report no. 93 (Bethesda, Md., 1987), fig. 8.1, p. 55.

Figure 12.7 "U.S. Nuclear Regulatory Commission, Information Digest, 1995 Edition," Report NUREG-1350, vol. 7 (Washington, D.C.: U.S. Nuclear Regulatory Commission, 1995), fig. 15, p. 32.

Figure 13.1 Adapted from J. H. Ausubel and H. E. Sladovich, eds., *Technology and Environment,* National Academy of Engineering, (Washington, D.C.: National Academy Press, 1989), fig. 2, p. 101.

Figure 14.1 Based on data in Lauriston S. Taylor, "History of the International Commission on Radiological Protection (ICRP)," *Health Physics* 1, no. 2 (1959), 97–104; and ICRP, *1990 Recommendations of the International Commission on Radiological Protection,* Publication 60, Annals of the ICRP, vol. 21, no. 1-3 (New York: Pergamon Press, 1991).

Figure 14.2 Based on data in D. C. Kocher, "Review of Radiation Protection and Environmental Radiation Standards for the Public," *Nuclear Safety* 29, no. 4 (October–December 1988), 463–475.

Figure 15.1 International Commission on Radiological Protection, *Principles of Monitoring for the Radiation Protection of the Population,* Publication 43, Annals of the ICRP, vol. 15, no. 1 (New York: Pergamon Press, 1985), fig. 3, p. 16.

Figure 15.2 Based on World Health Organization, "Guidelines on Studies in Environmental Epidemiology," *Environmental Health Criteria* 27 (Geneva, 1983), fig. 3.1, p. 109.

Figure 15.3 International Commission on Radiological Protection, *Principles of Monitoring for the Radiation Protection of the Population,* Publication 43, Annals of the ICRP, vol. 15, no. 1 (New York: Pergamon Press, 1985), fig. 2, p. 8.

Figure 16.1 U.S. Environmental Protection Agency, *Risk Assessment for Toxic Air Pollutants—A Citizen's Guide,* Report 450/3-90-024 (Research Triangle Park, N.C., 1991), unnumbered figures, p. 8.

Figure 16.2 D. F. Naugle and T. K. Pierson "A Framework for Risk Characterization of Environmental Pollutants," *Journal of the Air and Waste Management Association* 41, no. 10 (October 1991) fig. 3, p. 1301.

Figure 17.2 Based on data provided by the Council on Energy Awareness, Washington, D.C., and on data in Council on Environmental Quality, *Environmental Trends* (Washington, D.C.: Executive Office of the President, 1989), p. 15.

Figure 17.3 Council on Energy Awareness, Washington, D.C.

Figure 18.1 D. W. Moeller, "A Review of Countermeasures for Radionuclide Releases," *Radiation Protection Management* 3, no. 5 (October 1986), fig. 1, p. 72.

Figure 18.2 Atomic Energy Commission, *Radioactive Iodine in the Problem of Radiation Safety* (translated from USSR report) (Washington, D.C., 1972).

Figure 18.3 Federal Radiation Council, *Background Material for the Development of Radiation Protection Standards: Protective Action Guides for Strontium-89, Strontium-90, and Cesium-137,* Report no. 7 (Washington, D.C., 1965), table 2, p. 15.

INDEX